Sustainable Development and Renovation in Architecture, Urbanism and Engineering

Pilar Mercader-Moyano
Editor

Sustainable Development and Renovation in Architecture, Urbanism and Engineering

Springer

Editor
Pilar Mercader-Moyano
Department of Building Constructions I
University of Seville
Seville
Spain

ISBN 978-3-319-51441-3 ISBN 978-3-319-51442-0 (eBook)
DOI 10.1007/978-3-319-51442-0

Library of Congress Control Number: 2016963666

Printed on acid-free paper

This Springer imprint is published by Springer Nature
The registered company is Springer International Publishing AG
The registered company address is: Gewerbestrasse 11, 6330 Cham, Switzerland

Preface

The parts published in this book are taken from the III International Congress on Sustainable Construction and Eco-Efficient Solutions, held in Seville, Spain, in March 2017, this being the fifth edition at a national level. The chapters that are collected in this book are the best ones that have been selected from the 150 chapters presented at this event, by a double-blind peer-review performed by an International Scientific and Technical Committee.

The congress has been established as a forum bringing together academics, researchers, and professionals mainly from the construction sector, where available multidisciplinary environmental information is shared, participating from different areas of the construction process.

Its aim is the search for new alternatives to conventional construction solutions that minimize the environmental impact of the construction activity, improve energy efficiency of buildings, build or refurbish, being thus considered individually or at neighbourhood scale, always from a rentable and optimal cost in time. Therefore, the theme of this edition aims to extend the fields of action involved in the development of an ecological and sustainable society from all areas of knowledge, which is "Sustainable Development and Renovation in Architecture, Engineering and Urbanism" in response to the objectives, not only raised in the Horizon 2020 but from all the people who seek a more sustainable world.

Through three days, professionals, scientists, researchers, and public administration representatives worked together in thematic and parallel blocks, in round tables and debating sessions, in order to reflect on the decision-making that helps to improve the technique innovation in both the public and private building sectors. The content of the communications presented at the Congress is divided into thematic blocks, which continues with "re-" as the motto of previous editions and introduces new ones and of which this book is structured: Sustainable Renovation of Buildings and Neighbourhoods; Minimizing the consumption of material resources; Sustainable planning and urban development; Energy efficiency; Sustainable engineering; Eco-economy; Architecture and society; Sustainable building.

At the same time, students worked intensively in the workshops to achieve innovative building solutions, as alternatives to conventional solutions, which are capable of minimizing the environmental impact generated by construction, thereby improving the energy efficiency of the pre-existent buildings, with an optimal economic cost, and affordability in the long term.

These objectives have brought together more than 100 international researchers from eight different countries, which demonstrate the international scope of this call. Argentina, Chile, Ecuador, Mexico, Portugal, and Spain are the most representative countries. And also around 200 students have participated from the branches of Architecture, Engineering, Urbanism, and Environmental Sciences, among others.

There are involved institutions from Argentina (National University of La Plata, Consejo Nacional de Investigaciones Científicas y Técnicas, Experimental Production Center, University of Buenos Aires: Faculty of Architecture, Design and Urbanism and National University of San Juan); Chile (Bío Bío University); Japan (University of Shiga Prefecture); Ecuador (University of Cuenca and Private Technical University of Loja (UTPL)); Italia (Sapienza Università di Roma); Mexico (Autonomous University of Tamaulipas, Autonomous University of Coahuila and Autonomous National University of Mexico (UNAM)), Bolivia (Autonomous University Juan Misael Saracho), and Portugal (National Laboratory of Civil Engineering).

At national level, there is representation from the Polytechnic University of Cartagena, Jaume I University (Castellón de la Plana), University of Malaga, Polytechnic University of Catalonia, Polytechnic University of Madrid, La Salle Engineering and Architecture School of Ramon Llull University, University of Seville, University of Cordoba, University of Huelva, Polytechnic University of Valencia and University of Zaragoza.

The Congress also counts the participation of Spanish organizations such as "Centro Tecnológico de la Construcción de la Región de Murcia", Institute of Architecture and Building Science (IUACC), "Instituto Valenciano de la Edificación (IVE)", "Unidad de Investigación en Cuidados de Salud (Investén-isciii)", and the Eduardo Torroja Institute for Construction Science (CSIC).

The parts of this book provide a summary of the main debates and chapters that have taken place in this event and the results discussed in the sessions performed.

I would like to thank all the conference participants and especially the staff of the organization of the Congress for their valuable contributions and private and public companies and organizations that have contributed in the performance of the event.

Seville, Spain — Pilar Mercader-Moyano

Prologue

Following a thorough and lengthy procedure, we would like to thank all the contributors for their chapters, which are of the highest calibre, comprising this Special Springer Issue on "sustainable development and renovation in architecture, engineering, and urbanism".

In recent years, the building sector has been turning towards intervening in the existing city building stock. In fact, it is generally accepted that the refurbishment of buildings and the urban regeneration based on sustainability, must form the axis of reformulation of the building sector. Nowadays, achieving sustainable urban development inevitably involves improving existing buildings, thereby preventing the need for city growth, and for the emptying of established neighbourhoods. In order to meet this challenge, it is necessary to understand it in depth and thus break down any barriers that prevent its establishment. In this way, the topic of this Special Issue is framed; it encompasses the fields needed to guarantee sustainable renovation in architecture, engineering, and urbanism.

The research in the field of the rehabilitation of buildings and neighbourhoods provides the basis for a better formulation of the sector. Furthermore, it allows the design of short- and long-term strategies and policies for the sustainable development of cities. It is, undoubtedly, a broad subject that covers a multitude of aspects in this new way of addressing construction. In this Special Issue, we are proud to present 36 chapters authored by 91 researchers of various universities and companies in Argentina, Chile, Ecuador, Mexico, Portugal, and Spain. In Spain, several cities participate, including Barcelona, Cartagena, Castellon, Huelva, Madrid, Malaga, Seville, and Zaragoza. These chapters provide an overview of this issue, and delve into the most outstanding topics.

Part I, collects chapters related with Sustainable planning and urban development towards orderly and sustainable city growth. To this end, the development of synergies between sustainability and urban resilience provide the key for the adaptation of cities to future changes. The aim of Chapter "Model to Integrate Resilience and Sustainability into Urban Planning" is to clarify differences and synergies between these two approaches and to define a theoretical framework for the transition of sustainability and resiliency into urban planning.

Chapter "An Approach to Daylight Contrast Assessment in Mediterranean Urban Environments" proposes progress in the knowledge of urban daylight. The objective of this chapter is to attain a description of the visual environment in Mediterranean countries, in an urban context, in order to ascertain the range of outside luminance values and the expected contrast indoors.

Today, simulation software presents an important tool for the evaluation of the energy behaviour of buildings. The aim of Chapter "Analysis of the Influence of Variables Linked to the Building and Its Urban Context on the Passive Energy Performance of Residential Stocks" is, through the use of simulation software, to identify the level of influence in the passive energy performance of a set of covariates, linked to the urban scale and building scale, in an existing neighbourhood in the city of Castellon de la Plana (Valencian Community, Spain). Chapter "Urban Heat Island and Vulnerable Population. The Case of Madrid", framed inside the MODIFICA project, titled "predictive model for the energy performance of dwellings under the urban heat-island effect", contributes towards the continuous improvement of this software, and presents itself as a starting point for the integration of the urban heat island into the energy simulation process. This phenomenon, which increases the temperature in urban areas, is variable in both time and space, and implies a major lack of accuracy when simulating within an urban context.

Contemporary cities have existing buildings and land which are not in use, and these require recycling and refurbishment, both regarding the solution to their construction and the functional program offered to obtain a more sustainable city. Chapter "Opportunity Detection of Empty Architectonical Lands and Their Recycle for a More Sustainable City" carries out a study of existing tools in the city of Zaragoza, Spain to detect the needs and opportunities regarding the empty architectural voids.

The obsolescence of social housing built between 1950 and 1980 in Spain places these buildings as the principal objectives for regeneration. Chapter "Typological Analysis of H-Plan Social Housing Blocks Built in Spain Between 1957 and 1981" shows a study on 50 H-shaped blocks built in Spain between 1950 and 1980 as a contribution towards taking a step forward in the understanding of one of the most commonly used typologies in the construction of social housing in Spain.

In some cases, old buildings are protected by ordinances to maintain their historical character. These acts constitute a problem in building eco-efficient and sustainable constructions, since they impose a historic image that is not adapted to society's current demands. Chapter "The *Arrabal* of *Alcázar viejo* from Cordoba: Urban, Hereditary and Sustainable Regeneration of the Historic City Centre" proposes improvements to obtain sustainable regeneration in the established suburban areas, especially in the field of urban planning applied to the *Arrabal de Alcázar Viejo* of Cordoba, Andalusia, Spain.

Part II, titled *Architecture and Society*, shows the close relationships between people, architecture, and urban areas. In recent years, the relationship between people and urban areas has changed and it has become necessary to reflect on this change. In this way, Chapter "Heritage and Community Space as Contemporary

Housing Project Matters. Neighbors Courtyards." describes an approach to contemporary housing from the various ways of living and the appropriation of space. It allows the redefinition of a sociability that gives meaning to the urban space and favours the contribution of mechanisms that improve the city internally and rehabilitate community life through the revision of neighbours' courtyards. The aim of Chapter "The Empathic City. Towards a New Model of Urban Sociability" is the development of an analysis strategy and guideline proposals to incorporate urban empathy into the construction of free space in order to return its character of sociability to public spaces.

Chapter "Restoration of Pier-Dock of Clevedon. An Example of Involvement of Society in Defense of the Cultural Heritage" states the case of Clevedon dock, in England, where the political decision for its demolition was supplanted by a massive popular response that ultimately forced the structure to remain as part of their heritage.

Part III collects chapters about the Sustainable Renovation of buildings and neighbourhoods. Chapter "Steps Towards the Integration of Regeneration Processes Obsolete Buildings Envelope Spanish in the Paradigm of Sustainable Development" is a review of the current state of those aspects related to the approach of passive strategies on the energy rehabilitation of buildings in temperate climatic contexts, focusing on countries such as Argentina and Spain. The aim is the development of a model of sustainability assessment of potential regenerative actions on the envelope of types of buildings, in order to determine the suitability of such operations.

Quantitative studies of the improvement obtained in eco-efficient rehabilitations are shown in Chapter "Energy Retrofitting and Social Housing Instrumentation Attending Passive Criteria. Case Study in Winter" and "Environmental Assessment and Energy Certification for the Sustainable Restoration of a Traditional Residential Building". In Chapter "Environmental Assessment and Energy Certification for the Sustainable Restoration of a Traditional Residential Building", a comparative study is carried out between the sustainability and the percentage of improvement in terms of reduction of energy consumption in the case of the refurbishment of a residential building located in the city of Seville. Three states are evaluated: original, refurbishment of strict regulatory compliance, and refurbishment with sustainability criteria. Chapter "Energy Retrofitting and Social Housing Instrumentation Attending Passive Criteria. Case Study in Winter" proposes the adaptation of the Passivhaus standard for energy rehabilitation in existing buildings, and shows the case study of an envelope retrofit of a dwelling.

Chapter shows the results obtained in the I+D (Re)Programa research project carried out on 11 locations in the cities of Seville, Cordoba, and Jerez de la Frontera, Spain. This Chapter "(Re)Programa. Architectural Rehabilitation Incorporating Sustainability Criteria in an Andalusia Neighborhood" offers a reflection on the response that these residential compounds, built after the 40s, gives to its users, thereby clearly displaying the lack of energy efficiency and accessibility.

Social housing, which, predictably, has a greater risk of fuel poverty, is dealt with in articles 17, 15, and 18. The first article concerns current rental housing, while the other two refer to estates built in the mid-twentieth century.

Chapter "Social Rent Housing Refurbishment Demonstrator of LIFE Project "New4Old" (LIFE10 ENV/ES/439)" presents the experience and the result obtained in the refurbishment of two rental social housing buildings in the city of Zaragoza, and clearly shows that the proposed action improves the passive behaviour of the building and achieves greater thermal comfort and habitability without increasing the economic cost related to energy consumption.

Chapter "Assessment Method of Urban Intervention in Social Housing Developments: The Rehabilitation of Caño Roto (Madrid) Case Study" presents a graphic evaluation that enables the assessment of the quality and sustainability of social housing estates built in Spain following the Spanish War and during the period of Transition, in terms of both the state of the estates and the urban processes focused on their reactivation. To this end, the study of a concluded intervention, that of the Urban Rehabilitation of "Poblado Dirigido de Caño Roto" in Madrid, Spain is taken as a reference.

Article 18 carries out a study of existing mappable indicators for the prioritization of the refurbishment of social housing estates, for the case of Urban States in the city of Zaragoza, Spain. This study provides the available indicators and discusses how to combine these indictors for decision-making purposes.

Chapter "From Recovery Constructively Towards the Social Reactivation. The Integrated Knowledge of Traditional Architecture as a Sustainable Strategy" of this Part III proposes social reactivation in order to preserve local knowledgc for society. In this, challenge is for improvement of the knowledge of small hydraulic constructions, whose remains are in a deteriorated state or largely lost. This involves the enhancement of built heritage, considered here as part of the common heritage of a place that must be protected.

In order to achieve the goal of sustainability, it is necessary to minimize the consumption of natural resources and research on this topic is collected in this way in Part IV. A prominent field in achieving this goal is given by the reuse of construction and demolition waste, and related to this topic, Chapters "Study of Fine Mortar Powder from Different Waste Sources for Recycled Concrete Production," and "Concrete Sustainable Light and of High Performance". The first 21 performs a comparison between different sources of waste mortar, by applying dehydration processes of recycled material, and by obtaining a raw material that allows it to be experimentally observed whether there are significant differences between the source and condition of the recycled material and its relationship with the resistance of the recycled concrete on compression. The study carried out in the following Chapter "Concrete Sustainable Light and of High Performance" obtains results that allow recycled concrete to be manufactured that is light and of high strength with less clinker and quarry aggregate.

Natural fibres are regenerable materials that enable reduction of the extraction and use of conventional materials. Chapter "Assessment of the Relationship Between Diameter and Tensile Strength of Piassaba (*Aphandra natalia*) Fibers" presents the assessment of the relationship between diameter and tensile strength of one of the species of palm, which is considered one of the most economically important plant groups.

The management of water resources is dealt with in Chapter "Selection of Criteria for the Systematization of Technologies for a Sustainable Urban Water Cycle Management" of this part, whereby systematization of the information presently available is carried out, for the implementation of hydro-efficient strategies and technologies

Part V, titled *Sustainable Engineering*, collects chapters on the optimization and eco-efficient improvement of the facilities, structures, systems, and industrial processes that are part of contemporary construction. Chapter "Structural Refurbishment Projects. The Sustainability of Reinforcements Using Composite Materials." of this part studies the efficiency and environmental impacts of the structural refurbishment and reinforcement using composite materials, and states that reinforcement work using polymer reinforced with carbon fibre is of high efficiency. The objective of Chapter "NESS®, an Alternative System to Double Strand of Hot Water that Saves Water and Energy" is to establish the main differences between two water-saving systems: the double stream of hot water; and an alternative system called NESS. The results show that this alternative system is able to generate significant energy savings.

Ephemeral architecture is addressed in Chapter "Rethinking Ephemeral Architecture. Advanced Geometry for Citizen-Managed Spaces", whose aim is to lay the foundations to produce low-impact architecture, produced by digital processes, with low environmental impact and an innovative production cycle that minimizes assembly costs, thus rendering it more affordable.

Chapter "Project AURA: Sustainable Social Housing" presents the Aura Project, an award-winner of the XIII Biennial of Spanish Architecture, which is focused on very specific conditions: the tropical climate and problems of social housing and urban growth in the city of Cali, Colombia. The project is as a declaration of principles and intentions, leading to the implementation of social housing in a tropical climate, under the premise of the extrapolation of the results from the rest of Latin and trans-Mediterranean culture.

Certain mining cities in Mexico have expanded their territory in such a way that human settlements have been formed in places contaminated by deposits of mining waste. This causes open spaces, destined as green areas, to fail in their development of vegetation due to changes in the physical and chemical parameters of the soil. Chapter "Vegetation as a Design Element to Recover Green Areas in Settlements Developed on Contaminated Soils" presents a factorial experiment on the growth of seven plants that can be used for the design of such green areas.

Part VI, titled *Energy Efficiency*, collects chapters on research that can contribute towards the efficiency of energy consumption. Chapters "Update of the Urban Heat Island of Madrid and Its Influence on the Building's Energy Simulation" and "Urban Heat Island of Madrid and Its Influence over Urban Thermal Comfort" present studies funded by the aforementioned research project: the MODIFICA project. The first article 28 establishes the geospatial connection between the urban heat island and the most vulnerable population living in the city of Madrid. The study is aimed at understanding the influence of the urban heat island over this most vulnerable population. The aim of Chapter "Urban Heat Island of Madrid and Its

Influence over Urban Thermal Comfort" is to show the influence of heat island and the conformation of the urban fabric in the urban climate in order to identify the zones in which the worst conditions are produced, with the objective of improving comfort in public spaces.

Studies on buildings for tertiary use, such as hospitals, schools, and office buildings, are presented in Chapters "Method for the Implementation Of Active Solar Systems in Hospitals, in the Hospitalization Unit of the Hospital Clínico del Sur, Concepción, Chile", "Validation of a Dynamic Simulation of a Classroom HVAC System by Comparison with a Real Model", "Study on Envelope in Office Buildings Under the Influence of Climate Change in Santiago, Chile", and "Methodology for he Optimisation of Thermal Performance and Daylight Access to the Retrofit of Hospital Rooms in Mediterranean Climate". One of these articles 33 presents a study that includes both simulation and monitoring, performed in a hospital in Seville, Spain, to improve indoor thermal comfort and to reduce energy consumption of the hospital wards in the Mediterranean area, associated to the influence of the size and thermal properties of windows. Chapter "Method for the Implementation of Active Solar Systems in Hospitals, in the Hospitalization Unit of the Hospital Clínico Del Sur, Concepción, Chile" reviews current procedures for the installation of solar equipment in a hospital, in order to present a comprehensive methodology in the case of a hospitalization unit of a hospital in the city of Concepcion, Chile. Chapter "Validation of a Dynamic Simulation of a Classroom HVAC System by Comparison with a Real Model" defines and develops the validation process of a thermal dynamic simulation tool by means of comparison with a real enclosure, in medium-sized rooms with high internal loads. The study is carried out in a standard classroom of a school in the city of Lisbon, Portugal. The effect of the percentage and orientation of windows on the energy demand of office buildings is studied in Chapter "Study on Envelope in Office Buildings Under the Influence of Climate Change in Santiago, Chile". The research is focused on the effect of these variables on the energy demand of office buildings located in Chile.

The envelope is the most important part of buildings in terms of energy efficiency. Chapter "Thermal Energy Refurbishment of Envelope in Mass Neigbourhood Housing, Located in Semi-arid Climate of Argentina" presents technological variants to improve the thermal energy behaviour of the envelope in the case of a representative building in Argentina and the performance of each variant is evaluated, associated with its lifetime cost. Chapter "Threshold Values for Energy Loss in Building Façades Using Infrared Thermography" proposes a new methodology for the detection of the most significant sectors in energy transmission through the envelope of a building by means of infrared thermography. The main goal of this research work is to select building elements or zones of a thermal envelope according to their energy losses in order to avoid indiscriminate intervention on the opaque parts of envelopes. Chapter "Assessment of the Energy Efficiency of a "Cool Roof" for Passive Cooling. Comparative Study of a Case of Tropical Climate and a Case of Southern Spanish Climate" presents a study to determine, under tropical climate conditions, the energy efficiency of the ventilated façades that have proven their success in terms of passive cooling in those latitudes, and compares the results with those obtained for a typical climate in southern Spain.

In closing, on behalf of the Journal, we would like to express our sincere gratitude to all the authors of the chapters and to the reviewers for their systematic and innovative work.

Seville, Spain Pilar Mercader-Moyano

Contents

Part I Sustainable Planning and Urban Development

Urban Heat Island and Vulnerable Population. The Case of Madrid 3
Carmen Sánchez-Guevara Sánchez, Miguel Núñez Peiró and F. Javier Neila González

Analysis of the Influence of Variables Linked to the Building and Its Urban Context on the Passive Energy Performance of Residential Stocks 15
Marta Braulio-Gonzalo, Mª José Ruá Aguilar and Mª Dolores Bovea Edo

Typological Analysis of H-Plan Social Housing Blocks Built in Spain Between 1957 and 1981 27
Alfonso Guajardo

Model to Integrate Resilience and Sustainability into Urban Planning 39
Irina Tumini, Carolina Arriagada Sickinger and Sergio Baeriswyl Rada

Opportunity Detection of Empty Architectonical Lands and Their Recycle for a More Sustainable City 51
Santiago Carroquino Larraz

The *Arrabal* of *Alcázar viejo* from Cordoba: Urban, Hereditary and Sustainable Regeneration of the Historic City Centre 63
Rafael Cabello Montoro

An Approach to Daylight Contrast Assessment in Mediterranean Urban Environments 77
Judit Lopez-Besora, Helena Coch and Antonio Isalgue

Part II Architecture and Society

Heritage and Community Space as Contemporary Housing Project Matters. Neighbors Courtyards 91
María del Carmen Martínez-Quesada

Restoration of Pier-Dock of Clevedon. An Example of Involvement of Society in Defense of the Cultural Heritage 103
Concepción González-García de Velasco, Andrés Agudo Martínez and Miguel González Vilchez

The Empathic City. Towards a New Model of Urban Sociability 117
Isabel Jiménez-López, Angela Barrios-Padura and Marta Molina-Huelva

Part III Sustainable Renovation of Buildings and Neighbourhoods

(Re)Programa. Architectural Rehabilitation Incorporating Sustainability Criteria in an Andalusia Neighborhood 131
Begoña Blandón-González, Pedro Gómez de Terreros-Guardiola, Angela Barrios-Padura and Marta Molina-Huelva

Steps Towards the Integration of Regeneration Processes Obsolete Buildings Envelope Spanish in the Paradigm of Sustainable Development 143
Manuel Ramos Martín and Pilar Mercader-Moyano

From Recovery Constructively Towards the Social Reactivation. The Integrated Knowledge of Traditional Architecture as a Sustainable Strategy 153
Victoria Domínguez Ruiz

Energy Retrofitting and Social Housing Instrumentation Attending Passive Criteria. Case Study in Winter 165
S.G. Melgar, J.M. Andújar and M.A. Bohórquez

Assessment Method of Urban Intervention in Social Housing Developments: The Rehabilitation of Caño Roto (Madrid) Case Study 179
Noelia Cervero Sánchez

Environmental Assessment and Energy Certification for the Sustainable Restoration of a Traditional Residential Building 191
Carmen Rodríguez-Liñán, Mª Jesús Morales-Conde, Filomena Pérez-Gálvez, Paloma Rubio-de Hita and Silvia López-Alonso

Social Rent Housing Refurbishment Demonstrator of LIFE Project "New4Old" (LIFE10 ENV/ES/439) 203
Emilia Román, Gloria Gómez and Margarita de Luxán

Using Mappable Indicators for Prioritizing the Refurbishment of Social Housing. A Case Study of Zaragoza (Spain) 215
Marta Monzón and Belinda López-Mesa

Part IV Minimizing the Consumption of Material Resources

Assessment of the Relationship Between Diameter and Tensile Strength of Piassaba (*Aphandra natalia*) Fibers 227
Cristian Balcázar-Arciniega and Francisco Hernández-Olivares

Selection of Criteria for the Systematization of Technologies for a Sustainable Urban Water Cycle Management 239
Ana Prieto-Thomas

Study of Fine Mortar Powder from Different Waste Sources for Recycled Concrete Production 253
Eduardo Aguirre-Maldonado and Francisco Hernández-Olivares

Concrete Sustainable Light and of High Performance 263
Isabel Miñano, Francisco J. Benito, Carlos J. Parra and Pilar Hidalgo

Part V Sustainable Engineering

Project AURA: Sustainable Social Housing 277
Rafael Herrera, Paloma Pineda, Jorge Roa, Sebastián Cordero and Álvaro López-Escamilla

Structural Refurbishment Projects. The Sustainability of Reinforcements Using Composite Materials 289
Victoriano González, Angela Barrios-Padura and Marta Molina-Huelva

Rethinking Ephemeral Architecture. Advanced Geometry for Citizen-Managed Spaces 301
Andrés Martín-Pastor, Amanda Martín-Mariscal and Alicia López-Martínez

NESS®, an Alternative System to Double Strand of Hot Water that Saves Water and Energy 311
M. Carmen Ladrón de Guevara Muñoz, Eduardo J. Dueñas Ladrón de Guevara, Marcos Ortega Rodríguez and Luis Martín Martínez

Vegetation as a Design Element to Recover Green Areas in Settlements Developed on Contaminated Soils 325
Christhopher Contreras Lopez

Part VI Energy Efficiency

Update of the Urban Heat Island of Madrid and Its Influence on the Building's Energy Simulation 339
Miguel Núñez Peiró, Carmen Sánchez-Guevara Sánchez and F. Javier Neila González

Thermal Energy Refurbishment of Envelope in Mass Neighbourhood Housing, Located in Semi-arid Climate of Argentina 351
Irene Blasco Lucas

Method for the Implementation of Active Solar Systems in Hospitals, in the Hospitalization Unit of the Hospital Clínico del Sur, Concepción, Chile 363
Alberto Nope Bernal, Rodrigo García Alvarado and Ariel Bobadilla Moreno

Validation of a Dynamic Simulation of a Classroom HVAC System by Comparison with a Real Model 381
Miguel Ángel Campano, Armando Pinto, Ignacio Acosta and Juan J. Sendra

Study on Envelope in Office Buildings Under the Influence of Climate Change in Santiago, Chile 393
Carlos Rubio-Bellido, Alexis Pérez-Fargallo and Jesús A. Pulido-Arcas

Methodology for the Optimisation of Thermal Performance and Daylight Access to the Retrofit of Hospital Rooms in Mediterranean Climate 403
Juan Diego Pérez Téllez, Rafael Suárez Medina and Ángel Luis León Rodríguez

Urban Heat Island of Madrid and Its Influence over Urban Thermal Comfort 415
Emilia Román, Gloria Gómez and Margarita de Luxán

Threshold Values for Energy Loss in Building Façades Using Infrared Thermography 427
Juan José Moyano Campos, Daniel Antón García, Fernando Rico Delgado and David Marín García

Assessment of the Energy Efficiency of a "Cool Roof" for Passive Cooling. Comparative Study of a Case of Tropical Climate and a Case of Southern Spanish Climate 439
Carlos A. Domínguez Torres and Antonio Domínguez Delgado

Author Index 451

Subject Index 453

Contributors

Ignacio Acosta Instituto Universitario de Arquitectura y Ciencias de la Construcción, Escuela Técnica Superior de Arquitectura de la Universidad de Sevilla, Seville, Spain

Andrés Agudo Martínez Gabinete de Proyectos, Universidad de Sevilla, Seville, Spain

Eduardo Aguirre-Maldonado Departamento de Arquitectura, Universidad Tecnica Particular de Loja, Loja, Ecuador

J.M. Andújar TEP 192 "Control y Robótica" Research Group, Higher Technical School of Engineering, University of Huelva, Palos de La Frontera, Huelva, Spain

Daniel Antón García Higher Technical School of Building Engineering, University of Seville, Seville, Spain

Carolina Arriagada Sickinger Department of Planning and Urban Design, Universidad del Bío-Bío, Concepción, Chile

Sergio Baeriswyl Rada Department of Planning and Urban Design, Universidad del Bío-Bío, Concepción, Chile

Cristian Balcázar-Arciniega Department of Architecture and Arts, School of Architecture, Universidad Tecnica Particular de Loja, Loja, Ecuador; Department of Building Construction and Architectural Technology, School of Architecture, Polytechnic University of Madrid, Madrid, Spain

Angela Barrios-Padura Department of Architectural Constructions I., Higher Technical School of Architecture, University of Sevilla, Seville, Spain; Department of Architectural Technology II, Sevilla University, Seville, Spain; Department of Building Construction I, University of Seville, Seville, Spain

Francisco J. Benito Department of Architecture and Building Technology, Technical University of Cartagena, Murcia, Spain

Begoña Blandón-González Department of Building Construction I, University of Seville, Seville, Spain

Irene Blasco Lucas Instituto Regional de Planeamiento y Hábitat, Facultad de Arquitectura Urbanismo y Diseño, Universidad Nacional de San Juan, San Juan, Argentina

Ariel Bobadilla Moreno Universidad del Bío-Bío, Concepción, Chile

M.A. Bohórquez TEP 192 "Control y Robótica" Research Group, Higher Technical School of Engineering, University of Huelva, Palos de La Frontera, Huelva, Spain

Mª Dolores Bovea Edo Departamento de Ingeniería Mecánica y Construcción, Universitat Jaume I, Castellón de la Plana, Spain

Marta Braulio-Gonzalo Departamento de Ingeniería Mecánica y Construcción, Universitat Jaume I, Castellón de la Plana, Spain

Rafael Cabello Montoro Department of Urban Sciences and Territorial Planning, Seville University, Seville, Spain

Miguel Ángel Campano Instituto Universitario de Arquitectura y Ciencias de la Construcción, Escuela Técnica Superior de Arquitectura de la Universidad de Sevilla, Seville, Spain

Santiago Carroquino Larraz Universidad de Zaragoza, Zaragoza, Spain

Noelia Cervero Sánchez Department of Architecture, School of Engineering and Architecture, University of Zaragoza, Zaragoza, Spain

Helena Coch Architecture & Energy, Barcelona School of Architecture, UPC, Barcelona, Spain

Christhopher Contreras Lopez National Autonomous University of Mexico, Mexico City, Mexico

Sebastián Cordero Department of Building Construction, School of Architecture, Universidad de Sevilla, Seville, Spain

Antonio Domínguez Delgado Escuela Técnica Superior de Arquitectura de Sevilla, Seville, Spain

Victoria Domínguez Ruiz Departamento de Construcciones Arquitectónicas 1, Escuela Técnica Superior de Arquitectura. Universidad de Sevilla, Seville, Spain

Carlos A. Domínguez Torres Escuela Técnica Superior de Arquitectura de Sevilla, Seville, Spain

Eduardo J. Dueñas Ladrón de Guevara Métrica6 Ingeniería y Desarrollos S.L., Malaga, Spain

Rodrigo García Alvarado Universidad del Bío-Bío, Concepción, Chile

Miguel González Vilchez Department of Building Constructions I, Higher Technical School of Architecture, University of Seville, Seville, Spain

Concepción González-García de Velasco Department of Building Constructions I, Higher Technical School of Architecture, University of Seville, Seville, Spain

Victoriano González LABRUM Diagnosis and Specialized Consulting S.L, Seville, Spain

Alfonso Guajardo Universidad de Sevilla, Sevilla, Spain

Pedro Gómez de Terreros-Guardiola Department of Building Construction I, University of Seville, Seville, Spain

Gloria Gómez PhD Architect cc60 Estudio de Arquitectura, Research Group Sustainable Architecture and Urban Planning (GIAU+S), Universidad Politécnica de Madrid, Madrid, Spain

Francisco Hernández-Olivares Department of Building Construction and Architectural Technology, Superior Technical School of Architecture, Technical University of Madrid, Madrid, Spain

Rafael Herrera Department of Building Construction, School of Architecture, Universidad de Sevilla, Seville, Spain

Pilar Hidalgo Department of Quality and Environment, Cementos La Cruz, S.L, Murcia, Spain

Antonio Isalgue Architecture & Energy, Barcelona School of Architecture, UPC, Barcelona, Spain

Isabel Jiménez-López Department of Architectural Constructions I., Higher Technical School of Architecture, University of Sevilla, Seville, Spain

M. Carmen Ladrón de Guevara Muñoz Métrica6 Ingeniería y Desarrollos S.L., Malaga, Spain

Ángel Luis León Rodríguez Instituto Universitario de Arquitectura y Ciencias de la Construcción, Escuela Técnica Superior de Arquitectura, Universidad de Sevilla, Seville, Spain

Judit Lopez-Besora Architecture & Energy, Barcelona School of Architecture, UPC, Barcelona, Spain

Margarita de Luxán PhD Architect Emerita Professor at the Upper Technical School of Architecture, Universidad Politécnica de Madrid, Madrid, Spain

Silvia López-Alonso Universidad de Sevilla, Sevilla, Spain

Álvaro López-Escamilla Department of Building Construction, School of Architecture, Universidad de Sevilla, Seville, Spain

Alicia López-Martínez Department of Graphic Engineering, University of Sevilla, Seville, Spain

Belinda López-Mesa Universidad de Zaragoza, Saragossa, Spain

Luis Martín Martínez Hidrología Sostenible, Malaga, Spain

Amanda Martín-Mariscal Department of Architectural Design, Instituto Universitario de Arquitectura y Ciencias de la Construcción, University of Sevilla, Seville, Spain

Andrés Martín-Pastor Department of Graphic Engineering, Instituto Universitario de Arquitectura y Ciencias de la Construcción, University of Sevilla, Seville, Spain

María del Carmen Martínez-Quesada Department of Architectonical Projects, Higher Technical School of Architecture, University of Seville, Seville, Spain

David Marín García Higher Technical School of Building Engineering, University of Seville, Seville, Spain

S.G. Melgar TEP 192 "Control y Robótica" Research Group, Higher Technical School of Engineering, University of Huelva, Palos de La Frontera, Huelva, Spain; LAR Arquitectura, Huelva, Spain

Pilar Mercader-Moyano Departament of Building Construction I, University of Seville, Seville, Spain

Isabel Miñano Department of Architecture and Building Technology, Technical University of Cartagena, Murcia, Spain

Marta Molina-Huelva Department of Structures, University of Sevilla, Seville, Spain; University Institute of Architecture and Building Sciences, University of Seville, Seville, Spain; University Institute of Architecture and Construction Sciences, Sevilla University, Seville, Spain

Marta Monzón Universidad de Zaragoza, Saragossa, Spain

Mª Jesús Morales-Conde Universidad de Sevilla, Sevilla, Spain

Juan José Moyano Campos Higher Technical School of Building Engineering, University of Seville, Seville, Spain

F. Javier Neila González Grupo de investigación ABIO, Universidad Politécnica de Madrid, Madrid, Spain; ABIO Research Group, Universidad Politécnica de Madrid, Madrid, Spain

Alberto Nope Bernal Universidad del Bío-Bío, Concepción, Chile

Miguel Núñez Peiró Grupo de investigación ABIO, Universidad Politécnica de Madrid, Madrid, Spain; ABIO Research Group, Universidad Politécnica de Madrid, Madrid, Spain

Marcos Ortega Rodríguez Métrica6 Ingeniería y Desarrollos S.L., Malaga, Spain

Carlos J. Parra Department of Architecture and Building Technology, Technical University of Cartagena, Murcia, Spain

Juan Diego Pérez Téllez Instituto Universitario de Arquitectura y Ciencias de la Construcción, Escuela Técnica Superior de Arquitectura, Universidad de Sevilla, Seville, Spain

Alexis Pérez-Fargallo Department of Building Science, Faculty of Architecture, Construction and Design, Center for Research in Construction Technologies (CITEC), University of Bío-Bío, Concepción, Chile

Filomena Pérez-Gálvez Universidad de Sevilla, Sevilla, Spain

Paloma Pineda Department of Building Structures and Geotechnical Engineering, School of Architecture, Universidad de Sevilla, Seville, Spain

Armando Pinto Laboratório Nacional de Engenharia Civil, Núcleo de Acústica, Iluminação, Componentes e Instalações, Lisbon, Portugal

Ana Prieto-Thomas Department of Building Constructions I, Higher Technical School of Architecture, University of Seville, Seville, Spain

Jesús A. Pulido-Arcas Department of Building Science, Faculty of Architecture, Construction and Design, Center for Research in Construction Technologies (CITEC), University of Bío-Bío, Concepción, Chile

Manuel Ramos Martín University of Seville, Seville, Spain

Fernando Rico Delgado Higher Technical School of Building Engineering, University of Seville, Seville, Spain

Jorge Roa Department of Building Construction, School of Architecture, Universidad de Sevilla, Seville, Spain

Carmen Rodríguez-Liñán Universidad de Sevilla, Sevilla, Spain

Emilia Román PhD Architect cc60 Estudio de Arquitectura, Research Group Sustainable Architecture and Urban Planning (GIAU+S), Universidad Politécnica de Madrid, Madrid, Spain

Carlos Rubio-Bellido Department of Building Construction II, Higher Technical School of Building Engineering, University of Seville, Seville, Spain

Paloma Rubio-de Hita Universidad de Sevilla, Sevilla, Spain

Mª José Ruá Aguilar Departamento de Ingeniería Mecánica y Construcción, Universitat Jaume I, Castellón de la Plana, Spain

Juan J. Sendra Instituto Universitario de Arquitectura y Ciencias de la Construcción, Escuela Técnica Superior de Arquitectura de la Universidad de Sevilla, Seville, Spain

Rafael Suárez Medina Instituto Universitario de Arquitectura y Ciencias de la Construcción, Escuela Técnica Superior de Arquitectura, Universidad de Sevilla, Seville, Spain

Carmen Sánchez-Guevara Sánchez Grupo de investigación ABIO, Universidad Politécnica de Madrid, Madrid, Spain; ABIO Research Group, Universidad Politécnica de Madrid, Madrid, Spain

Irina Tumini Department of Planning and Urban Design, Universidad del Bío-Bío, Concepción, Chile

Part I
Sustainable Planning and Urban Development

Urban Heat Island and Vulnerable Population. The Case of Madrid

Carmen Sánchez-Guevara Sánchez, Miguel Núñez Peiró and F. Javier Neila González

1 Introduction

1.1 Relevance of the Research

The Urban Heat Island (UHI) is a well recognised effect that rises the ambient air temperature compared to surrounding less urbanized areas (Oke 1982). Recent studies of the UHI of Madrid has shown this temperature difference to be up to 8 °C (Núñez Peiró et al. 2016).

Projections from the Intergovernmental Panel on Climate Change point out that this effect will be exacerbated with temperature increase due to climate change (IPCC 2013). Besides that, estimations for the Iberian peninsula show an important increase in frequency and duration of heatwaves (Fischer and Schär 2010), which are expected to be more pronounced in dense urban areas combined with the urban heat island phenomenon.

The combination of all these effects will make population to be exposed to extreme temperatures. High daytime temperatures together along with warm night-time temperatures, for an extended duration of days, will have important impacts over people's health. High temperature health-related consequences have been already studied and some population groups have been identified as being more vulnerable towards this temperature raise. The age of the population has shown to be an important factor, but not only the elder and children are more susceptible to high temperatures; also people already suffering from a chronic disease or people living in high-rise dense areas of the city, are vulnerable towards these effects (Tomlinson et al. 2011; Culqui et al. 2013; Basu 2002; Díaz et al. 2002; Simon et al. 2005). In a recent report for the city of Madrid some districts

---

C. Sánchez-Guevara Sánchez (✉) · M. Núñez Peiró · F.J. Neila González
Grupo de investigación ABIO, Universidad Politécnica de Madrid, Madrid, Spain
e-mail: carmen.sanchezguevara@upm.es

P. Mercader-Moyano (ed.), *Sustainable Development and Renovation in Architecture, Urbanism and Engineering*, DOI 10.1007/978-3-319-51442-0_1

were already identified as being more vulnerable towards climate change due to their temperatures, social composition and building qualities (Tapia et al. 2015).

Derived from these differences in vulnerability towards extreme temperatures, researchers have started analysing whether social inequalities can be found as a consequence of urban temperature gradient distribution. Results of many of these studies have shown a correlation between urban location of more disadvantaged people and a higher exposure to high temperatures (Harlan et al. 2015). These studies are usually based on heat-related mortality data and point at some socio-demographic indicators as a key element in the measurement of this vulnerability: elderly groups of more than 60–65–70 years, secondary education attainment, people living alone as widowed, divorced and separated, low and middle incomes and poor housing conditions (Wong et al. 2016; Klein Rosenthal et al. 2014). Nowadays heat vulnerability indexes are being developed in different cities based on the correlation of high heat risk and socio-economic conditions (Johnson and Wilson 2009; Johnson et al. 2012; Wolf and McGregor 2013).

In line with these studies, Moreno Jiménez and Fernández García (2003) conducted a study in which urban thermal comfort was correlated to spatial distribution of population's income within the Region of Madrid, and the location of the poorer in less comfortable areas was confirmed. Most recent research has shown how this temperature increase can double cooling demand in dwellings of the centre of Madrid (López Moreno et al. 2015). Taking into account that almost 24% of households of Madrid face problems coping with their cooling and heating energy needs (Sánchez-Guevara et al. 2014) it is urgent to analyse whether these relations between disadvantaged people groups and extreme temperature exposure is taking place in the city of Madrid.

1.2 Aims and Objective of the Study

This paper explores the geospatial connection between the urban heat island of Madrid and the location of the most vulnerable population towards high extreme temperatures. The research is aimed at locating which are the neighbourhoods where those most in need live and hence, determining priorities for dwelling energy retrofitting and urban intervention in specific areas of the city.

2 Means and Methods

2.1 Summer Urban Heat Island of Madrid During a Heat Wave

Along 2015 and 2016 years, a proximity environmental temperature measurement campaign was conducted. Temperatures were registered by the method of urban

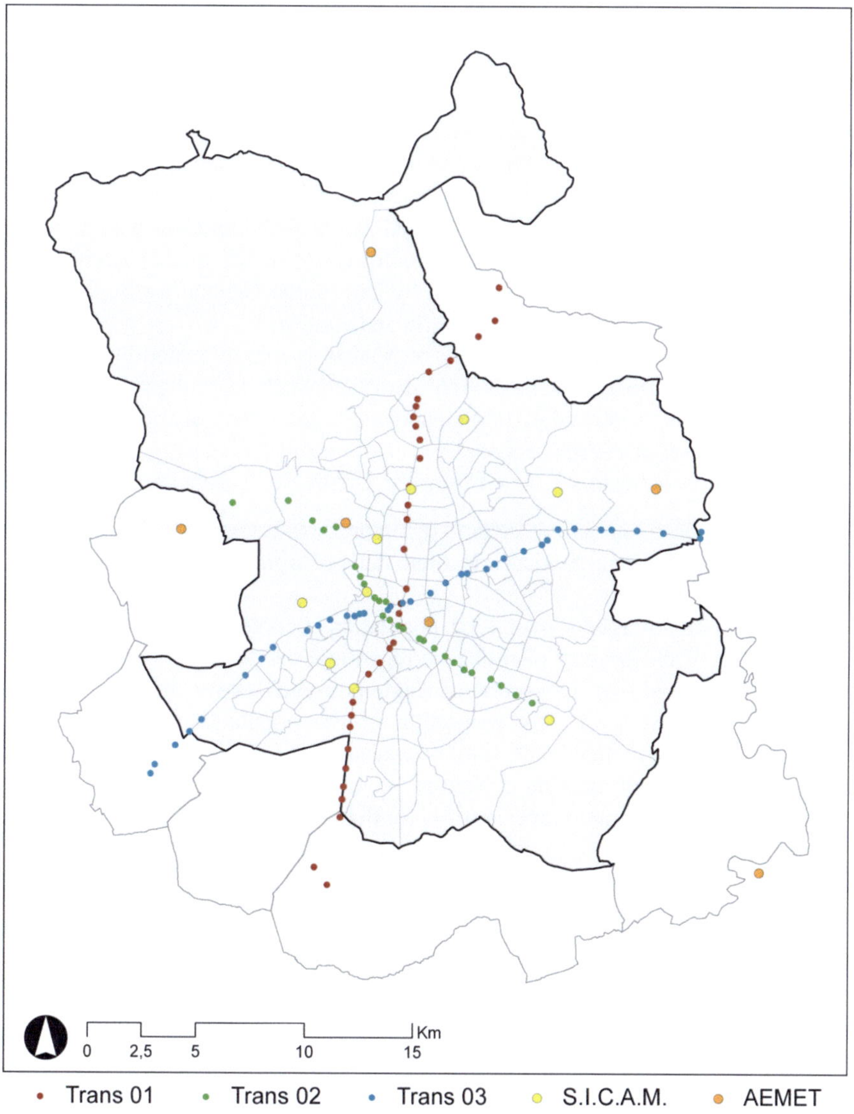

Fig. 1 Map of Madrid with the transects conducted on 2015, the 15th of July

transects, as shown in Fig. 1, already used in the first UHI existing study of Madrid (López Gómez et al. 1988). Results shown in present study were registered in the night of the 15th of July of 2015 under a heat wave event conditions and maximum intensity of the Urban Heat Island. Hence, this research has worked with an urban gradient temperature snapshot wherein the two phenomenon are combined.

2.2 Vulnerable Population Indicators Towards Extreme Temperatures

Previous research showed socioeconomic indicators as a key element to detect vulnerability to heat-related illnesses and mortality. Social indicators used for this study were selected from available data derived from the European project Urban Audit (Eurostat 2012). This project, started in the late nineties, was aimed at gathering statistical data that enabled the comparison of life quality among main European cities. The project is conducted by Directorate-General for Regional and Urban Policy of the European Commission and Eurostat.

Available statistical data for the city of Madrid is disaggregated by neighborhoods in what the project call *sub-city districts* which are urban areas with a population between 5000 and 40,000 inhabitants, smaller than districts and formed by the aggregation of census sections.

Four were the indicators selected from the Urban Audit database:

- Mean household annual net income (€): represents household's income and is calculated by National Statistics Institute derived from data of the Spanish Tax Agency.
- Proportion of foreigners over total population: people who do not have the nationality of the country of residence regardless the place of birth.
- Single household (%): proportion of single households over the total.
- Low educational level (%): proportion of the population between 25 and 65 years old with a maximum level of education ISCED 0, 1 or 2 according to the classification of the United Nations. Level 0 is for early education level, 1 for elementary education level and 2 is for the first stage of secondary education level.

All indicators were selected from 2011 data, except income values that were only available from 2013.

In line with existing literature that points at aged people and children as population at risk towards extreme temperatures, data for these groups was extracted from 2011 municipal census. Hence, another two indicators were incorporated to the study with the same disaggregation level as the sub-city districts:

- Elderly (%): proportion of the population over 70 years old.
- Children (%): proportion of the population under 4 years old.

Values of these indicators were studied for the 142 subcities or neighbourhoods of Madrid and different levels of vulnerability were established according to the range of these values. Hence, vulnerability threshold levels were established as the first or fourth quartile value of an indicator. Severe vulnerability was studied as well, and thresholds used in this case were the first or the ninth decile. Table 1 shows resulting thresholds for each indicator according to this method.

2.3 Geographic Information System Tool

All the information of this research was managed, analyzed and represented through a Geographic Information System (GIS) Tool. The software used was ArcGIS version 10.3.

Urban area disaggregation level was the *sub-city* previously defined, set by the indicators used from the Urban Audit. A reference temperature value was associated to each one of these areas as explained in Sect. 2.1, which is the arithmetic mean value of all contained points within each area. Calculations were made by means of the ArcGIS statistics package.

Finally, different data was overlap and vulnerability maps were obtained.

Table 1 Thresholds for vulnerability indicators

Indicator	Vulnerability threshold	Severe vulnerability threshold
Mean household annual net income (€)	<27,189.45€ (Q1)	<23,863.83€ (D1)
Immigrants (%)	>21.08% (Q4)	>25.31% (D9)
Low educational level (%)	>37.66% (Q4)	>47.69% (D9)
Single household (%)	>31.82% (Q4)	>35.71% (D9)
Elderly (%)	>17.16% (Q4)	>19.04% (D9)
Children (%)	>5.43% (Q4)	>7.15% (D9)

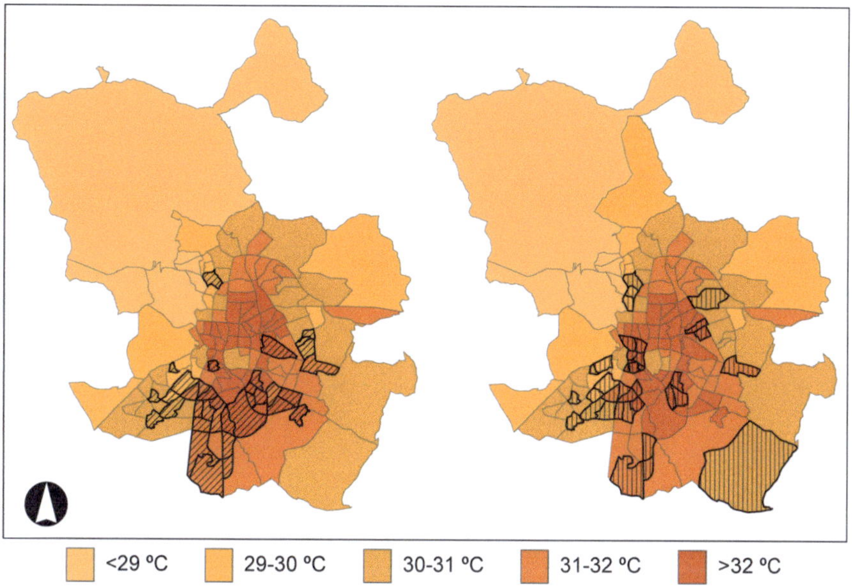

Fig. 2 UHI of Madrid during summer nighttime and neighbourhoods under the first quartile of income (*left*) and neighbourhoods with a proportion of foreigners over the fourth quartile (*right*)

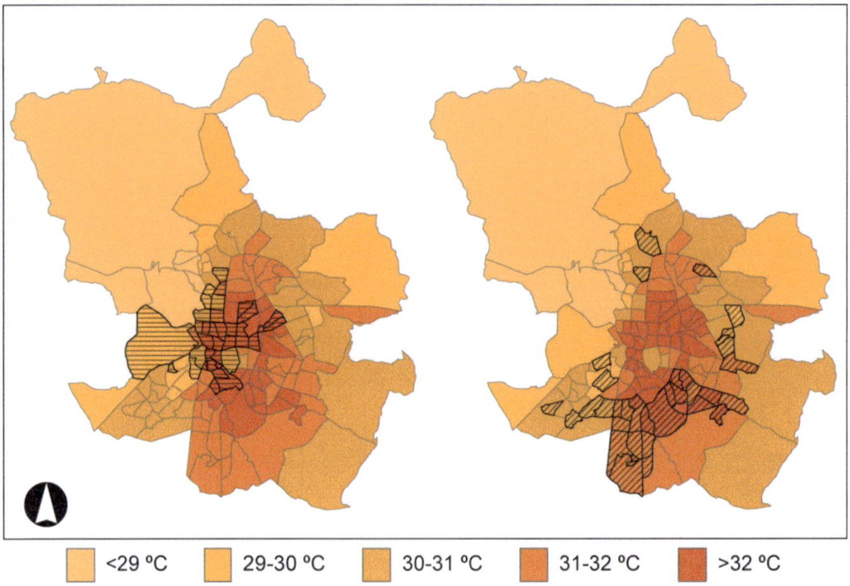

Fig. 3 UHI of Madrid during summer nighttime and neighbourhoods with a proportion of single households over the fourth quartile (*left*) and neighbourhoods with a population with low educational level over the fourth quartile (*right*)

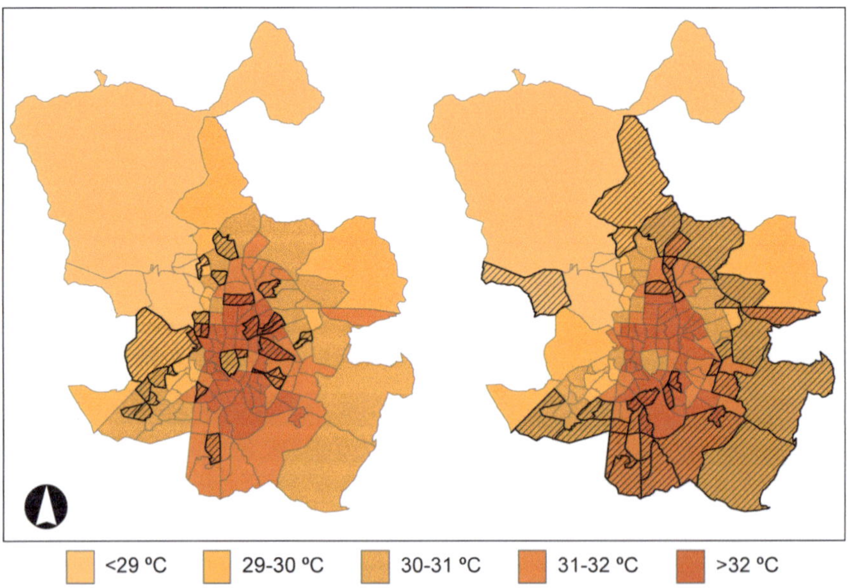

Fig. 4 UHI of Madrid during summer nighttime and neighbourhoods with a proportion of people older than 70 years over the fourth quartile (*left*) and neighbourhoods with a proportion of children below 4 years over the fourth quartile (*right*)

3 Results

3.1 Social Indicators and the UHI of Madrid

First part of the study consisted of analyzing the relation between temperature gradient of the UHI and selected socioeconomic indicators. For that purpose temperature gradient was calculated for each neighbourhood in order to set comparable geographical delimitations with the rest of indicators. Each indicator was separately analysed and compared with registered temperatures. Figure 2 shows neighbourhoods with a median income below the first quartile mainly located in the south of the city where higher temperatures were registered. This figure also enables locating some of the neighbourhoods with the highest presence of foreigners settled in hotter areas.

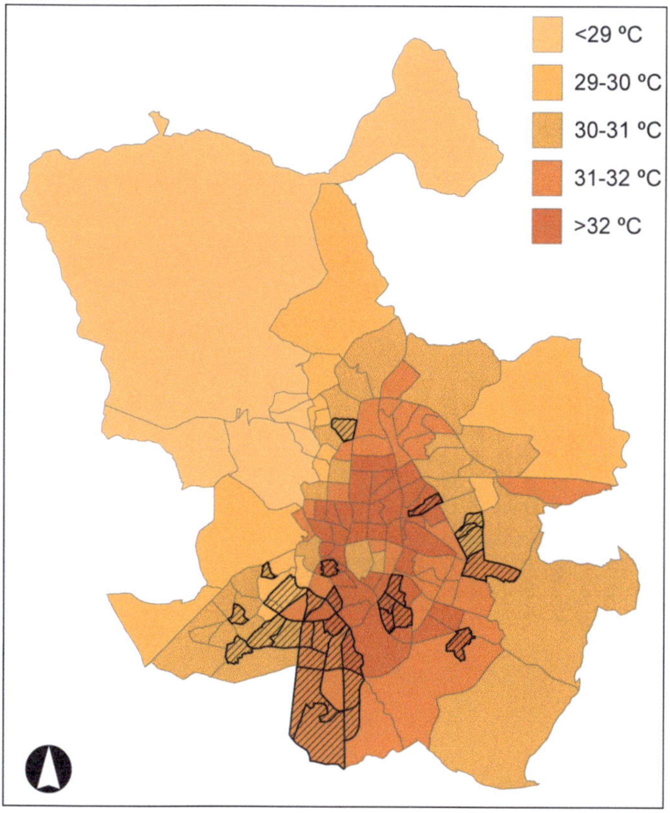

Fig. 5 UHI of Madrid during summer nighttime and vulnerable neighbourhoods according to selected indicators

Larger rates of single households can be found in the center of the city where extreme temperatures were recorded. By contrast, population with lower educational levels can be found in the south, as shows Fig. 3. Finally, Fig. 4 shows the relation of vulnerable population due to its age and the UHI of Madrid. It is possible to detect some of the most aged neighbourhoods in central areas where the highest temperatures are registered, while those with larger presence of children seem to be located in more peripheral areas and hence, with better thermal conditions. However, there can be found some areas in the south side of the city where the overlap between high temperatures and rates of children can be assessed.

From this first approach it can be stated the existence of an overlapping between the hottest areas in the city and the presence of neighbourhoods where those most in need live.

3.2 *Vulnerable Neighbourhoods and the UHI of Madrid*

After analyzing socioeconomic vulnerable indicators separately and comparing them with UHI gradient temperature, some areas of the city were delimitated. Neighbourhoods were defined as vulnerable when three or more of these indicators overlapped so most disadvantaged population could be located. Figure 5 shows these areas and their relation with the temperatures of the city. It can be seen that these neighbourhoods are mainly concentrated in the south of the city.

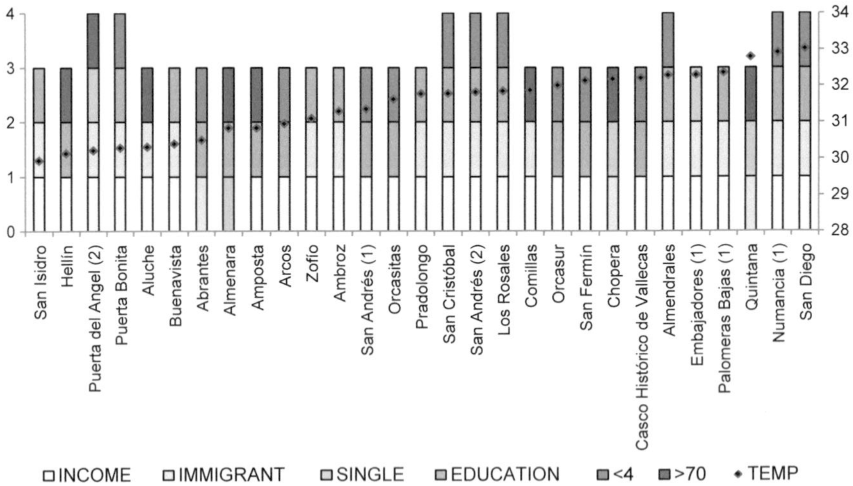

Fig. 6 Neighbourhoods' vulnerability indicators exceeding fixed thresholds and nighttime temperatures registered

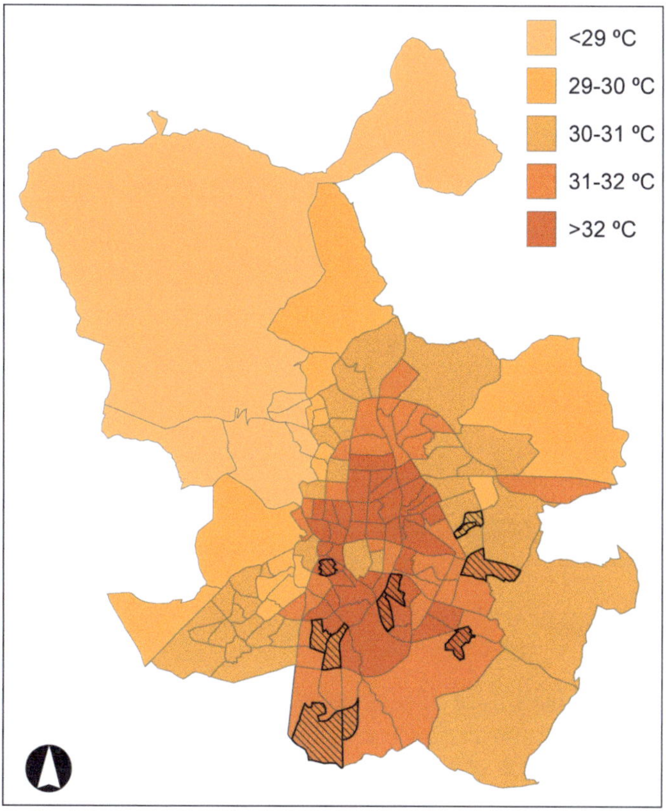

Fig. 7 UHI of Madrid during summer nighttime and severe vulnerable neighbourhoods according to selected indicators

Figure 6 plots this data detecting how many and which indicators were overpassed in each neighbourhood and the exact temperature registered. It can be stated that San Cristóbal, San Andrés, Los Rosales, Almendrales, Numancia and San Diego can be considered neighbourhoods that can be facing serious problems to cope with extreme temperatures.

Besides that, and in order to better understand neighbourhood vulnerability degree, those wherein at least too indicators exceeded the severe vulnerability thresholds were delimitated. Figure 7 shows the location of these neighbourhoods, mainly placed in the south of the city where also higher temperature levels are registered. Figure 8 gathers all vulnerable neighbourhoods and presents for which indicators severe vulnerability thresholds were exceeded.

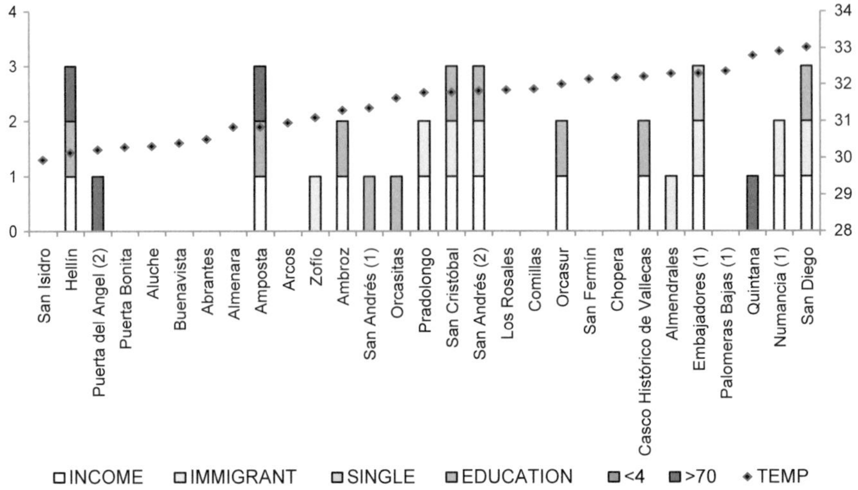

Fig. 8 Neighbourhoods' severe vulnerability indicators exceeding fixed thresholds and nighttime temperatures registered

4 Conclusions

This research has shown, for the city of Madrid, the existence of several neighbourhoods with an important presence of vulnerable population that are located in some of the hottest areas of the city, which poses important health risks for that population. Methodology developed poses a useful tool in order to establish different degrees of vulnerability and hence, prioritize interventions among urban areas.

Neighbourhoods delimitated as being vulnerable towards high temperatures are, in their majority, included in the 'Map of priority areas for the impulse of urban regeneration'—APIRU by its initials in Spanish—(Área de Gobierno de Desarrollo Urbano Sostenible. Dirección General de Estrategia de Regeneración Urbana 2016). This is mainly because this report is based on studies of deprived areas delimitation that take into account socioeconomic indicators of disadvantage of the population along with some others relative to the poor quality of dwellings.

Households settled in these are likely not to be able to cope with extreme temperature episodes either because of the poor quality of their dwellings or the lack of adequate cooling systems. As a conclusion it can be highlighted the importance of incorporating intercity temperature variations as another indicator that must play an important role in the urban regeneration decision-making process.

Acknowledgements This research is funded by *Programa de I+D+I orientada a los retos de la sociedad 'Retos Investigación'* of the Ministry of Economy and Competitiveness. Grant code BIA2013-41732-R, MODIFICA Project: Predictive Model For Dwellings Energy Performance Under The Urban Heat Island Effect.

Authors would also like to thank the State Meteorological Agency (AEMET) and Madrid Air Quality Integral System for weather data provided to carry out this research.

References

Área de Gobierno de Desarrollo Urbano Sostenible. Dirección General de Estrategia de Regeneración Urbana, 2016. *Mapa de áreas preferentes de impulso a la regeneración urbana (APIRU)*, Madrid

Basu R (2002) Relation between elevated ambient temperature and mortality: a review of the epidemiologic evidence. Epidemiol Rev 24(2):190–202

Culqui DR, Díaz J, Simón F, Linares C (2013) Impact of the effects of heat waves on mortality in the city of Madrid, Spain during the period 1990–2009/Análisis del impacto de las olas de calor sobre la mortalidad de la ciudad de Madrid durante el período 1990–2009. Revista Española de Salud Pública 87(3):277–282

Díaz J, Jordán A, García R, López C, Alberdi JC, Hernández E, Otero A (2002) Heat waves in Madrid 1986–1997: effects on the health of the elderly. Int Arch Occup Environ Health 75 (3):163–170

Eurostat (2012) Methodological Manual on City Statistics

Fischer EM, Schär C (2010) Consistent geographical patterns of changes in high-impact European heatwaves. Nat Geosci 3(6):398–403

Harlan SL, Brazel AJ, Jenerette GD, Jones NS, Larsen L, Prashad L, Stefanov WL (2015) In the shade of affluence: the inequitable distribution of the urban heat island. Equity Environ 15:173–202

Intergovernmental Panel of Climate Change (2013) Working group I. Contribution to the IPCC fifth assessment report of climate change 2013: the physical science basis

Johnson DP, Wilson JS (2009) The socio-spatial dynamics of extreme urban heat events: the case of heat-related deaths in Philadelphia. Appl Geogr 29(3):419–434

Johnson DP, Stanforth A, Lulla V, Luber G (2012) Developing an applied extreme heat vulnerability index utilizing socioeconomic and environmental data. Appl Geogr 35(1–2):23–31

Klein Rosenthal J, Kinney PL, Metzger KB (2014) Intra-urban vulnerability to heat-related mortality in New York City, 1997–2006. Health Place 30:45–60

López Gómez A et al (1988) El Clima urbano de Madrid: La isla de calor. CSIC, Madrid

López Moreno H, Sánchez-Guevara Sánchez C, Román López E, Neila González FJ (2015) Thermal characterization of urban heat island according to urban morphology of Madrid. In: Proceedings of the III international congress on construction and building research. Madrid

Moreno Jiménez A, Fernández García F (2003) El confort climático en los entornos residenciales de las capas altas, medias y bajas de la Comunidad de Madrid: otra forma de desigualdad socioespacial. La distribución espacial de la renta en la Comunidad de Madrid, Madrid

Núñez Peiró M, Román López E, Sánchez-Guevara Sánchez C, Neila González FJ (2016) Hacia un modelo dinámico para la isla de calor urbana de Madrid/Towards a dynamic model for the urban heat island of Madrid. Anales de Edificación 2(1):49

Oke TR (1982) The energetic basis of the urban heat island. Q J Royal Meteorol Soc 108(455):1–24

Sánchez-Guevara C, Sanz Fernández A, Hernández Aja A (2014) Income, energy expenditure and housing in Madrid: retrofitting policy implications. Build Res Inf 3218:1–13

Simon F, López Abente G, Ballester E, Martínez F (2005) Mortality in Spain during the heat waves of summer 2003. Eurosurveillance 10(7–9):168–171

Tapia C, Abajo B, Feliu E, Fernández JG, Padró A, Castaño J (2015) Análisis de vulnerabilidad ante el cambio climático en el Municipio de Madrid, Madrid

Tomlinson CJ, Chapman L, Thornes JE, Baker CJ (2011) Including the urban heat island in spatial heat health risk assessment strategies: a case study for Birmingham, UK. Int J Health Geographics 10(1):42

Wolf T, McGregor G (2013) The development of a heat wave vulnerability index for London, United Kingdom. Weather Clim Extremes 1:59–68

Wong MS et al (2016) Spatially analyzing the inequity of the Hong Kong urban heat island by socio-demographic characteristics. Int J Environ Res Pub Health 13(3):1–17

Analysis of the Influence of Variables Linked to the Building and Its Urban Context on the Passive Energy Performance of Residential Stocks

Marta Braulio-Gonzalo, Mª José Ruá Aguilar and Mª Dolores Bovea Edo

1 Introduction

Several studies have addressed the impact of buildings on energy demand and consumption in recent years. Urban areas are set to become greater energy consumers and the residential building stock has a great deal. Energy efficiency regulations in buildings are recent, and most of the housing stock is far from meeting present standard and is responsible of a high percentage of global energy consumption. Thus intervention in the consolidated city gain importance and the promotion of more sustainable urban development focus on reducing energy demand by adopting passive energy strategies.

This study is based on a previous work where a methodology for modelling the energy performance of existing residential stocks was developed (Braulio-Gonzalo et al. 2016). This model was based on a bottom-up approach (Swan and Ugursal 2009), where a representative sample of buildings was energy modelled to extrapolate conclusions to an urban scale.

In the present work, the influence of a set building and urban parameters on the energy performance of residential stocks is analysed. To this end, an existing neighbourhood in the city of Castellón de la Plana (Spain) (Fig. 1) was selected as a case study and the residential building stock was assessed by dynamic simulation with the DesignBuilder (2015) and EnergyPlus software (U.S. Department of Energy 2013) as a calculation engine. The selection of a set of covariates on both the building scale [shape factor (S/V), year of construction (Y)] and the urban scale [urban block (UB), street H/W ratio, and orientation (O)] allowed its influence on two response variables to be analysed: energy demand for cooling (ED_c) and for heating (ED_h).

M. Braulio-Gonzalo (✉) · M.J. Ruá Aguilar · M.D. Bovea Edo
Departamento de Ingeniería Mecánica y Construcción, Universitat Jaume I,
Castellón de la Plana, Spain
e-mail: braulio@uji.es

P. Mercader-Moyano (ed.), *Sustainable Development and Renovation in Architecture, Urbanism and Engineering*, DOI 10.1007/978-3-319-51442-0_2

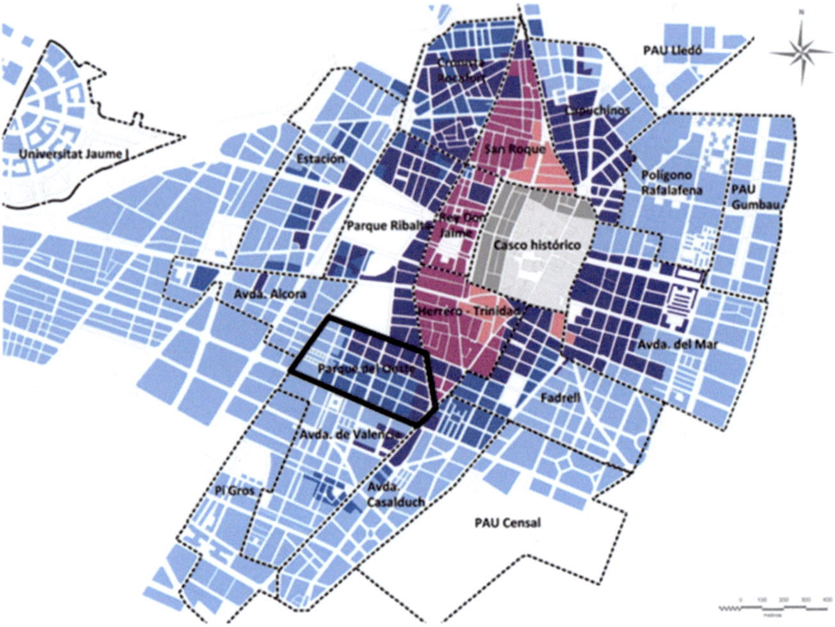

Fig. 1 Analysed neighbourhood (*Parque del Oeste*) in the context of Castellón de la Plana

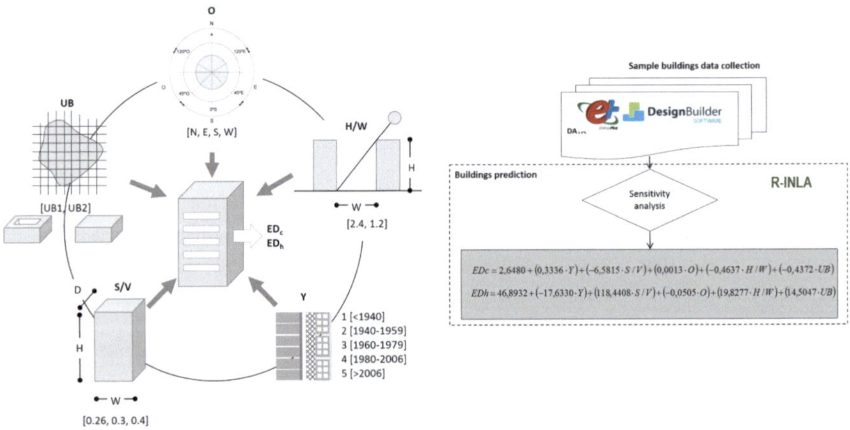

Fig. 2 Description of covariates (*left*) and prediction equations for ED_c and ED_h, adapted from Braulio-Gonzalo et al. (2016) (*right*)

The statistical data modelling through the Integrated Nested Laplace Approximation (INLA) (Blangiardo and Camaletti 2015) allowed one equation to be developed for each response variable (ED_c and ED_h) to predict the energy performance of every single building that make up the neighbourhood. Figure 2 graphically describes the covariates and the four prediction equations. INLA allowed a sensitivity analysis to be done and the level of the significance of the covariates to be identified in order to propose a set of passive design strategies, which implied energy savings in new urban developments.

Based on the conclusions drawn in the statistical analysis, a new urban layout is proposed, which is energy-assessed. The comparison made between the current scenario and the new one allowed the energy savings that can be achieved to be estimated by only integrating the use of passive energy strategies into urban planning. This study is presented in response to the growing need to develop comprehensive energy diagnosis tools to assist local authorities and other involved stakeholders in decision making during urban planning and regeneration processes.

2 Methodology

Drawing on the previous work described above, this study analyses the influence of the covariates (S/V, Y, UB, H/W and O) on the response variables (ED_c and ED_h). After conducting statistical modelling by INLA, every single covariate proved significant. So they all significantly influenced the energy demand of the residential stock. The order of covariates, according to level of significance, was: S/V, Y, H/W, UB and O, where Y and H/W were at the same level.

This analysis allowed a set of strategies for each covariate to be established in order to reduce the energy demand in a particular urban area. These passive energy strategies were used to propose a new urban layout as an alternative scenario to the current neighbourhood. The energy diagnosis of the new scenario allowed a comparison to be made with the existing neighbourhood and to then make an estimate of the energy savings that can be achieved when integrating energy efficiency criteria into urban planning design. Figure 3 presents the methodology followed in this study.

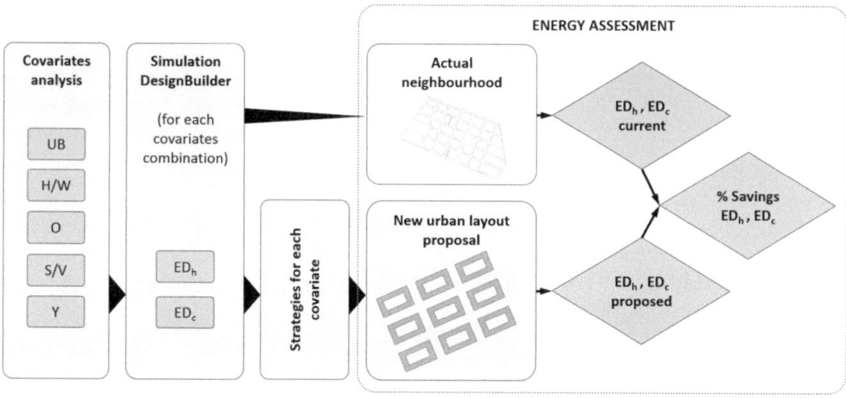

Fig. 3 Methodology

3 Analysis of Covariates

This section presents the covariates analysis results. Figure 4 offers the variation of energy demand for heating and cooling depending on each covariate. From the conclusions obtained, an array of urban design strategies is set to be applied to urban planning designs. The covariates analysis is detailed in the following sections by level of significance of the covariates.

3.1 Shape Factor (S/V)

The shape factor is the most significant covariate. The values adopted in the case study were 0.26, 0.30 and 0.40, which corresponded to the three building typologies found in the neighbourhood. It is observed that compact buildings, which mean lower S/V, have better energy performance for heating, but perform worse for cooling. However, the winter energy demand greatly exceeds the summer one; thus, lower S/V values are more convenient from the energy performance point of view.

3.2 Year of Construction (Y)

Secondly we found Y. In this case, year of construction is linked to the thermal transmittance (U-values) of the building thermal envelope since thermal characteristics, and whether thermal insulation is included or not, affect energy demands. In turn, envelope assemblies depend on year of construction. So five temporal periods were established and a set of envelope assemblies were assigned to each one for façade, dividing walls, roof, ground floor, windows and thermal bridges. These temporal periods included the buildings constructed in: before 1940, 1940–1959, 1960–1979, 1980–2006, and after 2006.

The results for Y showed that a general trend exists in reducing ED_h in recent buildings. Only a change in trend was observed in the buildings constructed during period 3 (1960–1979) when poorer quality construction techniques were employed. A general trend in increasing ED_c was seen in more recent buildings, which responded to loss of thermal inertia in building envelope elements.

3.3 Street Height-Width Ratio (H/W)

Street height-width ratio (H/W) defines the street proportion and is responsible for solar accessibility on the building's façade. According to the City Urban Plan of the neighbourhood under study, eight floors above the ground level are allowed and

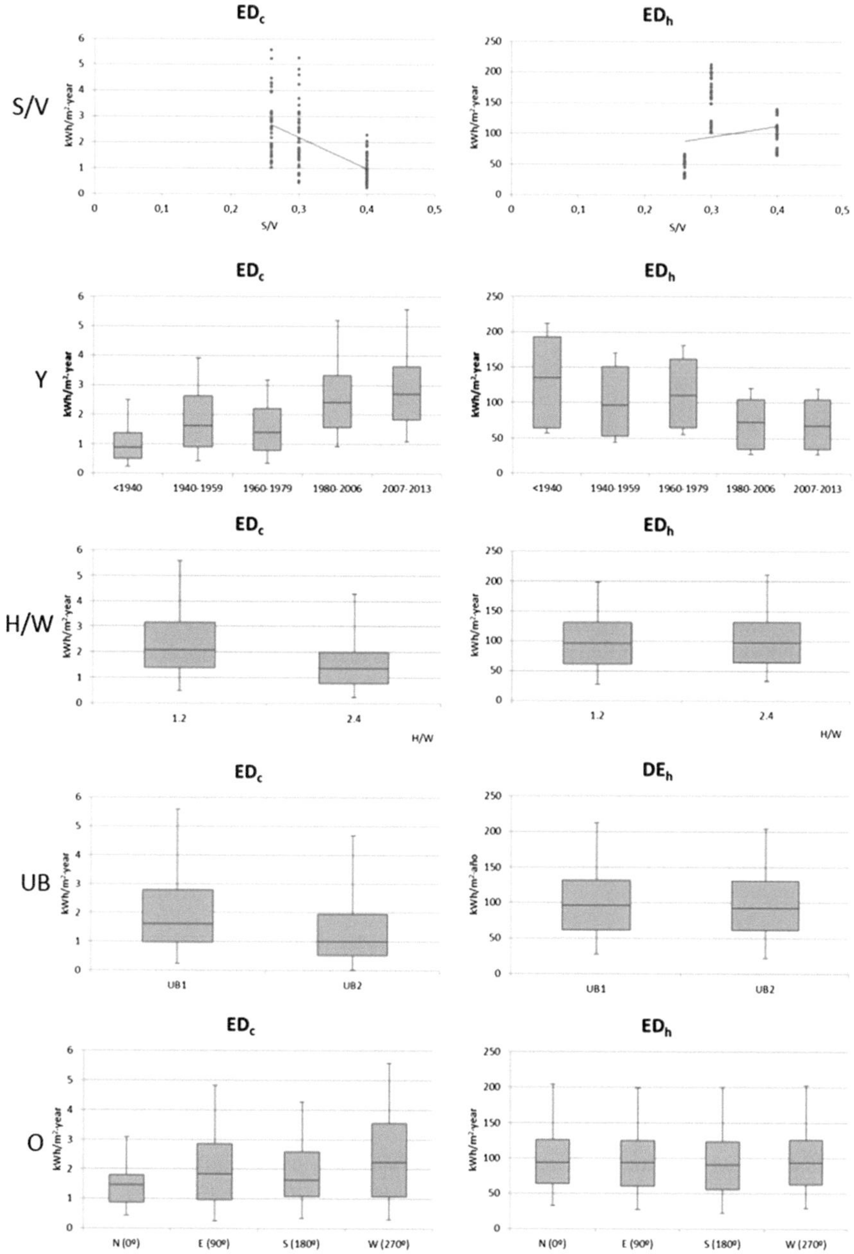

Fig. 4 Response variables results according to each covariate

two street widths were identified; 10 and 20 m. This results in two values for H/W, 2.4 in narrower streets and 1.2 in wider streets. The energy simulation results show that higher H/W ratios implied higher ED_h since narrow streets with high buildings limit solar accessibility. However, higher H/W ratios help decreasing ED_c.

In order to determine optimum H/W ratios, it was considered that an element had solar accessibility if it received at least 2 h of beam radiation during the winter solstice (Higueras 2006). Therefore for the latitude of Castellón de la Plana (39.98°) during the winter solstice (solar elevation 26.61°, at 12:00 p.m.), the H/W ratio should be 0.50 (Braulio-Gonzalo 2016), which implies that street width (W) must be more than twice the building height (H).

3.4 Urban Block (UB)

Two urban blocks are identified in the neighbourhood under study. UB1 has a big internal courtyard that allows solar gains on the south, east and west façades of buildings, with an inward orientation towards the courtyard. UB1 proves more favourable in the energy performance of buildings as it helps reduce ED_h, which is notably higher than ED_c. In contrast, UB2 does not have a big courtyard, but smaller light wells that act as internal building elements. It is observed that ED_c is lower in UB2 and that ED_h is almost the same in both blocks. The fact that UB1 has a big courtyard allows two opposite orientations on the building's façade, that confer passive solar gains benefits.

3.5 Orientation (O)

This covariate is the least significant of the model. It is motivated by the influence of the urban layout in the energy demand of buildings, which modifies the natural behaviour pattern of orientation. In the case study, UB1 provides two orientations to the building (street main façade and light wells façade). Therefore, little variation between data pairs S-N and E-O is observed. In UB2, the buildings have only one main façade (to the street), so the variability in ED_c among the four solar orientations is notable.

The streets with an N-S axis generate east- and west-oriented façades, which receive solar gains in the morning and the evening, respectively. But this option impedes façades having solar gains during central day hours, when a better benefit can be obtained. The streets with an E-W axis generate north- and south-oriented façades. An S façade is optimum in winter and is also quite favourable in summer as it does not yield high ED_c values and is easier to protect from extra sunlight than E and W façades. Therefore, an N-S orientation is the most favourable for façades. Nevertheless, Higueras (2006) and Olgyay (1963) suggest slightly rotating the façade towards an E orientation, which is considered ideal for temperate zones because solar gains are more beneficial in the morning, with lower temperatures than in the afternoon in W.

3.6 Design Strategies for Reducing Energy Demand

The above-conducted analysis for each covariate allows a set of suggestions for the urban planning design to be established for latitudes near 40° in the northern hemisphere. These are detailed below.

- **S/V**: Lower S/V values (more compact buildings).
- **Y**: Envelope assemblies with lower thermal transmittance values (U-values).
- **H/W**: The optimum value is 0.50.
- **UB**: The urban block shape should be rectangular [with length >1.5 width (Olgyay 1963)] and its longitudinal axis should be developed in an E-W orientation. Accordingly, most of the façade surface would be south-oriented. UB1 is the most appropriate because it allows two opposite façades to be oriented N-S.
- **O**: It has been suggested to orient the street axis in an E-O direction so that the streets shape is a rectangular grid with most building façades oriented to N and S. Buildings should be designed so they have two opposite façades, S-N or E-W. The orientation of the main façade of buildings to SE (with an azimuth angle of 18°) is, a priori, better than perfect S (azimuth 0°), but it should be verified experimentally. This influences covariate H/W because orienting the grid towards SE helps lower the W value, which leads to more compact urban structures. Thus in grids oriented to 18°E, H/W can be 0.53.

4 New Urban Layout Proposal

Following these criteria, we explored the optimum urban block by analysing the effect of dimensions and proportions. Each strategy should be applied in a logical order of the urban planning design from the urban scale to the building. The procedure to be followed is shown in Fig. 5.

Having identified UB1 as the most suitable, the parameter that conditions its dimensions is the maximum height of buildings provided by the City Urban Plan. In order to shape an urban block with an inner courtyard, which ensures sunlight on the S façades of buildings, the H/W ratio should c. However, considering the eight floors above ground level, set out in the City Urban Plan, would imply excessively big blocks. For this reason, it is proposed herein to reduce the maximum number of floors to six. This results in a street width of 28.30 m, which is more coherent with the actual urban layout in the city. The new urban layout is presented graphically in

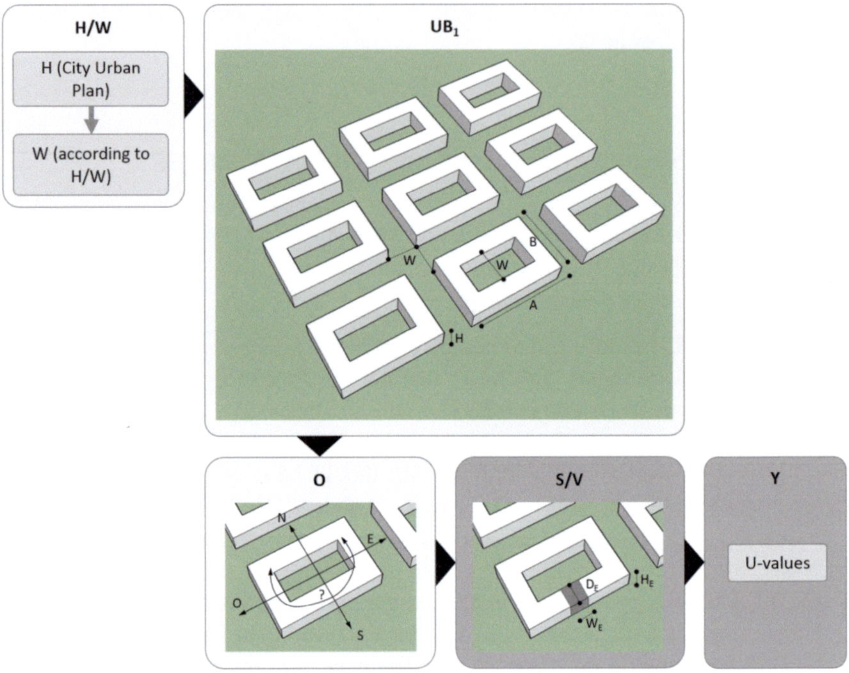

Fig. 5 Procedure for defining the new urban layout

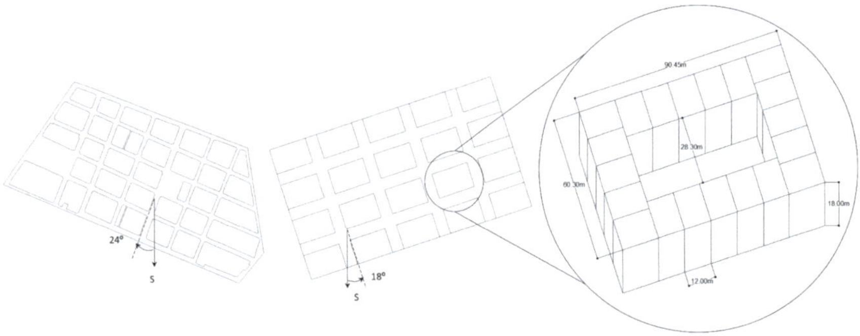

Fig. 6 Current neighbourhood planning (*left*) and new planning proposal (*right*)

Fig. 6. This can be considered when designing urban layouts in new urban developments.

At this point, new energy simulations were conducted to assess the energy performance of the buildings that would make up the new urban proposal. A multi-family building with low $S/V_{0.26}$ in UB1, $H/W_{0.53}$ and with building envelope assemblies from temporal period 5 (Year), in 24 possible solar

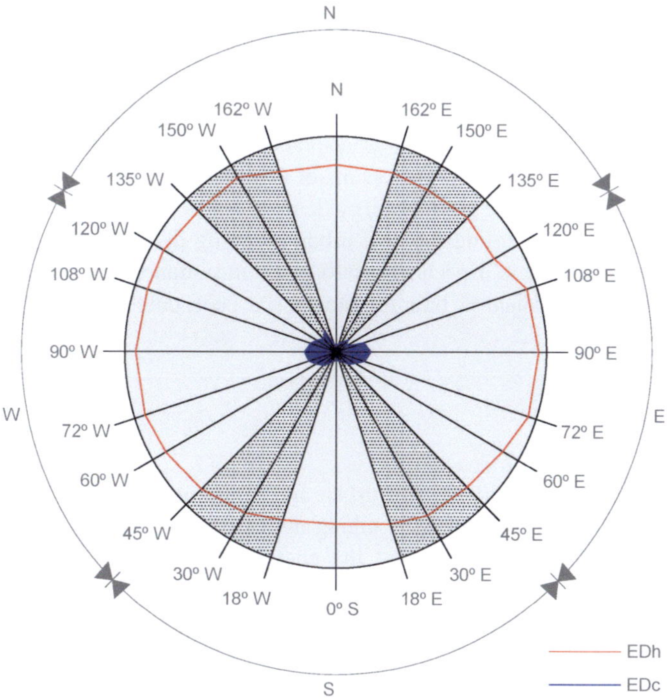

Fig. 7 Relationship between ED and orientation: optimum orientation

Table 1 Comparison between the existing layout and the new proposal (Braulio-Gonzalo 2016)

	Existing	New proposal	Variation %
Built area (m^2)	313,301.00	275,849.40	−11.95
ED_c (MWh/Year)	523.53	1073.88	205.13
ED_h (MWh/Year)	29,354.89	11,739.32	−60.01
ED_G^a (MWh/Year)	29,878.42	12,813.20	−57.12

[a] ED_G Global energy demand

orientations (O), was energy modelled. The results for ED_c and ED_h are shown in Fig. 7, which allow the optimum solar orientation to be identified. This optimum fell within a range of orientations between 18°E and 45°E, which yielded opposite façades between 135°W and 162°W. With these angles, lower ED_h, values are recorded and ED_c values are acceptable. The symmetric range (between 18°O and 45°O) could also be favourable, but the solar radiation in the morning is more beneficial than in the afternoon because of higher temperatures.

Table 1 presents the comparative results for ED_h, and ED_c between the existing neighbourhood and the new urban layout proposal.

The energy assessment of both scenarios allowed the energy demand for heating to be reduced by 60.01% to be estimated. The energy demand for cooling, however, significantly increased by 205.13%. In the existing neighbourhood, narrow streets in summer impede sunlight to reach building façades, which results in low energy demand for cooling. Yet narrow streets imply a notable increase in energy demand for heating. For this reason, the increase in ED_c compensates the more marked reduction in ED_h (60.01%), which is responsible for most of the global energy demand in the neighbourhood. The results of assessing both scenarios reveal that a reduction of 57.12% in the global energy demand for both heating and cooling can be achieved by implementing the new urban planning design, which implies considering only geometrical aspects. Therefore, the importance of urban morphology in the energy performance of building stocks has been demonstrated.

5 Conclusions

This work analyses the influence of urban and building parameters on the energy performance of residential stocks by taking into account energy demand. Based on a previous work done that analysed the energy demand in a neighbourhood in Castellón de la Plana with a statistical prediction model, this work identifies a set of passive design strategies which aim to reduce energy demand. These allow a new urban layout to be proposed, whose energy assessment outlines that important energy savings can be achieved (57.12%) by considering only passive urban and building design aspects.

The results of the analysis stressed that building-related aspects [shape factor (S/V) and year of construction (Y)] are the most influential for reducing energy demand. However, urban environment-related aspects should not be overlooked since they also notably affect the energy performance of the residential stock. In particular, H/W has the same level of significance as Y as it conditions solar gain possibilities in buildings, which greatly influence energy demand for heating and cooling. The urban block shape (UB), depending on it having a courtyard or not, also affects energy performance. Finally, O is the least significant covariate as its behaviour is strongly related to the other urban covariates (H/W and UB).

This study, which forms part of a wider methodology to assess the passive energy performance of residential building stocks, is a decision-making tool for local administrations when taking part in urban development interventions and regeneration projects. As acting on existing urban layouts is a complex process, it is crucial to provide tools that enable energy diagnoses to be made for exploring energy saving measures in existing stocks and their savings potential. Thus local administrations play a key role in integrating energy efficiency criteria into the City Urban Plan and also in prioritising retrofit actions, which have already been covered by Law 3R: urban rehabilitation, regeneration and renovation (Gobierno de España 2013). The integration of passive design criteria into urban planning in order to improve the energy efficiency of residential stocks is a key issue to achieve the energy savings targets set internationally to be met in the near future.

Acknowledgements The authors wish to thank economic support from the Spanish Ministry of Economy and Competitiveness, through Project BIA2013-44001-R, entitled: Protocolo de Diseño Integrado para la Rehabilitación de la Vivienda Social y Regeneración Urbana.

References

Blangiardo M, Camaletti M (2015) Spatial and spatio-temporal Bayesian models with R-INLA. Wiley, Chichester, West Sussex

Braulio-Gonzalo M (2016) Propuesta metodológica para la caracterización del comportamiento energético pasivo del parque edificatorio residencial existente considerando su contexto urbano. Universitat Jaume I

Braulio-Gonzalo M, Juan P, Bovea MD, Ruá MJ (2016) Modelling energy efficiency performance of residential building stocks based on Bayesian statistical inference. Environ Model Softw 83:198–211. doi:10.1016/j.envsoft.2016.05.018

DesignBuilder Software v.4 (2015) Retrieved from http://www.desingbuilder.co.uk

Higueras E (2006) Urbanismo bioclimático (vol 1ª edición). Gustavo Gili, Barcelona. ISBN 978-84-252-2071-5

Olgyay V (1963) Design with climate: bioclimatic approach to architectural regionalism. Princeton University Press, Princeton

Swan LG, Ugursal VI (2009) Modeling of end-use energy consumption in the residential sector: a review of modeling techniques. Renew Sustain Energy Rev 13(8):1819–1835. doi:10.1016/j.rser.2008.09.033

U.S. Department of Energy (2013) Energy plus software, Berkeley. Retrieved from http://apps1.eere.energy.gov/buildings/energyplus/energyplus_about.cfm

Typological Analysis of H-Plan Social Housing Blocks Built in Spain Between 1957 and 1981

Alfonso Guajardo

1 Introduction

The regeneration of mid-20th century residential estates is a major concern of the public Administrations (eu2010 2010; Ministerio de Fomento 2013; Jefatura de Estado 2013). The attention that city centers used to receive is now focused on peripheral residential areas. A number of problems of different nature can be identified in those neighborhoods (García Vázquez 2010). Firstly, from an urban point of view, they are commonly characterized by a certain level of isolation, a lack of facilities and the deterioration of their public space. Secondly, from a social perspective, many of them have also been identified as vulnerable neighborhoods, with high unemployment rates and low educational levels (Hernández Aja et al. 2015). Finally, architectural deficiencies come to light due to the inadequacy of their technical equipment, the low energy efficiency of these constructions and a lack of building maintenance. This situation has stimulated the interest of researchers to seek of solutions that can make cities more enjoyable and sustainable.

In concordance to what some authors, such as García Vázquez (2015), point out, one of the main causes behind the obsolescence of these buildings is the maladjustment that exists between the housing typologies and the needs and expectations of contemporary society.[1] These social housing projects were planned during Franco's regime to satisfy the basic needs of the Spanish working class of the 1950s. However, the advent of democracy a couple of decades later would see the rise of a new kind of house-demanding citizens, the profiles of which were very

[1]GarcíaVázquez has identified two quantitative maladjustments, size and functionality; and two qualitative ones, related to cultural changes and ways of life.

A. Guajardo (✉)
Universidad de Sevilla, Sevilla, Spain
e-mail: aguajardofajardo@us.es

P. Mercader-Moyano (ed.), *Sustainable Development and Renovation in Architecture, Urbanism and Engineering*, DOI 10.1007/978-3-319-51442-0_3

different. Accordingly, new social dynamics challenged previous architectural programs.

A deep knowledge of these housing typologies is a mandatory first step towards evaluation of their obsolescence. However, the study of the housing typologies built massively in Spain has not received enough attention and represents an unusual topic in contemporary architectural research. The main objective of this paper is to address that situation and provide potentially relevant information in order to: (1) identify housing unit typologies (2) analyze their characteristics. Specifically, this study is focused in the analysis of the H-plan block housing unit typology, given its wide-spread use in these public states.[2]

The H-plan block has been defined as a multi-story apartment building organized with 4 housing units per floor and a communication core. These buildings can be connected to others creating linear structures. The housing units have two facades: one towards the public space and another facing a small, inner courtyard.

2 Methodology

This research analyses the housing typologies of 44 H-plan blocks built in Spain between 1957 and 1981. The sample is mainly composed of buildings from western Andalusia, where a total of ten different archives were consulted.[3] For the identification of other cases in Spain, the complete collection of the journal "*Hogar y Arquitectura*" (edited between 1955 and 1977 by the national organism *Obra Sindical del Hogar*) was thoroughly reviewed. After the total sample collection by one or other means, the layout of each typology was redrawn in order to measure and analyze them. Three-bedroom units were identified in thirty six of the samples and four-bedroom units in the remaining eight. To carry out this typological analysis, only the blocks with three-bedroom units are taken into consideration consideration given that it is the only group with enough representative cases that would allow a size comparison analysis. The statistical analysis in Sect. 4.1 "Chronology of the H-plan block" includes them all.

The hypothesis to be tested is that quoted by Ignacio Paricio in regards to massive housing projects in Spain (Paricio Ansuátegui 1973a): "the width and depth of the plot have a significant influence in the housing layouts". Consequently, this study analyses the different typologies of the houses based on the relationship between the width and the depth of the blocks. The work "*Estudios de tipología de la vivienda: Entre Medianeras*" (Paricio Ansuátegui 1973b) regarding H-plan blocks from Barcelona was used as a reference. As part of this study's Conclusions

[2]Supported by a non-published study curried out by the author.

[3]Archivo Municipal de Cádiz, Archivo Histórico Provincial de Cádiz, Archivo de la Diputación Provincial de Córdoba, Archivo Histórico Provincial de Córdoba, Archivo Municipal de Córdoba, Archivo Municipal de Huelva, Archivo Municipal de Jerez, Archivo FIDAS, Archivo Histórico Provincial de Sevilla, Archivo Municipal de Sevilla.

a general characterization of the H-plan social housing block type is described, in the hope that it will serve as a reference for future regenerative initiatives.

3 Results: Typological Analysis

In this section, the typological analysis of 36 H-plan housing blocks types built between 1957 and 1981 has been carried out. An initial classification of the housing units into different groups is presented regarding how the day area (living room and kitchen) and the night area (bedrooms) are connected. These groups are named Type 1, Type 2 and Type 3. For each housing Type (1–3), a *characteristic layout* is defined based on the most abundant distribution. With that layout as a reference, the units within each Type are again analyzed and their differences regarding the "characteristic layout" are described. The dimensional relationship between the depth and the width of the blocks is taken into consideration for this last examination.

3.1 Type 1—Housing Units

Type 1—housing units (the main dimensional characteristics of which are summarized in Table 1) are defined by the dependence between the day and night areas, i.e., the access to the night area from the main entrance interrupts the circulation between the kitchen and the living room. 29 cases out of the 44 analyzed correspond to this type. Among them, 89.7% adhere to the characteristic layout of this Type, 6.9% are a variant that follow an irregular perimeter and 3.4% show a different solution. The first two groups will be described in this study.

3.1.1 Housing Units with a Characteristic Layout

The characteristic layout of Type-1 housing units fits the following scheme: a main entrance with a small hall, the kitchen located nearby and adjacent to the bathroom (comprising the wet area of the house), the living room across from of the kitchen

Table 1 Type-1 housing units. Basic characteristics

Average surfaces	Built	77.47 m^2	Structure	Reinforced concrete	89.65%
	Usable	63.58 m^2		Load-bearing walls	3.45%
Average depth		7.14 m		Mixed	3.45%
Average width		18.36 m		Prefabricated	3.45%
Width-depth ratio		2.7	Unique wet area		89.7%

and the night area organized around a small corridor. Terraces are associated to the living room or the kitchen (Fig. 1a).

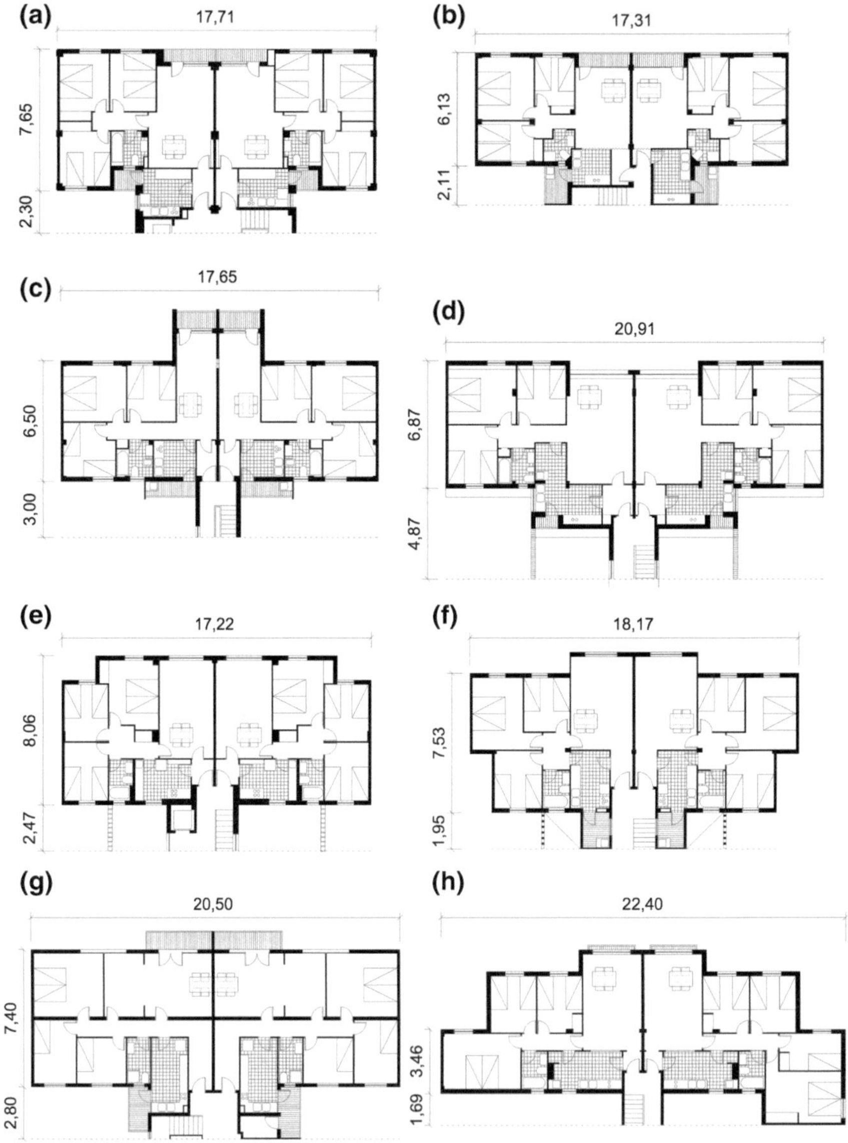

Fig. 1 Type-1 housing units. **a** J. Rebollo Dicenta (1974) La Fuensanta, Córdoba; **b** R. de la Hoz, G. Olivares (1965) Parque Cruzconde, Córdoba; **c** L. Recasens (1967) Pino Montano, Sevilla; **d** J. Álvarez et al. (1971) Moratalaz, Madrid; **e** J.E. Alba et al. (1976) Vega de Arriba, Asturias; M. Riquelme (1973) La Granja, Jerez; **g** P. Genovés (1975) Alcosa, Sevilla; **h** S. de la Fuente (1969) La Granja, Jerez

Extension towards the central courtyard A common strategy in the H-plan Type-1 group is the extension of the housing unit towards the inner courtyard, widening the space for any one of the rooms. This is achieved through more irregular shapes in the inner façade, in contrast with the basic rectangular shapes. That shape reconfiguration primarily affects the kitchen and the outdoor laundry area (terrace), which are placed closer to the communication core of the building. Of all the cases analyzed within this group, 55% of them make use of this scheme in their layout. This solution provides greater flexibility when it comes to locating the kitchen and living room. Locating the terrace on the exterior can also offer the advantage of serving as a ventilation space for the bathroom (Fig. 1b).

An inverse relationship between the depth of the unit and the use of this strategy was confirmed: 90% of the units with depths between 6 and 7 m partially or completely locate the kitchen in the inner courtyard of the building while this trend becomes less prevalent in units with larger depths. Among units with a depth of 7–8 m, for instance, only 55% adopt this solution and that percentage goes down to 20% in the group of units that are at least 8 m in deep.

The location of the kitchen in the inner courtyard is almost a must in units that are under 7 m deep. Figure 1c shows the case of a unit that is under 7 m deep that maintains the kitchen within the parallelepiped, forcing the longitudinal extension of the living room towards the façade, thus generating a very problematic geometry.

Flexible Circulation Flexible circulation is another characteristic of Type-1 housing units. This solution is found in a minority of cases (31%), and is only observed in blocks built after 1970. The most common strategy is to connect the entrance hall with the night-area corridor through the kitchen. This generates a circular itinerary around the wall that divides living room and kitchen.

Once again the layout of the unit depends on its depth. Among the units with a flexible circulation, two different configurations exist: those with short depth (around 6 m) have a small hall independent from the living room but with direct access to an L-shaped kitchen (Fig. 1d); while those with larger depths (around 8 m) maintain a rectangular-shaped kitchen. In the latter case, is possible to find two different ways of orienting the kitchen towards the inner façade: either parallel or perpendicular to it. The 90° rotation of the kitchen is an effective solution that allows flexible circulation while moving circulation away from the cooking area (Fig. 1f).

There is only one case in which a circular itinerary is not achieved around the kitchen but through the living room (Fig. 1g). This layout has two consequences: the mandatory shift of a second bedroom towards the inner façade and the need of a large corridor that goes through the entire unit. To be implemented, this layout requires large surface areas.

3.1.2 Housing Units with Irregular Perimeters

Figure 1h shows the case of Type-1 units whose layout differs from the characteristic distribution. Its main virtue is the rejection of the rectangular-shaped unit and a commitment to a more irregular contour. This novelty is not only a matter of a formal composition of the perimeter of the façade, but a reflection of the search for a more flexible architectural approach. This strategy could be understood as a new way of approaching the design of the H-plan block: rooms do not need to fit the rectangular shape of the house any longer, but instead, it is their aggregation what defines the shape of the block.

In these non-characteristic units, the main bedroom is placed at the end of the corridor and has a double façade, one towards the public space and another towards the inner courtyard. The other two bedrooms are next to the living room. This exceptional layout of the night area allows all the bedrooms to ventilate directly to public space (which is unique among all the Type-1 samples), it also provides independence and geometric proportion to the rooms and facilitates flexible circulations.

3.2 Type-2 Housing Units

Type-2 housing units (the main dimensional characteristics of which are summarized in Table 2) are defined in our study as those in which access to the night area can be done independently of the living room-kitchen circulation. No characteristic layout was found due to the few cases identified within this Type given the differences between them.

Figure 2a shows the floor plan of the smallest housing unit (60.22 m^2) found within this Type. We observe that the kitchen and living room comprise an architectural unit that is independent from the night area, making it possible to reach the main entrance directly from the bedrooms. This solution is achieved by placing the wet area across from the entrance hall and by opening the kitchen towards the living room. A central corridor organizes the circulation within the unit.

Two cases (Fig. 2b, c) are found in which areas are functionally separated by means of bays. One of them is designated to the bedrooms and the other to the living room, kitchen and bathroom. The main difference between these two blocks is the orientation of the load-bearing structure.

Table 2 Type-2 housing units. Basic characteristics

Average surfaces	Built	71.67 m^2	Structure	Reinforced concrete	40%
	Usable	57.08 m^2		Load-bearing walls	20%
Average depth		7.35 m		Mixed	40%
Average width		16.51 m		Prefabricated	0%
Width-depth ratio		2.3	Unique wet area		60%

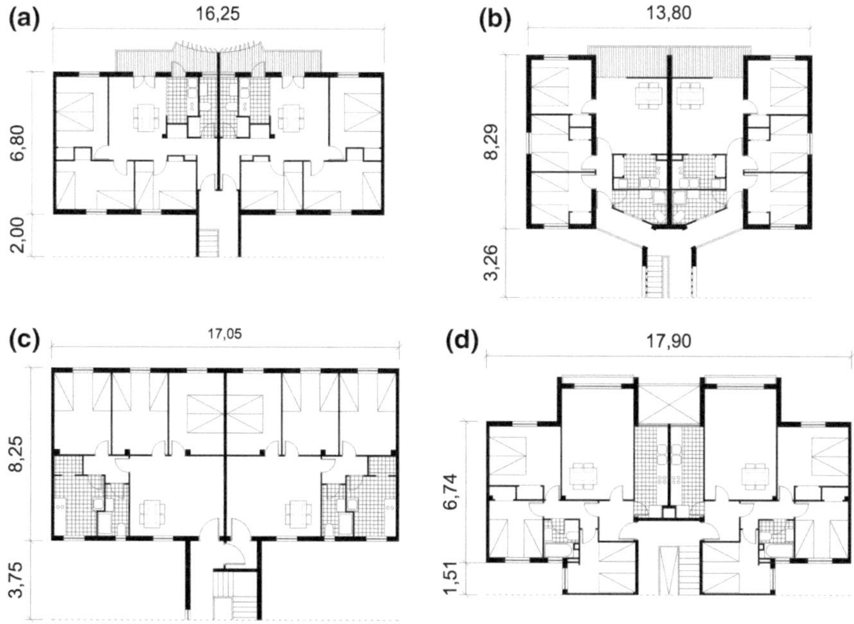

Fig. 2 Type-2 housing units. **a** R. Espiau (1958) Tiro de Línea, Sevilla; **b** A. Delgado Roig (1964) Tiro de Línea, Sevilla; **c** A. Delgado Roig (1957) Bami, Sevilla; **d** G. Olivares James (1970) Santuario, Córdoba

The unit displayed in Fig. 2b displays load-bearing walls following the same direction of the communication core. This solution is hard to implement in units that are less than 9 m deep and it also requires the existence of a window on the façade to ventilate the main bedroom. This last feature prevents the connection of additional buildings that would, in turn, create linear blocks. In this unit the wet area is located immediately beyond the communication core, a place generally reserved for the main entrance in this typology. Consequently, the access to the unit is moved to a middle point between the bedrooms and the kitchen, after crossing an outdoor open corridor. This layout allows a 4 m reduction in the average unit depth.

The unit shown in Fig. 2c, on the contrary, displays an orientation that is parallel to the main facades. Bedrooms are positioned facing the public space, while the living room, kitchen and bathroom ventilate facing a large, inner courtyard. This layout of the bedrooms forces these blocks to increase their depth.

The best-suited housing units among Type-2s are those exemplified in Fig. 2. Regardless of the increase in the unit's depth, an efficient organization of day and night areas is established, with the entrance hall serving as the connection space between them. The night area is planned around an autonomous hall that communicates the bedrooms and the bathroom. The kitchen and living room open directly onto the main entrance hall. This layout avoids the need to reach the bedrooms crossing through the living room but forces the orientation of two of the bedrooms towards a small inner courtyard.

3.3 *Type-3 Housing Units*

Type-3 housing units (the main dimensional characteristics of which are summarized in Table 3) are defined in our study as those in which the night area is disaggregated, i.e., access to some of the bedrooms, but not to others, must be done through the day area. Once again, no characteristic distribution was found for this Type given the low number of unit types (only two) that follow this layout.

The first (Fig. 3a) has a depth of 6.9 m and a surface area of 64.78 m^2. The disaggregation of the bedrooms is due to the central position that the living room takes in the unit. The wet area is planned with the bathroom facing the living room, placed so that it ventilates through the courtyard. The kitchen is located next to the main entrance.

The second example of this Type (Fig. 3b) is 7.44 m in depth and has a surface area of 88.49 m^2. Its larger dimensions allow a different organization of the disaggregated night area. A rare solution is applied for one of the bedrooms: it is located facing the central courtyard of the building. As a consequence, additional space is available for the functional division between the living and dining rooms. Two exceptional spaces can be seen in this unit type: a built-in closet in a common

Table 3 Type-3 housing units. Basic characteristics

Average surfaces	Built	76.64 m^2	Structure	Reinforced concrete	100%
	Usable	61.28 m^2		Load-bearing walls	0%
Average depth		7.17 m		Mixed	0%
Average width		17.80 m		Prefabricated	0%
Width-depth ratio		2.5	Unique wet area		100%

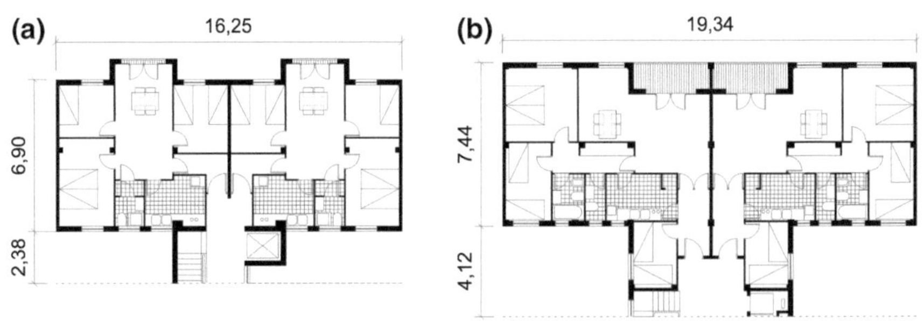

Fig. 3 Type-3 housing units. **a** J.L. Delgado Lejal (1967) Los Corrales, Cádiz; **b** M. AmbrósEscanellas (1966) Manoteras, Madrid

area and a second bathroom. The kitchen is open to flexible circulation, connecting the main entrance hall with the night area.

4 Discussion and Conclusions

4.1 H-Plan Block Chronology

The research carried out shows a gradual increase in the construction of H-plan blocks from 1950 to 1980: out of the 44 cases analyzed, only two (4.55%) were built in the 1950s, 18 (40.90%) in the 1960s, 22 (50%) in the 1970s and two (4.45%) during 1981. The explanation for this has traditionally been based exclusively on economic reasonings. Ignacio Paricio argues in his article "*Las razones de la forma en la vivienda masiva*" (Paricio Ansuátegui 1973a) that the reduction in building costs drove the typological evolution of the residential states. Paricio identifies two of the economic-based decisions that influenced this typological evolution of the blocks the most: a reduction in the length of facades and increase in the number of housing units per entryways. Indeed, the H-plan block typology produces significant savings in construction expenses in comparison to linear blocks, the most extensively used until then. The lower costs of H-plan blocks are based on these new traits: the implementation of four instead of two housing units per floor and the reduction of the thickness of the inner-courtyard façade to an average of 15 cm or even 10 cm. Paricio pointed out in his study that the construction of residential states in Spain in the seventies was monopolized by H-plan block typology. The chronological distribution shown in this study clearly reinforces that theory.

4.2 Generic Housing Unit of the H-Plan Block

It is possible to characterize the generic housing unit found in H-plan blocks built in Spain between 1957 and 1981. As demonstrated by the typological analysis, the most extensively used layout among the three-bedroom housing unit samples is that defined as Type 1 (80.55%). Type-2 and Type-3 account for a minority of layouts. Only five cases are identified within the Type-2 group and two as Type-3s.

The generic housing unit is comprised of a main bedroom (11 m^2), two secondary bedrooms (8 m^2), a living room (17 m^2), a kitchen (7 m^2) and a bathroom (3 m^2). Transition spaces are reduced to distribution areas. A second bathroom is only present in 11% of the cases.

This study has revealed that the average built surface area is 76.62 m^2, though an increase in the size of the units can be seen in the subsequent decades. In the 1950s, the average surface was 68.83 m^2, in the 1970s it rose to 79.38 m^2. Accordingly, the average usable surface area changed from 56.52 to 64.69 m^2 during those two decades. Their structure is basically made of reinforced concrete. Double bays are generally used, oriented along the main facades. Exceptionally, the load bearing structures can be found following in the same direction as the party walls. It can be stated that the units are wider than they are deep. An analysis of the width-depth ratios shows that the main facades are 1, 3 times larger than the side facades. Their average depth is 7.17 m^2. The average dimension of the inner courtyard is 5.44 m. The bathroom and kitchen are commonly built close to each other, so that 86% of the housing units have a comprehensive wet area. However, the use of a single wet area decreases slightly with time: from 93 to 78% between 1960s and 1970s.

4.3 The Architecture of the Housing Unit

This research has revealed the high level of homogeneity that characterizes the H-plan block social housing estate typology. It is remarkable that 72% of the housing units analyzed follow the exact same layout (described in this study as "Type-1 characteristic layout"). This high level of homogeneity becomes determinant when defining the obsolescence of H-plan block housing typologies in as much as it suggests that the study of a few cases can be representative of a larger sample and, therefore, solutions may be extrapolated. Even though some creative projects can be identified, the vast majority of the blocks contain a standardized-type house that fits within a parallelepiped. Housing unit design is reduced, consequently, to a matter of establishing its width and its depth and adapting a standardized layout to those parameters.

In fact, a relationship between those dimensions and the different layouts has been proven in this study: the geometry of the rooms, the location of the kitchen and terraces and the possibility of flexible circulations depend on the width and depth of the housing units.

The work presented here brings to light some important aspects of social housing architecture and aims to contribute to their regeneration by expanding the existing knowledge.

Acknowledgements This research has been supported by Universidad de Sevilla trough the V Plan Propio de Investigación (Ph.D. grant). We would like to thank Juan José Sendra and his team for their assistance in the data collection process.

References

eu2010.es/ Presidencia Española (2010, 22 de junio) Reunión Informal de Ministros de Desarrollo Urbano. Declaración. Toledo: eu2010.es/ Presidencia Española 2010

García Vázquez C (2010) Obsolescencia Urbana. El caso de las barriadas residenciales. La ciudad Viva 3:4–10. Junta de Andalucía

García Vázquez C (2015) La obsolescencia de las tipologías de vivienda de los polígonos residenciales construidos entre 1950 y 1976. Desajustes con la realidad sociocultural contemporánea

Informes de la Construcción67(EXTRA-1). doi:10.3989/ic.14.045

Hernández Aja A et al (eds) (2015) Atlas de Barrios Vulnerables de España: 12 Ciudades 1991/2001/2006. Instituto Juan de Herrera, Madrid

Jefatura de Estado (2013) Ley 8/2013, de 26 de junio, de rehabilitación, regeneración y renovación urbanas. Boletín Oficial del Estado, nº 153

Ministerio de Fomento (2013) Real Decreto 233/2013 que regula el Plan estatal de fomento del alquiler de viviendas, la rehabilitación edificatoria y la regeneración y renovación urbanas para 2013–2016. Boletín Oficial del Estado, nº 86

Paricio Ansuátegui I (1973a) Las razones de la forma en la vivienda masiva. Cuadernos de Arquitectura y Urbanismo 69:2–18

Paricio Ansuátegui I (1973b) Estudios de Tipología de La Vivienda: Entre Medianeras. La Gaya Ciencia, Barcelona

Model to Integrate Resilience and Sustainability into Urban Planning

Irina Tumini, Carolina Arriagada Sickinger and Sergio Baeriswyl Rada

1 Introduction

The cities of the future must reach a higher level of sustainability and that is only possible if they are capable of being more resilient against crises produced by natural events and global change (UNISDR 2015). Because of this, it is fundamental that governments and responsible entities develop planning tools that allow developing interaction or synergies between resilient design and sustainability.

In spite of the damage that natural disasters can produce in cities, these always offer an opportunity so that an urban system is reinvented and evolves towards a new status, improving and promoting changes that strengthen its reaction capacity. This process is always accompanied by innovation.

Currently, there are numerous studies which link sustainability with resilience (Ahern 2011; Childers et al. 2015; Brand 2009), however, difficulties still persist in the definition of terms and the provision of clear examples which serve as reference. This limits the possibility of designing integration tools and strategies between both concepts when planning the city.

This research project intends on advancing towards the integration of the concepts of resilience and sustainability in planning tools. It presents the case of post-disaster reconstruction implemented in the town of Dichato in Chile, after the earthquake and tsunami in 2010. By evaluating this case study, it can be shown that there are synergies between sustainable planning and urban resilience against natural disasters. In the conclusions, some suggestions are presented for the definition of tools that will allow transforming and adapting cities to face future changes.

I. Tumini (✉) · C. Arriagada Sickinger · S. Baeriswyl Rada
Department of Planning and Urban Design, Universidad del Bío-Bío, Concepción, Chile
e-mail: irina.tumini@gmail.com

P. Mercader-Moyano (ed.), *Sustainable Development and Renovation in Architecture, Urbanism and Engineering*, DOI 10.1007/978-3-319-51442-0_4

2 Definition of Urban Sustainability

During the last 3 decades and after the publication of the Brundtland Report (1987), sustainable development has directed international and domestic policy, leading to multiple programs, agendas and tools that are focused on urban sustainability. This is something however, that has not always generated an effective change in the development models of the cities.

To understand urban sustainability, the city can be considered as a Socio-Ecological System (SES), which must look for balance with the surroundings that support it and in its internal structure using the following three pillars: environmental, economic and social (Naredo 2003).

Starting from the ecological approach, several authors propose an approach towards sustainability of the SESs through the evaluation of the urban metabolism (Li et al. 2016; Robinson 2011). This is evaluated as the relationship between the input flow of resources, energy and information in the system, demanded to develop its functions, and the output flow in the form of heat, waste and contaminants which are finally emitted into the environment.

In the search for a city model which balances urban metabolism with its surroundings, the compact European city is acknowledged as a good example of sustainable cities, aided by its contained structure, the mix of uses and services, and the dynamism of the public spaces which facilitate efficient transportation systems. Starting from this model, the definitions and action directives in urban planning have been strengthened, both at a city level and on a neighborhood scale (Messerschmidt et al. 2008; Fariña Tojo 2008).

2.1 Sustainability Indicators

Currently, there are numerous systems of indicators developed by different entities, some of which have allowed defining directives or evaluation systems which have been incorporated into urban design and planning. For the compact European city, one of the most relevant systems is the CAT-MED sustainability indicators system, developed through a European research project of the "Climate Change and associated natural risks program" (Marín Cots 2012). This system understands the average European city as a consolidated morphological structure contrasted against urban sprawl; a model which has increased the imbalance between built space and the environment (Marín Cots 2012; Turégano Romero 2009; Tumini 2016). These indicators are organized around four key concepts:

Compactness is the parameter which deals with the physical reality, being directly related with the building density, the soil use, the amount of green spaces and the existing roads. This parameter looks to evaluate the proximity between urban uses and functions, seeing the built volume associated to the provision of equipment and public spaces as a whole.

Complexity deals with urban organization for the mix of uses and functions in the area. Urban complexity reflects the interactions that exist in the urban space and that mirror the city's vitality. This parameter is linked to the concept of urban diversity, shows the maturity of the urban fabric and the wealth of the economic, social and biological capital.

Metabolic efficiency is a concept related to the flow of materials, energy and information that the system exchanges with its surroundings. The sustainable city must reach efficient management of the incoming resources and reduce the emission of contaminating products as much as possible.

Social cohesion refers to the capacity of the cities to satisfy their role as a motor of social progress, economic growth and as a space for the development of democracy. For this, it is necessary to maintain the social balance, both at an urban and interurban level, protecting cultural diversity and co-existence between the players. In this sense, the success of the urban space is in creating the conditions which promote opportunities for meeting and exchange, facilitating co-existence, thus making the reduction of conflicts possible. In the urban design, social cohesion can be fostered using the concept of *proximity* as an expression of the city's vicinity. Fariña (200) defines proximity as that where the urban surrounding has a domestic nature, close to the home, well distributed into urban grids, and multi-functional. An urban "proximity" design promotes a different management of the space, pedestrian movement, local stores, contact with the people and the proximity of equipment and roles (Marín Cots 2012; Tumini 2016; Rueda 2012; Fariña Tojo 2009; Robinson 2011).

3 Definition of Urban Resilience

In the literature, references of the concept of resilience can be found in different areas, each one providing a more suitable definition for its application. For the urban setting, two main approaches can be recognized: engineering and ecological. From the engineering point of view, resilience is the capacity of a system to resist a disturbance, mitigate the effects and return to the point of stability once the event has ended. This definition refers to the system's "resistance" and "elasticity" (Brand 2009).

The ecological approach is based on observing the response of the natural systems under the action of a disturbance, of how certain structures mutate, sometimes some species disappear and are substituted by others, reorganizing roles and relationships between them (Holling 2001; Folke 2006). Applied to the SES, resilience is defined as "the capacity of the systems: cities, communities or societies exposed to threats to efficiently resist, absorb, adapt or recover from the effects of the threats in a reasonable time, including the maintenance and recovery of their basic structures or functions" (Jabareen 2013:221).

Susan Cutter (2003), in her studies on vulnerability, relates resilience of the SES with the geographic conditions. Her research is based on the hypothesis that it is possible to associate vulnerability to spatial patterns and that once identified, these form the directives of the adaptability of the urban system (Cutter et al. 2003; Cutter et al. 2014); Allan and Bryant (2011) in their studies about post-disaster recovery, acknowledge that the urban setting offers a series of resources that can be used in the emergency phase and that help the recovery. Therefore, it is possible to measure urban resilience as the capacity of the built environment to adapt to the changes caused by the natural events, facilitating useful resources for the early recovery of the functionality (Allan and Bryant 2011; Bryant and Allan 2013).

The extent that sustainable urban structure can contribute to the resilience depends on its capacity of resisting and mitigating the impact of the events and its flexibility towards change, taking advantage of working in a network and the capacity of reorganizing the structures and resources available. For this, the existing models must consider the risk factors and unpredictability of the events, the interaction between levels and dimensions. In this way, tools will be generated that can evaluate the status of the system and warn about the critical aspects as well, thus aiding in making decisions about preventative actions (Milman and Short 2008).

3.1 System of Indicators to Evaluate Urban Resilience

The approach to resilience by studying the SES, proposes that the urban system is capable of providing resources for innovation, to be adaptable and/or redundant. The phase immediately after the crisis, i.e. the emergency, is where the biggest changes are produced, because the system must reorganize itself as quickly as possible to start working again.

As a result of this, resilience analysis is organized around four main attributes:

- *Diversity* in terms of structures and functions, characteristic which allows guaranteeing a diversity in the response to the disturbance, in the options and in the resources available for the recovery (Walker et al. 2015).
- *Flexibility* which describes the capacity of the urban elements to adapt to the changes and to assume new roles (Allan et al. 2013; Walker et al. 2015; Villagra et al. 2014).
- *Connectivity*, essential to maintain the functionality. In the case of the SES, it is important to maintain both the physical connectivity and the exchange of information that is produced among the individuals of the same community or between different communities.
- *Modularity* is the fourth requisite which is related to Diversity and Connectivity. Resilient urban systems must be capable of shaping themselves as a modular system, in which the modules can work independently and simultaneously be connected in a network with the others. In this way, if one module collapses due to the crisis, the others can provide resources and services for the recovery.

The evaluation of these attributes can be done through the urban design indicators where we can recognize the synergies with the key concepts of sustainability.

4 Research Methodology

The research work proposes an approach to sustainable urban planning for the cities exposed to natural risks and that due to this need to improve and increase their resilience and simultaneously assure the quality of life for their citizens.

As a method, it is proposed to relate the concepts of the CAT-MED compact European city (compactness, complexity, metabolic efficiency and social cohesion) with the resilience attributes (diversity, flexibility, connectivity and modularity), setting out a system of urban design indicators that will be evaluated in the case of the post-disaster reconstruction of the town of Dichato (see Fig. 1).

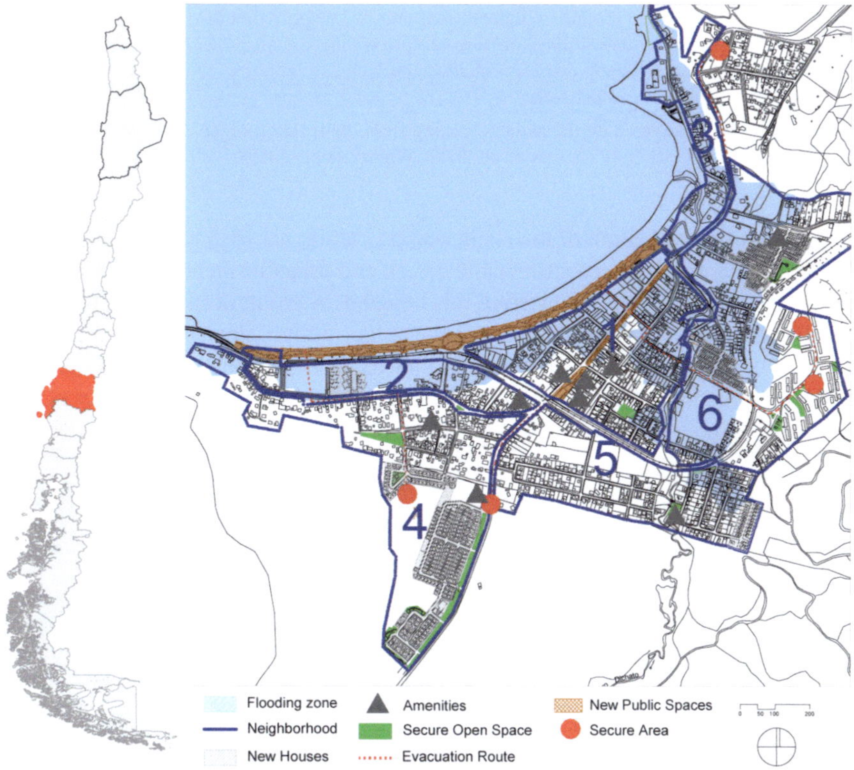

Fig. 1 Reconstruction of Dichato. Neighborhoods: *1*. Centro, *2*. Litril, *3*. Villarica, *4*. Posta, *5*. Santa Alicia, *6*. Villa Fresia. *Source* own preparation from the cartography of Tome Council and from PRBC18

Table 1 Table of resilience attributes and indicators

Resilience attributes	Synergy with sustainability	Urban design indicators
Diversity	Is mainly related with the concept of urban *complexity*. Assures *metabolic efficiency* on reducing displacements	Open spaces in safe areas: m^2 open spaces/inhabitants
		Public buildings in safe areas: m^2 built/inhabitants
Flexibility	*Compactness* in the city defines the balance between the spaces built and the open spaces, to avoid sprawl and congestion. Assures *metabolic efficiency* on reducing the displacements	Population density: inhabitants/hectares
		Urban Compactness (corrected): m^3 built/m^2 public spaces
Connectivity	Proximity between persons and services facilitates connectivity and *social cohesion* in the community Guarantees *metabolic efficiency* on promoting a more efficient mobility	Proximity Index: percentage of inhabitants with access to at least 1 utility
		Walkability Index: percentage of spaces and sidewalks over the total number of roads
Modularity	Organization in neighborhoods allows that every module is independent and can provide resources in case of crises. Improves the *metabolic efficiency* by operation in network	N° of independent and resilient neighborhoods

The table has been prepared by the authors starting from the revision of: (Allan et al. 2013; Cutter et al. 2003; Marín Cots 2012; Norris et al. 2008; Walker et al. 2015)

Through the evaluation of the resilience indicators, we wish to evaluate whether there are synergies between sustainable design and urban resilience and whether the current planning tools allow meeting this demand. A group of indicators is defined below, orientated to the evaluation of the urban design elements which have an impact on the resilience. This work is limited to the evaluation of the physical resilience dimension, because, among the different dimensions that make up the resilience of the community, this is the one that depends directly on urban planning (Table 1).

4.1 Case Study: Dichato

On February 27th 2010, an earthquake of 8.8 on the Richter Scale and a tsunami generated by this, affected five regions of the central-southern zone of Chile. Many of the coastal cities were severely damaged by the event, along with small fishing villages and touristic areas, which were the zones that saw the greatest destruction.

After the catastrophe, the Government set a reconstruction process in motion in these coastal areas. In the case of the Biobio Region, the process was organized and handled by the Regional Government, which proposed an integrated management program with the idea of rethinking the cities from a more complex and cross-sectorial view (Baeriswyl 2011). The city of Dichato experienced major intervention, becoming a reference of the Coastal Reconstruction Plan (GSAPP 2015; Bio-Bio Gobierno Regional 2010) (Map).

Dichato is part of the district of Tome and is set in Coliumo Bay. Its geography makes it especially vulnerable to tsunamis due to a combination of geographic factors, including the low altimetry of the town center and the form of the bay which amplifies the hydrodynamic effects of a tsunami. Likewise, the low resistance of the buildings, mainly consisting of wooden one story houses with no foundation, increased the town's vulnerability.

The losses were substantial, both in number of homes and in urban services and infrastructure. Along with the material losses, the production sector, mainly associated to fishing and tourism, was seriously compromised. The town's reconstruction was a greater challenge due to the urgency of acting in an integrated manner in the urban system, mobilizing the resources needed for the economic and social recovery of the affected communities (Cartes Siade 2013; GSAPP 2015).

The reconstruction project proposed the replacement of the existing equipment and a significant increase of the green areas and public spaces. Different mitigation measures were prepared as a coastal defense, a promenade and a mitigation forest, with the purpose of reducing the hydrodynamic force of a possible future tsunami and thus, its impact on the city. The Master Reconstruction Plan proposed changes in the soil use, relocating critical equipment into safe areas (schools, fire station, police station and health services). In the case of the residential areas, when their relocation to a safe height was not possible, resilient homes were proposed which comply with higher structural design standards to facilitate their reconstruction (Baeriswyl 2013).

For the evaluation of the case study, the indicators in six neighborhood units of the city have been analyzed: Centro, Litril, Villarrica, Posta, Santa Alicia and Villa Fresia. The data used only considers the infrastructure above the flood zone, as all the infrastructure below this height cannot be considered as a useful resource for recovery (see Fig. 1).

5 Results and Discussion

The values collated are presented in Table 2 and refer to the configuration set out in the Master Reconstruction Plan. The data provided by the 2002 census, information collected onsite, cartography of the Tome Council database and the information provided by the PRBC18 Coastal Urban Reconstruction Plan were used to prepare these values. The data has been prepared using the GIS program.

The data shows that after the reconstruction process of the Centro neighborhood, it is still the one with the highest urban density. Litril, Centro and Villarrica, located in the flood zone and greatly affected by the 2010 tsunami, in spite of not seeing

Table 2 Collation of indicators in the dichato reconstruction phase

Indicador	Litril	Centro	Villarrica	Posta	Santa Alicia	Villa Fresia
Population Density (inhab/h)	18.71	41.13	82.94	18.89	35.84	50.18
Open Space (m^2/inhab)	0.00	0.00	3.89	38.84	19.80	11.28
Public Buildings (m^2/inhab)	0.00	0.00	0.00	0.72	0.00	0.00
Urban Compactness (m^3/m^2)	0.00	0.00	0.02	0.37	0.18	0.18
Walkability Index (%)	0%	0%	0%	100%	100%	100%
Proximity Index (%)	0%	0%	0%	86%	23%	0%
Population Density (inhab/h)	18.25	91.84	21.58	10.12	29.17	38.34
Open Space (m^2/inhab)	0.00	3.36	3.89	46.63	1.04	3.53
Public Buildings (m^2/inhab)	2.60	1.10	0.00	46.17	0.80	1.70
Urban Compactness (m^3/m^2)	0.00	9.07	0.00	9.13	101.10	12.35
Walkability Index (%)	100%	100%	100%	83%	100%	100%
Proximity Index (%)	100%	100%	100%	100%	100%	100%

Pre-tsunami and post-reconstruction
Own preparation from the data provided by the Master Plans and the Tome Council. The pre-tsunami values are presented only as references because the objective of paper is assessed the synergy between sustainability and resilience in the reconstruction project

major changes in the population density, have not improved in terms of equipment. In the neighborhoods located at higher levels, like Posta, Santa Alicia and Villa Fresia, these received new buildings and thus increased their population density.

In terms of the amount of open spaces and public buildings that are useful for the emergency, the values are very low and insufficient still for the resident population in five of the six neighborhoods analyzed. The data shows that the location of the new equipment has been set up in the Posta neighborhood. The highest number of open spaces and services is concentrated in the Posta neighborhood; this is because the neighborhood is in the safety area above the flood zone.

In the evaluation of the corrected urban compactness, Litril has a complete lack of open spaces and Villarrica a lack of public buildings. Santa Alicia has a very high urban compactness value, because it has a high volume of constructed buildings and very little open space. In a post-disaster scenario, the population of Santa Alicia would have great difficulty to organize the emergency. From the point of view of sustainability, the green space greatly reduces the environmental impact produced by the urbanization, while it provides meeting places and spaces for social cohesion.

In regards to connectivity, the neighborhoods comply with the proximity and pedestrian route requirements, demonstrating good connectivity both inside the neighborhood and in the rest of the system. The Posta neighborhood has limitations in terms of pedestrian connectivity when compared to other neighborhoods of the system.

Thus, the assessment shows that the only Posta is sustainable and resilient neighbourhood, because it provides open spaces and basic services useful for emergency activities and recovery. Despite, the lack of connectivity with the other neighbourhoods could limit the networking capacity by providing spaces and services to citizens that live in less resilient neighbourhoods.

In terms of the objectives of the Master Reconstruction Plan in regards to environmental improvements, this has been partially met from the perspective of this analysis matrix. Although the reconstruction process implied a considerable increase of the amount of services and public spaces, the locating of these was not balanced in the whole urban area, producing big contrasts between neighborhoods and something that without a doubt affects the city's resilience and sustainability conditions.

The neighborhood which has the best behavior in terms of its resilience is Posta. In fact, in case of emergency, this neighborhood could continue working and provide useful resources for the recovery of others with fewer equipment; however, the lack of connectivity may be an important limitation. In terms of the modularity and operation in the system's network, the town of Dichato maintained its original fabric after the reconstruction process, as such it did not improve this aspect, which could have been provided a greater resilience.

6 Conclusions

The work presented proposes a quantitative approach to analyze sustainability and resilience in the urban space. Through the analysis of the post-disaster reconstruction of Dichato, the idea is to demonstrate that sustainable urban design also contributes to improving the resilience of risk-exposed towns, placing value on synergies between the two approaches.

From the results obtained, it can be concluded that there are synergies and differences between the two approaches, as such it is necessary to adapt the tools and models orientated to sustainable design in order to integrate the concepts of resilience and adaptability to facing natural events.

In the case analyzed, a general improvement of the sustainability and of the resilience of the urban whole is obtained, underlining some disagreements in concrete aspects. The first is the imbalance in locating public spaces and green areas, which are mainly concentrated in the Posta neighborhood. On considering these spaces as useful resources for the recovery (Bryant and Allan 2013; Allan et al. 2013; Villagra et al. 2014), it is necessary to even out the amount of equipment in all neighborhoods. The second is related with the deficient connectivity in this neighborhood, which could make it difficult for them to provide services and resources to the others in case of emergency. This leaves it clear that, to reach the resilience goals, sustainable design processes must also incorporate the vision of modularity and operation in the urban system network (Walker et al. 2015).

References

Allan P et al (2013) The influence of urban morphology on the resilience of cities following an earthquake. J Urban Design, 18(October 2014), pp 242–262. Available at http://www.tandfonline.com/doi/abs/10.1080/13574809.2013.772881

Allan P, Bryant M (2011) Resilience as a framework for urbanism and recovery. J Landscape Arc 6:34–45

AHERN J (2011) From fail-safe to safe-to-fail: sustainability and resilience in the new urban world. Landscape Urban Plan 100(4):341–343. ISSN 01692046. doi:10.1016/j.landurbplan.2011.02.021. Disponible en: http://dx.doi.org/10.1016/j.landurbplan.2011.02.021

Baeriswyl S (2011) Plan de Reconstrucción del Borde Costero—Región del Bío Bío. Revista CA, No 145:62–68

Baeriswyl S (2013) La experiencia de Reconstrucción Urbana del Borde Costero de la Región del Bío Bío. Edición especial Revista Escala, Bogotá Colombia

Bio-Bio Gobierno Regional (2010) Plan de Reconstrucción del Borde Costero. Plan Maestro Dichato

Brand F (2009) Critical natural capital revisited: ecological resilience and sustainable development. Ecologi Econom 68(3):605–612. ISSN 09218009. doi:10.1016/j.ecolecon.2008.09.013. Disponible en: http://dx.doi.org/10.1016/j.ecolecon.2008.09.013

Bryant M, Allan P (2013) Open space innovation in earthquake affected cities. Available at: http://www.intechopen.com/books/approaches-to-disaster-management-examining-the-implications-of-hazards-emergencies-and-disasters/open-space-innovation-in-earthquake-affected-cities

Cartes Siade I (2013) Dichato desde la crisis a la reconstrucción. Un modelo de gestión de riego y resiliencia. Urbano: 33–40

Childers D et al (2015) An ecology for cities: A transformational nexus of design and ecology to advance climate change resilience and urban sustainability. Sustainability, 7(4):3774–3791. ISSN 2071–1050. doi:10.3390/su7043774. Disponible en: http://www.mdpi.com/2071-1050/7/4/3774/

Cutter SL, Boruff BJ, Shirley WL (2003) Social vulnerability to environmental hazards. Soc Sci Q 84(2):242–261

Cutter SL, Ash KD, Emrich CT (2014). The geographies of community disaster resilience §. Global Environ Change, 29: 65–77. Available at: http://dx.doi.org/10.1016/j.gloenvcha.2014.08.005

Fariña Tojo J (2008) Ecocity, manual para el diseño de ecociudades. http://www.elb.com/

Fariña Tojo J (2009) Parques de proximidad en el siglo XXI. blogspot. Available at: http://elblogdefarina.blogspot.cl/2009/03/parques-de-proximidad-en-el-siglo-xxi.html Accessed 29 Nov 2015

Folke C (2006) Resilience: the emergence of a perspective for social-ecological systems analyses. Glob Environ Change 16(3):253–267

GSAPP (2015) Learning from 27F: a comparative assessment of urban reconstruction processes after the 2010 eathquake in Chile, Santiago

Holling CS (2001) Understanding the complexity of economic, ecological, and social systems. Ecosystems 4(5):390–405

Li Y et al (2016) Evaluating urban sustainability potential based on material flow analysis of inputs and outputs: a case study in Jinchang City, China. Resour Conserv Recycl 110:87–98

Marín Cots P (2012) Modelos urbanos sostenibles. Metodología de trabajo y resultados, Malaga

Messerschmidt R et al (2008) Proyecto ECOCITY. Manual para el diseño de ecociudades en Europa. Bakeaz, ed., Bilbao

Milman A, Short A (2008) Incorporating resilience into sustainability indicators: an example for the urban water sector. Glob Environ Change 18(4):758–767

Naredo JM (2003) Instrumentos para paliar la insostenibilidad de los sistemas urbanos. Boletín CF S, (24)

Norris FH et al (2008) Community resilience as a metaphor, theory, set of capacities, and strategy for disaster readiness. Am J Community Psychol 41(1–2):27–150

Robinson D (2011) Computer modelling for sustainable urban design: physical principles, methods and applications routledge. ed., Earthscan. Available at http://books.google.es/books?id=SiLOSAAACAAJ

Rueda S (2012) Certificación del urbanismo ecológico

Tumini I (2016) Acercamiento teórico para la integración de los conceptos de Resiliencia en los indicadores de Sostenibilidad Urbana. Revista de Urbanismo 34:4–19

Turégano Romero JA (2009) Arquitectura bioclimática y urbanismo sostenible de Zaragoza PU (ed)

UNISDR (2015) Marco de Sendai para la Reducción del Riesgo de Desastres 2015–2030, Available at: http://www2.ohchr.org/spanish/bodies/hrcouncil/docs/gaA.RES.60.1_Sp.pdf

Villagra P. et al (2014) A GIS-base exploration of the relationships between open space systems and urban form for the adaptive capacity of cities after an earthquake : the cases of two Chilean cities. Appl Geo 48: 64–78. Available at: http://dx.doi.org/10.1016/j.apgeog.2014.01.010

Walker B et al (2015) General resilience. A discussion paper based on insights from a catchment management area workshop in south eastern Australia: 1–14

Opportunity Detection of Empty Architectonical Lands and Their Recycle for a More Sustainable City

Santiago Carroquino Larraz

1 Reasons for the Recycling of Empty Architectonical Lands

1.1 Biocapacity and Productive Sectors

The awareness that the energy and resource consumption rate exceeds the capacity of regeneration of the planet marks a fundamental milestone which should affect all kinds of human activity. The planning and management of the territory therefore, as support system and matrix of human relationships on the planet, must be strongly influenced by such knowledge. Inefficient and unsustainable architecture and urbanism appear between the reasons why biocapacity has been surpassed.

The measurements of the whole planet's ecological footprint realized in the year 2000 show a result of overconsumption which vary between 31 and 20% according to the indicators used (Wackernagel et al. 1997). The scientific confirmation of the planet over-exploitation and the social perception of such reality have created consequently a sustainability demand, understanding this as the restriction of the contaminating capacity and the optimized control of energetic resources in the development. From the result of the new mentality different international commitments were born, with the Kyoto Protocol of 1997 being the first legally-binding agreement. This marks the environmental target of not exceeding in more than 2 °C the increase of global temperature with respect to preindustrial levels.

Within the European Union, the Europe 2020 strategy defines the triad of 20/20/20, consisting of the reduction of greenhouse gases at least 20% compared to 1990, the increase of the percentage coming from renewable energy sources in the final consumption more than a 20% and the improvement of the energy efficiency

S. Carroquino Larraz (✉)
Universidad de Zaragoza, Zaragoza, Spain
e-mail: scarroq@unizar.es

P. Mercader-Moyano (ed.), *Sustainable Development and Renovation in Architecture, Urbanism and Engineering*, DOI 10.1007/978-3-319-51442-0_5

over 20%. Regarding this guidelines, for the year 2020 Spain must have achieved a reduction of 10% in the emissions with respect to 2005 in the diffuse sectors. Considering such as: transport, residential, commercial, institutional, agrarian, waste and fluorinated gases (Asunción et al. 2009).

The strategy Europe 2020 includes various measures in the area of social inclusion with the aim of reducing the number of people at risk of poverty and social exclusion at least 20 million. In the case of Spain the expected drop is 1,500,000 people, reaching an employment of 75% for the people between 20 and 64 years old. Likewise, some recommendations are made from the European Commission to Spain after confirming that social exclusion is increasing and social protection for reducing poverty is of limited efficacy.

1.2 Need to Reconsider Habitability

The progress towards a lower ecological footprint in the construction sector which assures a sustainable growth trying to balance the planet biocapacity involves the redefinition of habitability concept, issue of the sector, concerning sustainability. It became necessary to rethink and redefine the group of activities intendend to produce and maintain such habitability, necessary on the other hand, to accommodate the development of social and individual activities (Casals-Tres et al. 2013). All of this involves adjusting the resources used in obtaining it, recovering and optimizing the ones already consumed and the constructions already built. Considering "Urban unsustainability" as the existence of buildings and building sites without user or led to deterioration due to obsolescence, marks a new satisfactor to solve for an efficient city.

The need of sustainability affects the competitivity of services and the pretension to provide cities with habitability. Apparently the only possible solution to be able to increase sustainability is increasing the efficiency, and this involves making more things work with less. To do this we should not waste energy and land, nor pollute with unnecessary or expensive displacements, or with the warming up or cooling down of bioclimatically ineffective buildings. In the global scene today a more efficient operation in our urbanized areas is urgently needed (Fariña Tojo 2013).

1.3 Energy Costs of Construction

The main component of the energy consumption in construction is due to the use of the building itself, in other words, because of the Operational Energy necessary to the regular functioning of the property. It is therefore necessary to improve and renovate energy wasteful buildings to achieve savings in the regular operation. The second major factor is the one due to the integrated energy in the building construction, the so called Embodied Energy for being the energy cost that materials include in their development. This concept not only includes the installation of materials but also every previous process: the obtaining of the raw material, its transformation and subsequent transport to the construction site.

The studies carried out by López-Mesa and others have demonstrated for the case of social housing, how after an appropriate renovation the post_renovation operational energy in 25 years is below 5% of the operating energy before the renovation for the same period of time. Being the energy costs of construction and renovation an 8.4 and a 9.6% respectively. That it to say, the operational energy pre_renovation means a 77% of the total, which shows the energy waste in this life phase of the building. While in the curriculum of the building other potential energy costs exist associated with renovation, life of the renovated building and final demolition, the ones associated with the Operational Energy and Embedded Energy are of greater importance.

With regard to the embodied energy because of the construction, this can reach a 40% of the total. This is, however, the case of buildings with a high construction cost and low energy consumption, more likely to appear in construction conceived with "Passive House" concepts or in others with a short service life, than in buildings realized with regular construction standards and common service life. In any way it becomes clear that extending the life of buildings is a sustainable practice due to the reuse of the embodied energy while increasing their life cycle (López-Mesa et al. 2013).

1.4 Effects of Population Access in Energy Costs

Being the previous parameters typical of every construction, in the case of urban services and facilities it becomes necessary to weigh the impact of energy consumption per number of served population. One facility that serves double population reduces its energy consumption per person by half.

The concept of population access, becoming clear from the point of view of urban installations whose optimization is proportional to the number of users, is easy to assimilate in facilities. A service building used by a wide sector of society is an optimized architecture. By contrast, a not inhabitated construction and not valued by use, or utilized by few inhabitants, under-uses the embodied energy. The energy savings are therefore directly proportional to the number of potential served users, defining this as population access.

A sustainable development should be proposed with equality criteria in its social aspects so the benefits reach a higher number of citizens, specially those residing in vulnerable areas and the most economically disadvantaged (Gallegos Ferrer 2009).

It is possible to extrapolate this criteria to urban renovation, since when having an effect on a higher number of population it becomes more efficient. In the words of Tejedor Bielsa the processes of densification which lead to a more efficient urban transformed land, properly regulated could be a magnificent path to urban, social and economic recovery, optimizing the use, generating quality housing, and with services and urban provisions integrated in the urban space (Tejedor Bielsa 2013).

1.5 Urban Revitalization as Improvement of Habitability

Leal Maldonado differences between two types of interventions in residential space in terms of public space: the transformation of the public space itself and the transformation of the residential conditions in themselves. In the first case the effects on residential space are indirect, to the extent that the improvement of the environment affects the condition of the housing, in its image, in its value and, specially, in the life conditions of its residents. In the second case the improvements are established in three intensity and consequence grades (Leal Maldonado and Sorando Ortín 2013). Analyzing the different virtues of all cases and subcases, we can conclude that urban renewal processes which act on public spaces and facilities, while from the point of view of general energy savings represent a lower amount than the ones applied to wide sectors of residential fabric, from the perspective of social cohesion are clearly more satisfactory and in the end they improve energy efficiency in the city. They are therefore of high interest for the sustainability in the city, provided that its population access is appropriated.

In any case, the existence of urban facilities without a use causes a general rejection from the population in the surroundings being possible that it degrades into vulnerability situations, acting on them will be therefore of benefit for society. The contemporary city has multiple service buildings in the need of recycling and renovation in its constructive solution as well as in the offer of the program. Recycling, understood as the repeated subjection of a material to the same cycle, bring us to a new use of obsolete architectures. Renovation, understood consistently with sustainability implies the adaptation of constructive solutions to a greater efficiency. Urban sustainability, consequently, will lead to the activation of this architectures promoting their use and inverting the degradation of the empty spaces to optimize the urban fabric.

2 Study of Tools for the City of Zaragoza

Once established the convenience of recycling empty of use architectures it becomes necessary to review the tools we have to detect opportunity spots. The methodology will consist of a study of the urban areas with needs and the possibility to resolve them with reoccupation and starting up of constructed voids. To that end three groups of documents will be used: Planimetry obtained from the

Atlas of Urban Vulnerability of Spain implemented on the city of Zaragoza, Mapping of empty of service islands and accessibility to such services, obtained from the programm *Zaragoza manzana a manzana*; and finally the Map of Municipal Spaces with Obsolescence of use, generated according to the information obtained from the City Council of Zaragoza through the program SIARQ.

2.1 Vulnerability Indicators

The Atlas of Urban Vulnerability of Spain, which is a part of the project from the Vulnerability Observatory carried out by the Ministry of Development, allows to analyze Urban Vulnerability in the level of census section in every municipality of Spain, this being done from the data present in the population and housing census. Thanks to the possibilities of the tool, the study has been focused in the city of Zaragoza, ensuring that the comparative contexts, when possible, refer to the municipality itself or, failing that, to the Autonomous Community of Aragón. For each census section it offers a total of 96 vulnerability indicators and indexes organized in 4 domains: Inticators of Urban Vulnerability, Contextual Analyse of Urban Vulnerability, Indexes of Urban Inequality and Synthetic indexes of Urban Vulnerability. The planimetry of the Contextual Analyse of Urban Vulnerability compares such indexes of the census section with the municipal, regional and national context. In our case we will use the municipal context with the purpose of developing proposals according to the immediate environment.

2.2 Mapping of Empty of Service Spaces and Accessibility to Them

For the detection of empty of service islands in the city of Zaragoza the cartography made by EBRÓPOLIS within the framework of the Urban Observatory of Zaragoza and its environment has been used, developing an own in which the areas in the city without the analysed service or facility will be emphasized. The planimetry is generated considering ranges of accessibility in accordance with immediate access and pedestrian movement distances for facilities of daily or occasional use. This planimetry is also complemented with the implementation of a plan of the city of Zaragoza in which the density of population data is provided according to the darker or lighter gradation of tone in the block. On this topography, and through self elaboration, the existence of equipping voids is highlighted. To that end, a shading of the islands out of reach for accessibility to services according to the relationship type of facility-needed accessibility, is realised. This offers a better visualization of the empty spaces allowing a more detailed analyse when combined with others.

2.3 Map of Municipal Buildings with Obsolescence of Use

With the purpose of a practical application of the case of study in the present work, counting with a database of urban facilities with obsolescence of use becomes necessary. To that effect there has been a collaboration with the Architecture Conservation Service of the City Council of Zaragoza, which from October 2009 has been developing the Director Plan of Facilities, starting with the analyse of existing services and facilities. The work with the tool SIARQ, outcome from such works, shows a result of 63 municipal spaces without identified use. Even though installations are of different format and importance, remarkably is however the amount of potential occupancies whose starting up, properly renovated, would make the city of Zaragoza more sustainable. After conducting a short study of the data provided by SIARQ, the number of recyclable municipal spaces is reduced to 56.

3 Methodological Proposal

Once presented the documents to use, the work methodology is divided in two phases: Analyse of the demand and Case Studies. In the first phase, by virtue of the vulnerability plans, specially delicate areas in the city will be detected according to the indicators. Continuously, the accessibility to facilities will be studied, detecting the empty of service islands. Finally, once both data is combined and the need determined, the possibilities of starting up close constructions contemplated in SIARQ will be reviewed. Of this last aspects will consist the case studies in the city of Zaragoza.

3.1 Analyse of Vulnerability Indicators

From the multitude of vulnerability indicators, or combinations, which the Atlas offers it is summarized, as an example of the whole consulted, the results of four of them in the area of the city of Zaragoza.

The IDSM Municipal Socioeconomic Inequality Index is the weighted sum of the Basic Urban Vulnerability Indicators of % of population out of work and % of population without studies. In the analyse, various specially vulnerable areas are detected. Highlighted are: the census section of the surroundings of Conde Aranda street; the adjoinings of Francisco Rallo and Jerónimo Cáncer streets; the surroundings of Orense street, the surroundings of Madrina Salinas street and the ones of Robert Baden-Powell street. In this areas social facilities are specially necessary, such as Civic Centres, Municipal Councils and libraries.

The ISVUR-R Synthetic Urban Vulnerability Index according to Residence Criteria is a weighted sum of all Residential vulnerability indicators: housing area, provision of toilets, year of construction, etc. The analyse shows a devastating landscape. Problems of vulnerability exist in the east zone as well as in the west one, in the historic centre, and in the districts of Vadorrey, La Jota and Picarral. The poor condition of the residential stock in the districts of Oliver, Valdefierro and Torrero-La Paz is pressing. By contrast, the good conditions of the residential stock are clear in the central strip North-South, cut by the historic city. In the vulnerable areas, what is needed is a residential renovation, to that end municipal councils are needed as public management instigators.

The ISVUR-S Synthetic Urban Vulnerability Index according to Subjective Criteria is measured by the weighted sum of the five subjective vulnerability indicators. This values the perception of the inhabitants of Zaragoza from the census section where they live with regard to topics such as cuantity of Green areas, delinquency, noise, etc. As those are subjective values, the dissatisfaction of a high percentage of people with its immediate environment is significant. The coincidence of satisfied areas in the city with the areas where exhibitions rooms, cinemas and theaters are located is curious. Even though the indicator is submitted to the subjectivity of the surveyed, and therefore the result appears to be a little arbitrary, perhaps the lack of this facilities is a clue of the uses to locate.

Finally the ISVUR-G Synthetic Urban Vulnerability Index according to Global Criteria is measured by a weighted sum of the twenty Urban vulnerability indicators. Being a global indicator it serves to detect vulnerability areas in the city in general and without an enhanced criteria. As this is a summary of all previous indicators, the previous vulnerable areas appear again: Historic Centre and Oliver, Valdefierro, Torrero-La Paz, Vadorrey, Picarral and Actur Norte districts.

3.2 Analyse of Accessibility to Facilities

The website of the program *Zaragoza manzana a manzana* allows to visualize the accessibility to diverse facilities for the city of Zaragoza. It also allows some study options of the accessibility in combination with various. In our case, to analyse the lack of facilities, we should obtain the negative of the accessibility plan. To that end the areas of the city out of the different accessibility radius to facilities have been highlighted, that is to say: The Empty of Service Islands. Finally, as in the indicators case and to collate the recycling options, I have placed the list of reciclable municipal spaces obtained from SIARQ listing. The process has been developed for 10 types of facilities selected between the range of possibilities that the program *Zaragoza manzana a manzana* offers. The following have been chosen: Health Centres, Children Educational Centres, Civic Centres, Sport Centres, Libraries, Exhibition spaces, Museums, Cinemas and Theaters, Municipal Councils and Small Bussinesses.

4 Case Studies in the City of Zaragoza

In the case studies of the city of Zaragoza the process will be from the general to the particular, addressing the recycling of empty architectonical lands for a more sustainable city.

In the first one we will establish the area with greater needs according to the concentration of vulnerability indexes. Subsequently, for that area, the accessibility to facilities of interest will be studied. As a conclusion, the two values will be collated with the list of recyclable municipal facilities, establishing its suitability according to the potential of the construction to reuse.

In the second one, the recycling potentiality of an architectural emptiness will be detected, in accordance with the coincidence of this space in various empty of service islands. Afterwards, its inclusion in vulnerable areas will be studied, proposing the type of facility more suitable. At last, the adequacy of the program will be reviewed in accordance to the the construction to start up.

Finally other options and possible lines of investigation, consequence of the study of Vulnerability Indicators and Void Islands, will be indicated.

4.1 Case of Study According to Vulnerability Indexes

Being the more general indicator the ISVUR-G (Fig. 1) when reviewing the 20 Vulnerability indicators, we will use it to develop a first scan of the city of Zaragoza detecting greater vulnerability areas. From this ones, the districts of Hustorical Centre, Delicias, Oliver, Valdefierro, Torrero-La Paz, Las Fuentes Este, Vadorrey, Picarral and Actur Norte stand out. In all of them a big number of vulnerable census sections appear. The Indexes ISVUR-R and ISVUR-S repeat the same areas as the most vulnerable. By contrast, the analyse of more specific indexes exclude some of the previous districts. The IDSM, excludes as vulnerable the districts of: Delicias, Valdefierro, Vadorrey and Actur Norte; keeping as delicate: Historical Centre, Oliver, Torrero-La Paz, Las Fuentes East and Picarral. El IDUM excludes as vulnerable almost every census section with the exception of the ones included in the districts of Historical Centre, Oliver and Las Fuentes East. The ISVUR-SD excludes also a big number of districts maintaining the attention in: Historical Centre, Delicias, Torrero-La Paz, Picarral and Actur Norte. Finally the ISVUR-SE reiterate the districts of the previous indicator adding Oliver and Valdefierro disctricts. As a summary of the observation of the more general vulnerability indexes it is inferred the reiterated appearance of the Historical Centre district and specially the census sections of the surroundings of San Blas street.

Regarding specific Vulnerability Indexes concerning the percentages of: foreign population, child foreign population and popoulation out of work; the area's vulnerability is reiterated. By contrast, the percentage indicator of single households of

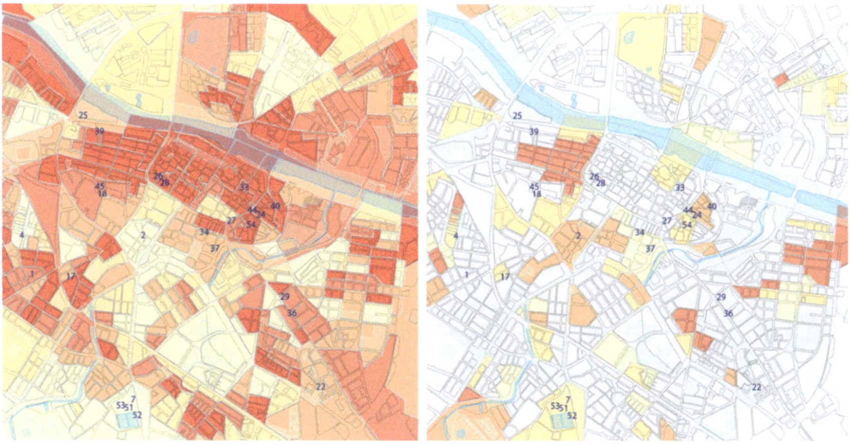

Fig. 1 Detail of the plans ISVUR-G and IDSM. *Source* Atlas Vulnerabilidad Urbana

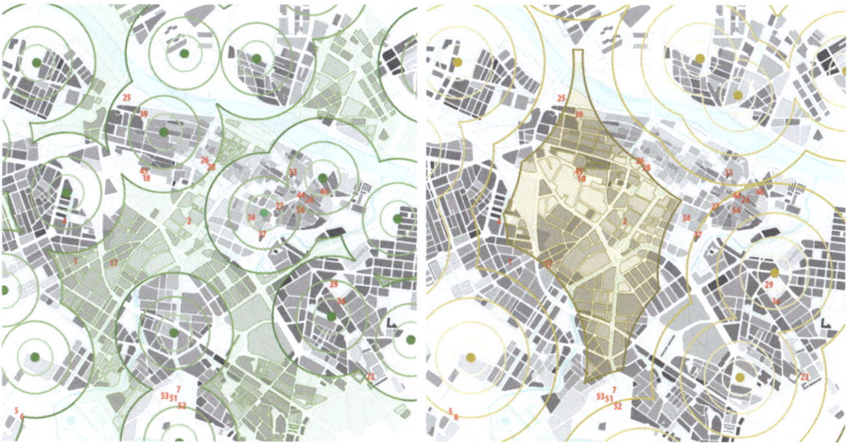

Fig. 2 Void Islands of Primary Care Centres and Civic Centres, surroundings of Historical Centre. *Source* own elaboration according to *Zaragoza manzana a manzana*

people older that 64 years old, does not contemplate the Historic Centre as a vulnerable area, this being one of the rare cases where it is not included.

Moreover, the study of Empty of Service Islands (Fig. 2) clearly shows that the district is relatively well equipped with municipal facilities, with the exception of Civic Centres and Primary Care Centres. In general, the centre of the city is a big empty island. That is why the recycling and starting up of a municipal space in the Historic Centre would resolve the shortage not only for the district but also for a greater area, and highly dense, as it is the centre of the city, a sector with high Population access.

We are finally left with the issue of checking the existence of recyclable facilities in the area. Five possible municipal spaces are highlighted: the Old Sangenis Pontoneros Barracks, the House of Bombs in the district Almozara, the Old Imprenta Blasco, the Palace of Fuenclara and the Luis Buñuel Institut. The study of all of them offers a wide range of possibilities due to the size of the buildings as well as the interest in recovering the memory of the city, the interest associated with the depreciation of the embodied energy and the different accessibility in terms of the functional program proposed.

In any case, it becomes clear that the facility demand oriented in solving an equipping emptyness and its inclusion in a city area with high levels in the vulnerability indexes, as well as the achievement of a more sustainable city claims for recycling and occupation of some of the empty architectonical lands detected.

4.2 *Case of Study According to Empty of Service Islands*

In the second case of study we will carry out the reverse process. Analysing empty islands, the emergence of Valdespartera district is repeated (Fig. 3). The bad accessibility to almost every facility of this area is partly a consequence of its recent creation. The lack of the following items is detected, Civic Centres, Libraries, Municipal Councils, Exhibition Spaces, and Cinemas and Theaters. In other words, with the exception of Primary Care Centres the district has a lack of practically every municipal facility. From all of this facilities, Exhibition Spaces and Cinemas and Theaters are considered with less priority. That is why reoccupation of empty architectonical lands intended to be used as Civic Centres, Libraries or Municipal Councils is considered suitable.

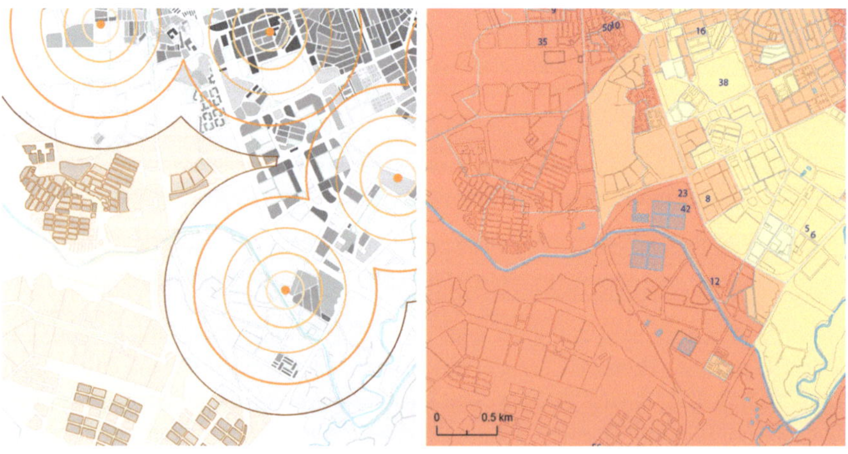

Fig. 3 Detail of planimetries from Empty Islands of IDSE, Valdespartera surroundings. *Source* Elaboración propia en base al programa Zaragoza manzana a manzana y Atlas Vulnerabilidad

Moreover, the study of the vulnerability indicators only highlights the district of Valdespartera in the Municipal Socioeconomic Inequality Index. This consist of the weighted sum of the percentage of population out of work and the percentage of population without studies. For this reason, to reverse the vulnerability process, the creation of any of the previous facilities is considered convenient. Furthermore, as a major part of the city, the district also stands out negatively in the Subjective Synthetic Urban Vulnerability Index, therefore every operation which means the equipping of a facility will be much welcomed.

Finally, the study of Recyclable Municipal Facilities shows one only possibility: the Old Militar Pavillon of Valdespartera. After an inspection of the construction, it is noted that the empty architectonical lands could be recycled with any of the three potential uses, leaving the optimization of the embodied energy, the recuperation of the building's memory, and in general the achievement of quality architecture to the expertise of the Architect.

5 Conclusions to the Case Studies

Even though a first analyse shows the previous evidences, a more detailed study of the specially vulnerable areas recommends to densify the number of facilities in specific areas, solving with a greater offer a greater vulnerability. This is the case of the districts of Historical Centre and Delicias, where, despite having in general a good accesibility to facilities, the existence of a high number of construction without use, its population density and its reiterated appearance in the vulnerability indicators, recommend the starting up of the maximum number of municipal spaces. These should be renovated and reused with programs suitable for the context, with the intent to reduce specific vulnerability. Moreover, the existence of great vulnerable areas in the districts of Oliver, Valdefierro, Actur North, Picarral and las Fuentes east sector is highlighted. In such areas the absence of municipal constructions keeps us from thinking about the recycling of empty architectonical lands. However, the accesibility study of the facilities highlights the need of their existence.

6 Future Investigation Lines

As a conclusion, the study offers a large list of opportunity spots where added value can be generated, acting aware of the sustainability demand and proposing a more sustainable city. Nevertheless, the list of recyclable buildings used is limited. Staying consistent on the concept of a more sustainable city, the list should include every construction potentially reciclable into a facility. Moreover, the program *Zaragoza manzana a manzana* carries out the study about existing facilities according to accesibility rate, being desirable its consideration in accordance to population access. On the other hand, while working with facilities of regular programs it is not possible to investigate about alternative programs, or hybrids from the existing ones. These could be created intentionally to solve concrete shortages in specific areas.

References

Asunción M, Segovia E, Willstedt H (2009) La ruta de España hacia Copenhague. Propuestas de WWF para reducir un 30% las emisiones de CO_2 en los sectores difusos en España 2005–2020. WWF España, Madrid

Casals-Tres, M, Arcas-Abella J et al (2013) Aproximación a una habitabilidad articulada desde la sostenibilidad. Raices teóricas y caminos por andar. INVI 28(77):193–226

Fariña Tojo J (2013) Ciudad sostenible, rehabilitación arquitectónica y regeneración urbana, Issue XV:15–27

Gallegos Ferrer G (2009) La lógica del proyecto urbano y la transformación del espacio no construido, dentro de la renovación urbana de los contextos históricos

Leal Maldonado J, Sorando Ortín D (2013) Rehabilitación urbana y cambio social en las grandes ciudades españolas. Issue XV:205–237

López-Mesa B, Palomero Cámara JI et al (2013) La Rehabilitación y la mejora de la eficiencia energética de la vivienda social a examen. Issue XV:283–319

Tejedor Bielsa J (2013) Nuevos instrumentos de planificación y gestión de la rehabilitación y la regeneración urbana. Issue XV:27–73

Wackernagel M et al (1997) Ecological footprints of nations: how much nature do they use? How much nature do THEY have? Comisionado por el foro Rio+5, Toronto

The *Arrabal* of *Alcázar viejo* from Cordoba: Urban, Hereditary and Sustainable Regeneration of the Historic City Centre

Rafael Cabello Montoro

1 Introduction

Cities are cultural landscapes between two realities: territorial growths and socioeconomic dynamics. The treatment of different urban scales leads to reorganization problems, especially when we have to improve the relationship between a historic city centre, which contents the city's heritage identity along the centuries, and their outskirts. In the same way, historic city centres are heavily rigid owing to their material and urban structure. Likewise, this character has increased during the last years because of strict and protective regulations, which have frozen the actual urban image of a part of the city. That is an irony, since the historic city centre has always been an historical and changeable city according to the different civilizations that have dwelled them. This character is essential to a sustainable and eco-efficient historic city centre (a city from its beginnings to Industrial Revolution). Nowadays there is a breakdown between the new outskirts and the historic city centre, which is less attractive to the contemporary society because of their new necessities, causing negative effects such as excessive outsourcing, depopulation and loss of identity, thematic tourism or obsolescence of the urban and public spaces.

Current historic city centre plannings belongs to the new stream of urban renovation that carried out in Europe from 1960s to the beginning of the 21st century. This new urban trend was in favour of revaluating and recovering of the inherited city. In other words: monuments (defined as an element with great singular relevance) and architectural ensembles, whose value is collective. In this way, the Modern Movement, whose supporters' ideas of "new city" imposed on the existing city without any sensitivity, finished. In its place, a new stream appeared whose goal was the protection and conservation of urban heritage. Thanks to that trend, most of the ancient city centres of the old European capitals have been preserved.

R. Cabello Montoro (✉)
Department of Urban Sciences and Territorial Planning, Seville University, Seville, Spain
e-mail: rafael.cabello@coacordoba.net

P. Mercader-Moyano (ed.), *Sustainable Development and Renovation in Architecture, Urbanism and Engineering*, DOI 10.1007/978-3-319-51442-0_6

Nowadays, forty years after, we can assure that this perspective in urban regulations has been consolidated. However, in many cases the regulation is extremely protectionist and prevents to renovate buildings formally and functionally. Because of that, several urban scientists think that this is the cause of our serious contemporary problems like material urban deterioration, obsolescence, thematic tourism and depopulation of the old city centres. So like this, they agree to a solution that should present from a point of view according with the urban renovation and requalification morphologically and functionally between the different zones of the historic city centre and the connection with the outskirts. We might ask ourselves: Is this a new urban wave? To Falini (1985) the idea of an united and homogenous historic city centre is in crisis, while the sustainably regeneration allows to recuperate the residential heritage, in other words, to preserve the old city alive.

Nowadays, a new urban and heritage cycle is beginning in Europe: urban scientists believe that regeneration is the key to find an equilibrium between two focus: society-territory versus tourism-heritage. Likewise, it is fundamental to obtain a real and sustainable integration between the historic city centre and the outskirts. Our challenge consists on finding the resources to protect and rebuild the relationship among the local population, whose social and economic demands are changeable, and its territory. To Busquets (2000) urban planning in the twenty-first century must guarantee that historic city centres keep being alive.

2 The Historic City Centre of Cordoba

Cordoba is considered one of the biggest cities of the Middle Ages, after Roma and Constantinople (so much that Christian conquerors related it after the heroic deeds of the siege of Cordoba). However, this city begun to decline after the Christian conquest, given that it was the first time that Cordoba is not a headquarter of a state or national government. Progressively, Cordoba became an ordinary city of the Christian Kingdom of *Castile*. As a consequence, the historical suburbs or *arrabales* (in other words, neighbourhoods outside the historical wall) were destroyed, except from the Roman founding city. Of course, this city was inside the Wall and have been called *Medina* by the Arabs and *Villa* by the Christians. An other Moorish and walled *arrabal* (called *Axerquía*), linked and placed in the East of the *Medina*, survived too. One of the Christian's first decisions was to fortify the southwest side of the city increasing the surface belonging to the Moslem *alcázar* (old Arab castle) and creating the last historical *arrabal* inside it. They designed a Christian and urban setting, that today it still exists and we know as *Alcázar viejo* neighborhood.

Beginning of the past century there were settlements outside Cordoba's walls that consolidated along the years later. However, it caused the non-interconnection between all the outskirts and the historic city centre. Because of this, general plans, approved by local governments after Spanish Civil War, tried to link the different parts of the city in order to transform Cordoba into an only and compact town.

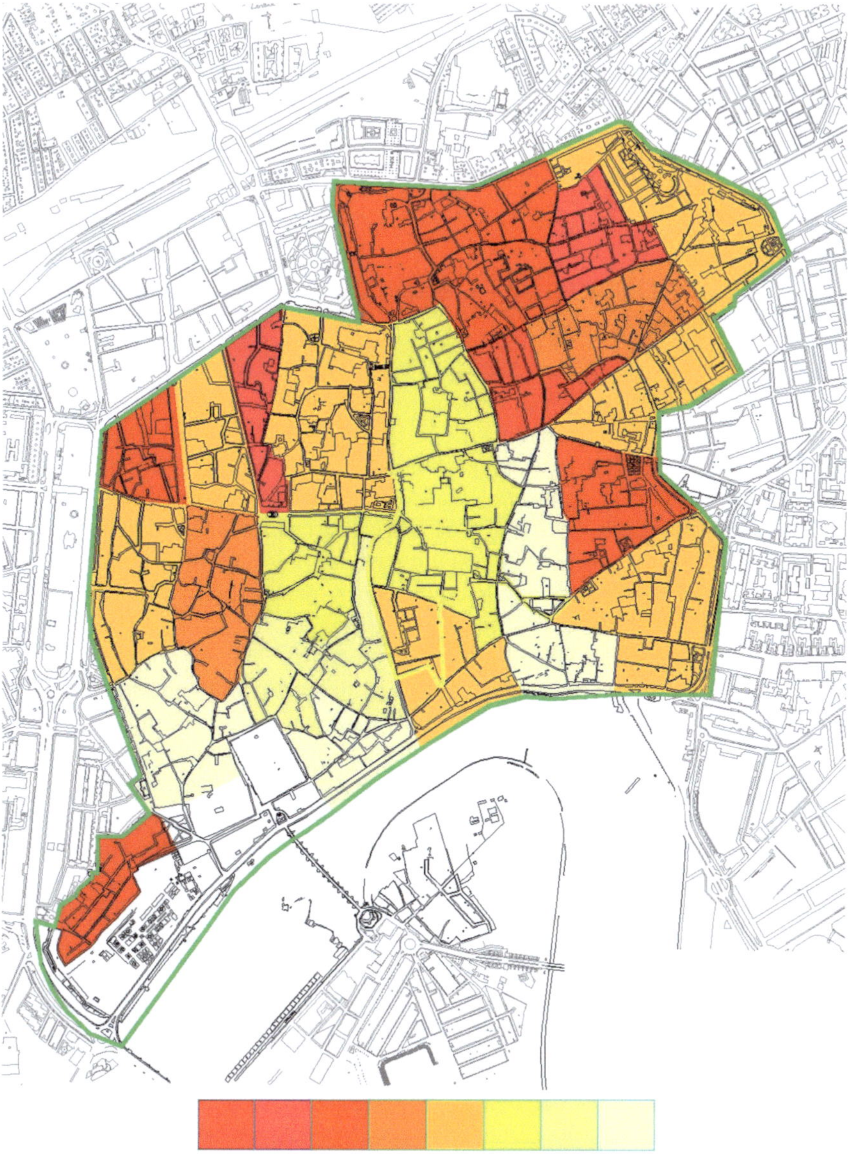

Fig. 1 Depopulation of the historic city centre

However, nowadays there are still unconnected urban areas that constitute an opportunity to work on.

The historic city centre of Cordoba contains its population's heritage identity. Inside it, the most important civilian or religious monuments and the most popular and traditional houses can be found. Because of that we can affirm that the main character of the historic city centre is its heterogeneity. However, Cordoba's main value keeps on being the conservation of its ancient use as residential area (for 50,000 inhabitants, the 15% of the actual total population). However, it can be seen in Fig. 1, currently the population is moving from the centre to the outskirts since the end of the last century. Not surprisingly, Castilla del Pino (1973) recommends: hurry up if you want to see Cordoba.

3 Cordoba's Historic City Centre Protection Special Plan (PEPCH-2003)

The 8th of May, 2003, the government approved definitely Cordoba's Historic City Centre Protection Special Plan (PEPCH in Spanish). Its author Daroca Bruño (2003) recognizes the difficulties to reach PGOU-1986's objectives (previous plan), which consisted in matching up "downtown" with the historic city centre. However, he tries to avoid the depopulation of the historic city centre. In this way, he pretends to reinforce its central position, its historic past, its general public facilities and its ring-shaped accessibility. The ring city centre's main quality is balancing the relationship between historic city centre and outskirts.

As its name points out, PEPCH is a more protective than reforming instrument, since it "lays down three different building ordinances from maximum to less protection, this is according to its possibility of being renovated: (a) Monuments, buildings and architectural ensembles; (b) Zone of protective building typology and (c) Renovated Zone". Subsequently, the entire traditional old town is under a strong urban discipline.

The ordinance of protective building typology affects in the whole of the historic city centre except for the already renovated zone (North *Villa*). Effects of imbalance and a possible depopulation can be seen because a single ordinance imposes the same rules to preserve the same *patio* in the very heterogeneous ancient city. Therefore, it turns out that an ideal and historical building typology that does not allow to construct housing according to new generations' necessities. Apart from rules about maximum building height, projections, façade's composition and buildings' covers, the main and characteristic element of the ordinance of protective building typology is the heritage courtyard o *patio* and the adjacent gallery. In this way, all the building must be organized around that singular element that pretends to evoke the traditional *patio* of Cordoba. It can be seen in the Fig. 2, the *arrabal*'s only currently ordinances are "Monuments, buildings and architectural ensembles" and "Zone of protective building typology".

Daroca Bruño (2015) thinks that is dangerous to lace the historic city centre because of an overly strict general interpretation of Urban Management of Cordoba, who does not study each particular case.

Fig. 2 Impact of Cordoba's HCC protection special plan on *Alcazar viejo*

4 Cordoba's *Patios* Festival and Its Traditional Housing

Solano Márquez (2014) affirmed in his announcement of Cordoba's Patios Festival 2013 that the beginning of this traditional contest could be the neighbours' disputes over having the most flowery or well-decorated *patio*. Little by little, the upper and middle class had a good impression of this popular competition because of its traditional and folkloric atmosphere. Soon after, the local government legalized *Patios* Festival's setting ups which *patios* could participate. Likewise, this also applied to *rejas* (bars) and *balcones* (balconies) following contests.

The housing involved in the contest use to be the home of several families (between 8 and 16). Each one of them lived in single rooms of maximum 20 m^2. These little "flats" surrounded a big and shared popular courtyard, where the common services such as the kitchen, the bath-room, the laundry-room and the well were located.

In the fifties, Cordoba's *Patios* Festival found its consolidation as a party of great interest for the city, the province and the rest of Spain. Different local governments think that *patios* are a singularity of Cordoba that differentiates it from other cities, thus relevant to attract national and international tourism.

From the second middle of the 20th century, the city councils followed politics of renovation, subsidies and supports to *Patios* Festival and the house's owners such as a single lower class family with limited resources to participate in the festival.

In 21st century, all these efforts are giving results and Cordoba's *Patios* Festival reappears strongly, especially thanks to an international recognition: in December 2012, *Patios* are declared World Immaterial Heritage by UNESCO. Because of this, in 2013, there has been a touristic boom in Cordoba which has caused that some *patios* have been opened not only in May (the month of the festival), but also the rest of the year—although with less intensity. As a result, tourism has increased considerably in the city, which is implying the emergence of more bars, restaurants and all kind of business regarding tourism. In the next figure, it can be seen a touristic road-map where historic city centre's *patios* are split into groups according to its close proximity to neighbours (Fig. 3).

Fig. 3 Map of Cordoba's *Patios* festival 2016

5 The *Arrabal* of *Alcázar viejo*

The *arrabal* of *Alcázar viejo* is a new and Christian historical suburb that knocked into shape after the reconquest of *Qurtuba*. In the elderly Middle Ages, Christians' new ideas about urban hygiene and health standards were embraced thanks to a planned urban setting, which was more regular than the tortuous and narrow streets of the Muslim *Medina*.

Despite of its youth compared to the rest of the city, the *arrabal* played a very relevant role, historically recognized. As an example, in an ancient civil war that took place in 1367, Solano Márquez (1986) affirmed that the King Peter I of Castile, called the Cruel, left Córdoba after a great defeat because he could not penetrate the city because of the brave women that defended the *Alcázar viejo* while men were fighting in the *Campo de la Verdad* (nowadays, a neighbourhood in the outskirts at the other side of Roman Bridge Gate).

There were few important reforms in the *arrabal* in the Modern Age, except for the establishment of the residential use after losing the military character. However, Christian Monarchs' Alcazar was a military dungeon until Spanish civil post-war period. Nevertheless, the Christian urban setting consolidated facing the traditional Muslim typology. To illustrate this, Nieto Cumplido and Luca de Tena (1980) explain that originally there were three main streets and a little square in the centre of the *arrabal*, where the Church of the Ancient Convent of *Basilios* was located, even though the religious city centre kept on being the Cathedral-Mosque.

5.1 *The Rural Housing of the 19th Century*

While the transformation of Northern Córdoba (present historic city centre) into a modern and industrial city, the working and popular class, who belonged to the rural scene, remain at the south and the east of the city. In this way, the *arrabal* of *Alcázar viejo* was rebuilded without planning, causing the buildings to be lined up along the street, a single storey, with an only bay and a rear courtyard that in fact was the rest of the plot (see Fig. 4). They were like the three kinds of elderly medieval houses described by Pino García (1999): tenement housing, gates housing and palace-housing.

Fig. 4 *Arrabal*'s housing for the 19th century

5.2 The Rural Housing of the 20th Century

During this past development period, *Alcázar viejo* was overfilled by housing and evolved: houses grew from one to two floors. Façades were renovated more healthily, with a better proportion among openings (windows and doors), like it can be seen in Fig. 5. Inside the houses, owners constructed a new bay which was joined to neighbours' party walls. In this way, the rear courtyard was reduced and became a central element with a more functional character respect to the rest of the house. These transformations follows Sierra's thesis (1980): a poor, regional and timeless architecture, objectifying an urban space very similar to others that you can find in Andalusia.

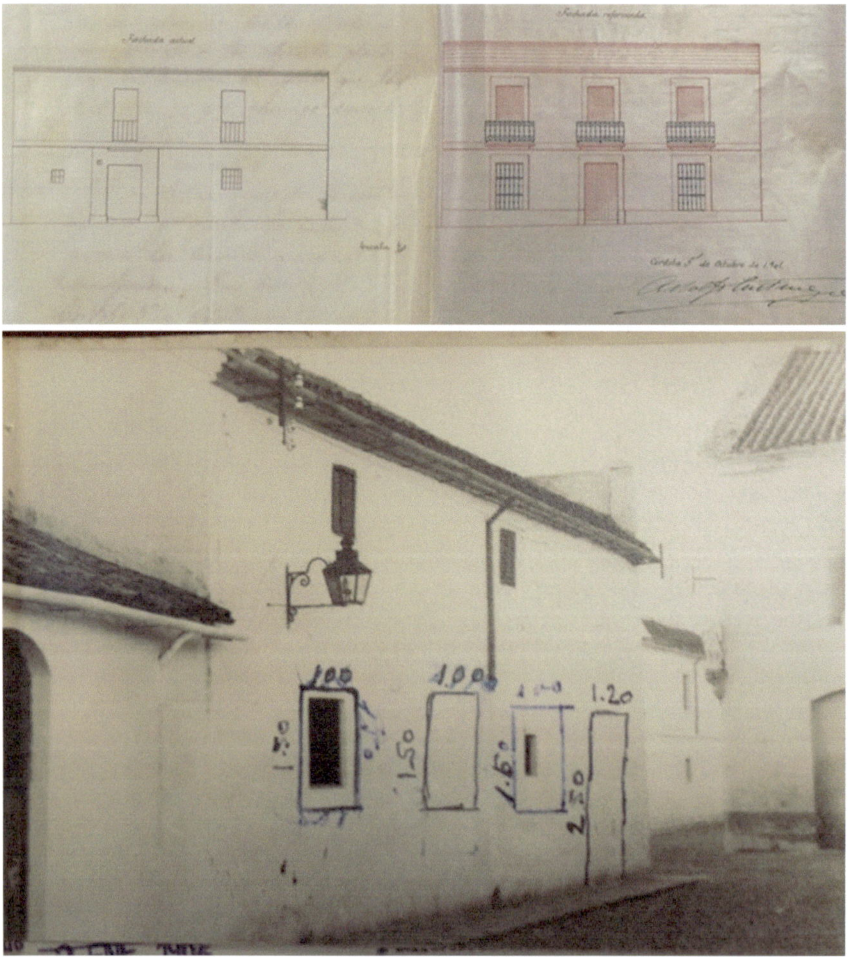

Fig. 5 Housing's transformation at the beginning of the 20th century

However, the people living in these buildings still belonged to the working class, who transformed single-family homes into multifamily housing. In fact, they were like little *familisterios* without planning, where 10 or 15 families lived in single rooms of 20–30 m^2 and shared common services around the central courtyard or *patio*: kitchens, laundries, bath-rooms, even water supply from an Andalusian well.

5.3 The Urban Boom of the Last Years of the 20th Century

In the seventies and eighties, the arrabal grew one floor more, from two to three storey buildings, thanks to local urban rules. Buildings such as the one showed in Fig. 6 were allowed. Furthermore, the lacking of protection for the private *patio*-housing favored that new property were built in that years. In this way, old *arrabal*'s residents did not need to emigrate to the outskirts, because the number of houses increased inside *Alcázar viejo*. Nevertheless, *arrabal*'s population decreased approximately two thirds since the old patio-housing were too crowded and there were not enough single family homes. In addition, residents' purchasing power increased and they could afford better housing, even though they had to move outside *arrabal*.

For this reason, lots of *patio*-housing were demolished. However, something of them remains preserved thanks to some families, whose generations have always been there, and to popular initiatives that succeed at saving them. This is the case of *Patio*-Housing number 50 of *San Basilio* Street, which was kept thanks to the intervention *Patios*' Friends Association.

Fig. 6 A real estate promotion in the elderly 20th century

UNESCO declared in 1994 South *Villa* and the *arrabal* of *Alcázar viejo* as World Heritage Site, which allowed the decrease of the voracious property development. However, the present protective urban tools have not become true until the beginning of the 21st century.

5.4 The Imposition of an Idyllic Model in the 21st Century

The hard urban protectionism that is applied in the old *arrabal* comes from Cordoba's Historic City Centre Protection Special Plan, approved definitely the 8th of May, 2003.

Like it can be seen in Fig. 7, the ordinance of protective building typology was not strictly able to apply to *Patio*-Housing number 42 of *San Basilio* Street, specially because of the striking absence of perimeter galleries.

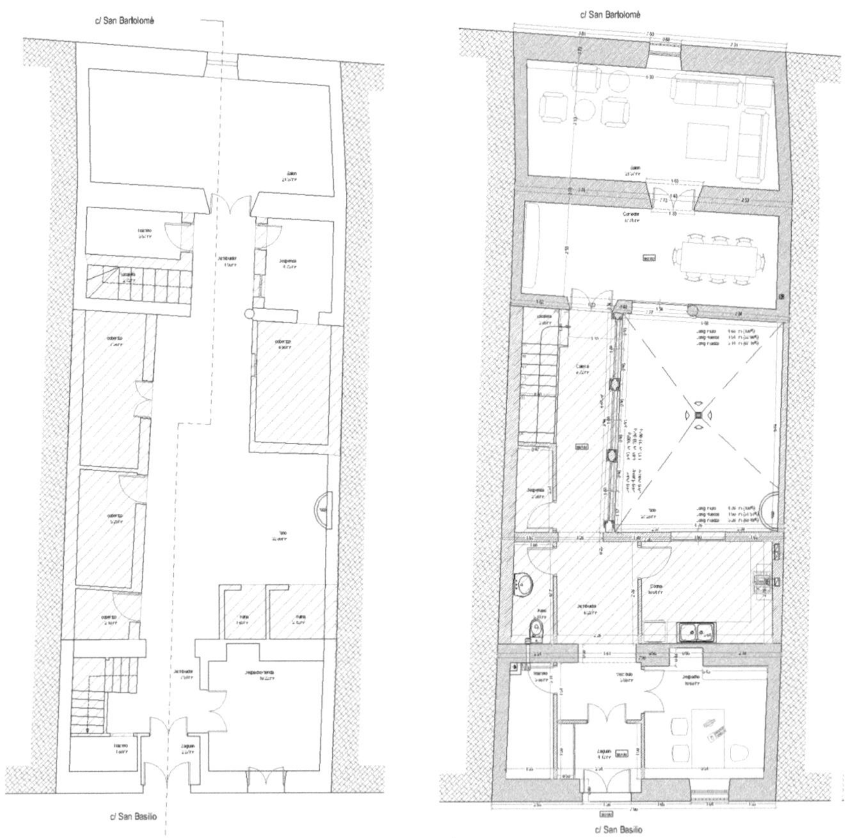

Fig. 7 Real implementation of the ordinance of protective building typology

This only current ordinance is not considered appropriate to renovate housing—taking account of dimensions, forms and areas of the *arrabal*'s characteristic parcel. Moreover, the imposed typology does not correspond with *Alcázar viejo*'s traditional one. Consequently nor PEPCH helps to resolve the problem of depopulation neither keeps the essence of the ancient *arrabal* in the new constructions. In the long term, this will imply the lacking of constructive, environmental and efficient sustainability.

The big central courtyard with adjacent gallery is more appropriate for a knight's ancestral home than a popular neighborhood community's Cordoban *patio*-housing. Therefore, the model that the PEPCH imposes in the old *arrabal* does not seem the most adapted to its identity.

To Barrionuevo (2005) it is a very interesting detail to study the evolution of the *patio* as principle of growth and construction of the city. It can be seen in Fig. 8, there are two ancient patio-housing just as 100 years ago and 2–3 families are still living here. These are *Patio*-Housing number 7 and 9 of *Martín de Roa* Street; curiously, both *patios* are strongly linked by the same *zaguán* (little hall). This implies that there is an only entrance from the public street, in other words, it can be come in them through the house number 7's door.

The *arrabal*, today *Alcázar viejo* neighborhood, is still an urban piece with human life and activity. It is not a dead museum or historical theme park. Its wealth lies in the fusion of urban sciences and sociology, in this way its total depopulation should cause its decline.

Fig. 8 Present *patio*-housing of *arrabal* of the beginnings of the 20th century

6 Conclusions

The regeneration of historic city centre's population is an objective of Cordoba's PEPCH. However, this has still to be met, in fact, the historical district is empting from the centre to the historical walls such as it can be seen in Fig. 1 on page three (municipal population census's files of 2011, eight years after the approval of the Special Plan, nowadays the same census reveals a decrease of total population). It should therefore be borne in mind that an excess of building protectionism and rigid constructive rules directly affect to people's lives and a sustainable renovation of the historic city centre.

It is true that, in the end of the 20th century there were abuses against the heritage and new typologies, which were not in conformity with the past, were implemented. However, there also were generations of families which, as their level of income was increased, renovated their homes according to their hygienic, light, technologic and energetic necessities. Like it can be seen in Figs. 4 and 5, housing grew one floor and more doors and windows were opened with a better proportion. It is told to the tourists that lots of *arrabal* of *Alcázar viejo*'s houses are 200–300 years, but in fact the majority of them are not centenaries. If the *arrabal*'s housing sustainably evolved during the two first thirds of the 20th century, why are the urban planning of 21st century freezing an heredity and idealized image even though the house has not so much singular value.

This research does not pretend to override the current ordinance to implement a new one, but establishing an action programme to renovate sustainably the residential houses from the planning and without losing environmental values and heritage. The first point to consider is: the historic city centre is an heterogeneous set alive. This building diversity implies that the population were an also heterogeneous set but with a strong territorial identity. However, everything have to keep an equilibrium with other kind of uses that it can be developed in the historic city centre, specially domestic trade, tertiary use, education and tourism. To Ximeno (2007) the functional and aesthetic homogenization promotes people's decoupling with the territory.

A new law about touristic accommodations speeds up administrative procedures to obtain touristic licenses. The ordinance of protective building typology also seems to promote the creation of this kind of homes instead of the traditional ones. Nowadays, private promoters are buying houses in the historic city centre in order to bring them into touristic flats which are organized around an idyllic *patio*. While the historic city centre's population is decreasing, the number of the temporary inhabitants or tourists increases without the balance point has still been achieved. Nor it has been expected by the urban planning, neither there are programmes or rules to control that,

Maybe some politicians pretend to transform the historic city centre into the biggest thematic park of the world. However, how can sustainably a building park maintain like four times Disneyland Park Paris' area? To Cervellati (2007) without a respect project for history and work, for environment and art, the present can be built; then the future cannot either. If he calls "*Chalépolis*" to actual cities' outskirts, Cordoba's historic city centre can be called "*Patiópolis*" in a near future. Urban planning has not only to protect the historical buildings, but people who are living inside them. The real identity, the maintenance and the sustainable building renovation must originate from these people.

Acknowledgements **Thesis Tutor and Director** Victoriano Sainz Gutiérrez, **Contributor** Isabel Parras Pancorbo.

References

Barrionuevo A (2005) Seville, forms of growth and construction of the city, 2nd edn. University of Seville, Secretary's office of publications, Seville

Busquets J (2000) Toledo and its future. District Council of Toledo and Saving Bank of Castile La Mancha, Toledo

Castilla del Pino C (1973) Hurry up to see Cordoba. Triunfo, 538 year XXVII, pp 20–23

Cervellati PL (2007) The low density city, Provincial Council of Barcelona, From the city-garden to "chalépolis", pp 185–200

Daroca Bruño F PEPCH, www.gmucordoba.es

Daroca Bruño F (2015) Personal interview, 15th of September, 2015. Architect and Professor of ETSA's Project Department of the University of Seville, Andalusia, Spain

Falini P (1985) Project's new forms of the existing ones in Italy (Chapter 2) Evaluation of the necessities of the renovation. M.O.P.U, Seville

Nieto Cumplido M, Luca de Tena C (1980) The Alcázar viejo, a repopulation from Cordoba in the 14th century, Axerquía Journal of studies from Cordoba

Pino García JL (1999) The Cordoban house in the elderly of the Middle Age, Cordoba in the history: the construction of the city, Conference proceedings, 20th–23rd of May, 1997, Cordoba, pp 249–262

Sierra JR (1980) Introduction to the formal analysis of the popular and domestic architecture in Seville. University of Seville (thesis)

Solano Márquez F (1986) Cordoba's squares from A to Z. no 29. The *Corredera*, an unpublished novel. Daily paper of Cordoba, Cordoba

Solano Márquez F (2014) Cordoba is *patio*. Editions Buendía, Cordoba

Ximeno F (2007) The low density city, Provincial Council of Barcelona, Free spaces and urban centres to restrain the low density, pp 427–452

Figures

Figure 1: Cabello, R. (2011) The trace of the *Corredera*. (PhD thesis)

Figure 2: Daroca, F. (2003) Building Plan from Cordoba's PEPCH

Figure 3: Cordoba Town Hall, (2016). Cordoban *Patios* Festival and Contest Plan, from 2nd to 15th of May, 2016

Figure 4: Building Projects from Cordoba's Historical and Municipal Archive (1880–1900)

Figure 5: Building Projects from Cordoba's Historical and Municipal Archive (1920–1950)

Figure 6: Plans from Project of multi-family residential building on *San Basilio* Street, 47, Cordoba´s Historical and Municipal Archive (1984)

Figure 7: Crespo, J. E. (2010) Plans from Project of renovation of a single-family home between Mediatrix on *San Basilio* Street, 42

Figure 8: UPM Research Group New Techniques, Architecture, City. 2009–2010 (2010) Study about collective-patio housings from Cordoba. VIMCORSA

An Approach to Daylight Contrast Assessment in Mediterranean Urban Environments

Judit Lopez-Besora, Helena Coch and Antonio Isalgue

1 Introduction

It is widely known that light and Mediterranean countries are strongly related. The availability of light in this environment is immense because, most of the time, the clear sky condition provides the buildings with a large amount of exterior daylight (Baker et al. 1993). It means that city and building designers have to cope with more light than needed, and with high contrasts between light and shadow. This situation is especially extreme during summertime, when the incidence of sun is strong and the white and pale architectural surfaces reflect a large amount of light. As a consequence, urban visual scenes are featured by a high luminance contrast within the scene itself and among different ones. It has been pointed out above that the most frequent sky type in Mediterranean environments is the standard clear sky (ISO, 15469:2004). This sky type do not show a high value of luminance, but lower than other surfaces of the environment flooded directly by the sun. However, in places with a higher occurrence of overcast skies, the luminance values of the sky vault are higher than those of the clear skies; these values, in comparison with the luminance of urban surfaces under diffuse light are not so different, and therefore the luminous contrast is lower. Whereas information about illuminance values in horizontal surfaces in hot climates such as the Mediterranean (Baker et al. 1993), or the spectral analysis of daylight (Hernández-Andrés et al. 2001) are present in literature, there are few studies about the values and distribution of luminance in urban environments, since the vast majority are focused on the study of diffuse light, much easier to predict in terms of brightness.

The presence of direct sunlight in the urban environment influences the vision at the time of entering a building. In these cases, the observer experiences a kind of sudden blindness that can be explained with the visual adaptation process.

J. Lopez-Besora (✉) · H. Coch · A. Isalgue
Architecture & Energy, Barcelona School of Architecture, UPC, Barcelona, Spain
e-mail: judit.lopez.besora@upc.edu

P. Mercader-Moyano (ed.), *Sustainable Development and Renovation in Architecture, Urbanism and Engineering*, DOI 10.1007/978-3-319-51442-0_7

According to that, the visual system needs a few seconds, even minutes, to get a complete vision adapted to the new level of illumination, considerably lower than the level outside. The vision in these spaces, called transition spaces, is object of study (Lasagno et al. 2011; Araji et al. 2007; Lopez-Besora 2015) because it is possible to soften the effects of adaptation through the architectural design. However, the most extended solution is the increase of illuminance levels and the use of more lighting fixtures, with the consequent waste of energy.

Along the history, Mediterranean architecture has integrated filters, large elements of shadow, and other control mechanisms so as to soften the effects of direct sun (Ruggiero et al. 2009) in interiors. It helps to reduce the heat load and to create more uniform lit environments with a gradual transition (Araji et al. 2007). But today, the situation is different. As the huge stream of natural light is very difficult to handle, artificial lighted and controlled environments are becoming the most extended solution. It leads to poor and monotonous atmospheres excessively lighted in which the dynamism of natural light is not present and the cost of energy is high. One of the things that can be done to tackle this situation is to progress in the knowledge of the lit environment in cities as a previous step to estimate the possible contrast with the interior.

The objective of this paper is to establish a methodology to get a description of the Mediterranean visual environment in an urban context so as to know the range of luminance values outside buildings and, with this information, help lighting and building designers in their projects. It is expected to determine the luminance values in a visual scene in a Mediterranean urban environment, and how these values are distributed within the scene. At a later stage, it will be possible to compare these values with those from an interior scene and determine the zones where a higher contrast will be present. To attain this objective, a geometric description of the visual scene has been made and luminance values from real locations have been measured to put into practice the methodology. The work consisted on collecting information from a case study and the subsequent data processing following the proposed methodology.

Here, a description of a representative Mediterranean visual environment has been defined. Further work should be done in other locations in order to have more data and contrast the values with those obtained here. Apart from this, the methodology proposed can be applied in other climates making possible to extract the features of other visual environments.

2 Case of Study

The methodology was first put into practice in the exterior of a building located in Tarragona (Spain). It was chosen for the ease to identify the pavement, façade and sky zones in a low density urban environment. The colours of the environment are pale, characteristic of the Mediterranean architecture. It allows extrapolating the results to other environments located in this climate. With this specific case of study

Fig. 1 Situation of the building and entrance view

it is expected to illustrate the procedure and show the luminance values in this specific context.

The building houses private offices in two parallel blocks connected by a reception area. The blocks are oriented east-west and are composed of ground floor and one storey with large white surfaces (Fig. 1). The observation and measurements took place in July of 2013, at noon (between 13:00 and 14:00 pm) in order to coincide with the moment of greatest sun incidence in the Mediterranean climate. It consisted on a series of pictures and luminance measurements in specific points facing the entrance of the building.

3 Methodology

In order to tackle the luminous characteristics of the visual scene in a Mediterranean urban environment it was decided to base the study in luminance, since this parameter is the most suitable to determine the brightness of a scene. With the aim of checking the distribution of luminance in the visual scene, it was divided in zones with similar features, which would permit to make the analysis in each of them, as explained further. The methodology has two parts: the first one consists on analysing the visual urban scene; in the second part, data from measurements and pictures are evaluated and compared through three different procedures, as explained below.

3.1 Visual Scene

Analysing the visual scene in urban areas is a complex task which can be simplified if an abstract approach is outlined. It means that the essential parts (or surfaces) of a city that define the visual scene must be included in a geometric framework. Once these parts are clear, all urban environments can be described in terms of composition and proportions, allowing the analysis of any urban scene.

The vision of an urban scene is framed by the human visual field. In broad terms, the human visual system is made of two eyes located in front of the head with a superimposed visual field which covers most of what is in front of him. The most sensitive area is located around the focus point, and vanishes away from this direction. At the periphery, only movements can be detected, as well as other changes such as flashing lights or isolated stimulus. In the graphic description of the horizontal and vertical visual field (Panero and Zelnik 2007), a static observer is supposed; however, when the observer moves, the horizontal visual field makes scarce sense because the observer is constantly changing position in this plane, so the scene is framed by upper and lower limits. In this situation, the visual field becomes a horizontal stripe with the line of sight approximately in the middle, and different sensitivity zones depending on the angle related to this line (Fig. 2). The central stripe, corresponding to the zone of chromatic discrimination, is considered the most important zone in terms of vision. This zone is where the sensitivity of the eye is higher. But, when the observer moves in an urban environment, what elements are present in this stripe, and to what extent?

This brings us back to the definition of the visual scene. In an attempt to simplify the parts of a visual scene, it has been considered that a built environment can be

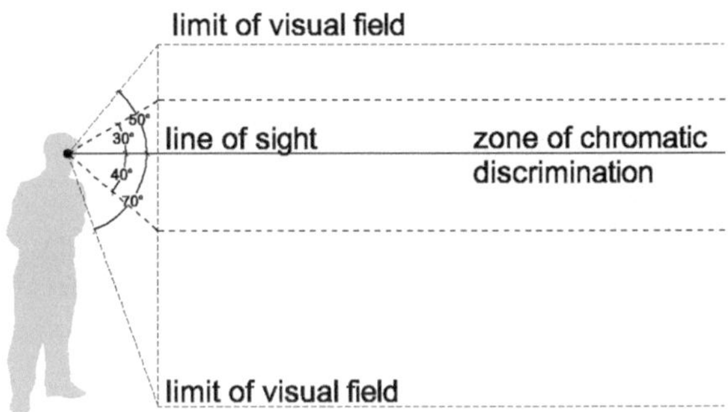

Fig. 2 Visual field of a moving observer

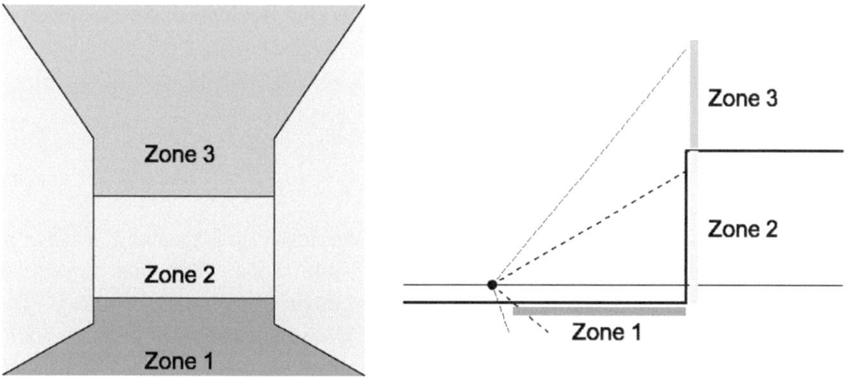

Fig. 3 Graphic description of the visual scene zones

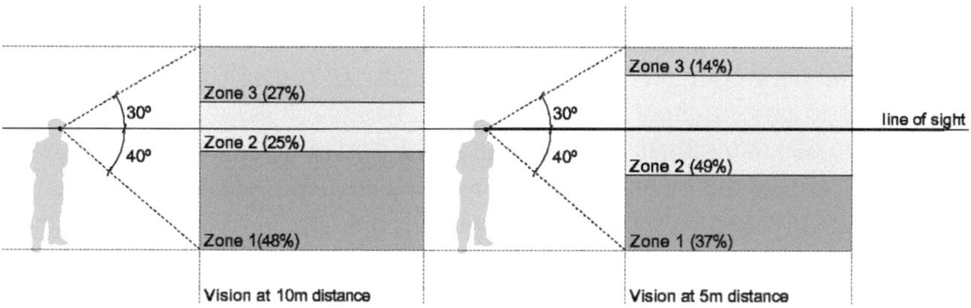

Fig. 4 Visual scene zones in the case study

reduced to three categories that make it up, which are surfaces in different positions and therefore, with different sun incidence. The first category comprises horizontal surfaces located at the bottom of the visual scene. It usually corresponds to the pavement. The second category includes vertical surfaces, such as façades, in the middle. The third category is the sky dome and horizontal surfaces seen from below, at the top. The resulting zones are named 1, 2 and 3 from bottom to top (Fig. 3). Their proportion in a visual scene depends on the characteristics of the environment (geometry) and the distance of vision (the position of the observer).

In the case study, the proportion of each zone in the visual scene was deduced from a geometrical approximation taking the height of the building and different view distances combined with the visual field of a moving observer. Then, two scenes were defined, at 5 and 10 metres distance from the entrance (Fig. 4). The scenes included horizontal and vertical surfaces and the sky dome with the proportions shown below. Besides their proportion in the visual scene, it has to be considered the luminosity of each zone, which is strongly linked to the incidental light, the orientation of the surface, its position in the visual scene, and other architectural or morphological features such as the typical materials used in buildings.

3.2 Light Analysis

Once defined the geometry of the environment the next step is to obtain a range of luminance values for each zone and for the whole scene. With this purpose, a methodology to getting and using data from real environments was proposed. The source data are pictures taken in the moment of maximum radiation (next to noon) and a series of measurements taken at the same time.

To start with, standard pictures of the visual scene were taken with a digital camera. The series of pictures included the outside of the building from the entrance point. It is important that the position of the camera reproduces the position of a standing observer in front of the building and that the parameters of the camera are adjusted to the actual light conditions. In addition, another series of pictures was taken with the same point of view but with different exposures. The purpose of these pictures was to process them as High Dynamic Range (HDR) images using a program so as to obtain luminance maps (http://www.jaloxa.eu/webhdr/index.shtml). It is necessary to take the pictures with the parameters given in the program instructions in order to obtain a precise result. In both cases, the pictures included lit and shaded zones and reproduced what an observer would see in a standing position. The pictures were taken with a digital camera Canon EOS 350D (resolution used 3456 × 2304 pixels). Apart from the pictures, a list of points in light and shadow from each zone (pavement, façade and sky) were detected and its luminance was measured using a Konica Minolta LS-110 luminance meter. The information was processed using three procedures, which are different depending on the nature of the given source.

The first procedure was a picture luminance analysis to obtain luminance data from photographs, carried out with a straightforward software developed by authors. The software extracts RGB light data from pixels, which are converted in luminance values using a reference value taken during the measurement campaign. In each picture, Zones 1–3 were identified and processed independently from each other and all together as a whole. The interest of the study are the values obtained individually for each zone, since they make possible to characterize pavements, façades and sky in a Mediterranean urban environment. The results obtained with this software are average luminance, maximum and minimum luminance values, and standard deviation as shown in Fig. 5 for the case of study.

The second procedure consisted on HDR processed images in order to obtain a luminance map. A series of three pictures with different exposures was taken and processed with a specific program (http://www.jaloxa.eu/webhdr/index.shtml). As a result, a distribution of luminance values in the image was obtained. The value for a particular point with its coordinates can be extracted directly from the program in a

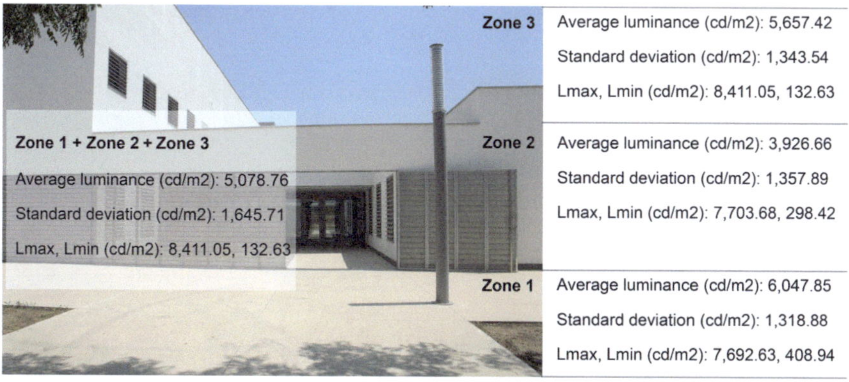

Fig. 5 Standard picture analysis

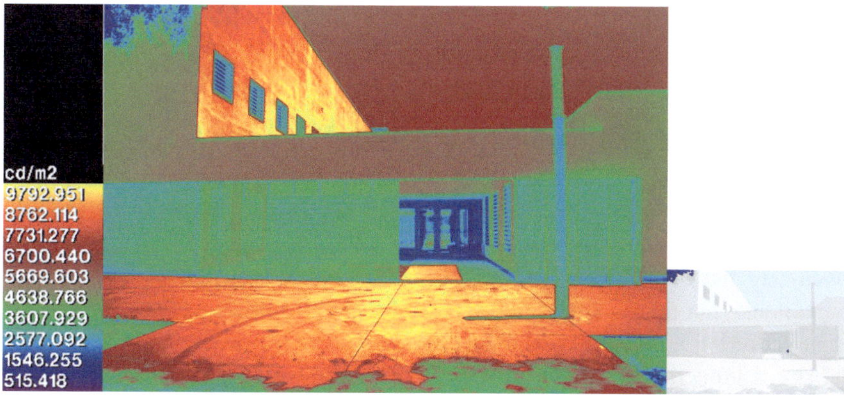

Fig. 6 HDR luminance map and 'heat map' with the pixels whose photometric information is not reliable (*in blue*)

dynamic luminance map. The resulting map for the picture used in the standard picture analysis is shown in Fig. 6.

In the third procedure real measurements were taken at different parts of the wall, pavement and sky dome. It corresponds to Zones 1–3 explained in the process of determining the visual scene. The measurements were taken in lit and shaded surfaces, and were superimposed in the same picture used in the other procedures (Fig. 7). The values from the measured points can be compared with those from the pictures; however, as the points were chosen from a visual test in situ, they cannot be considered as extreme luminance values of the scene.

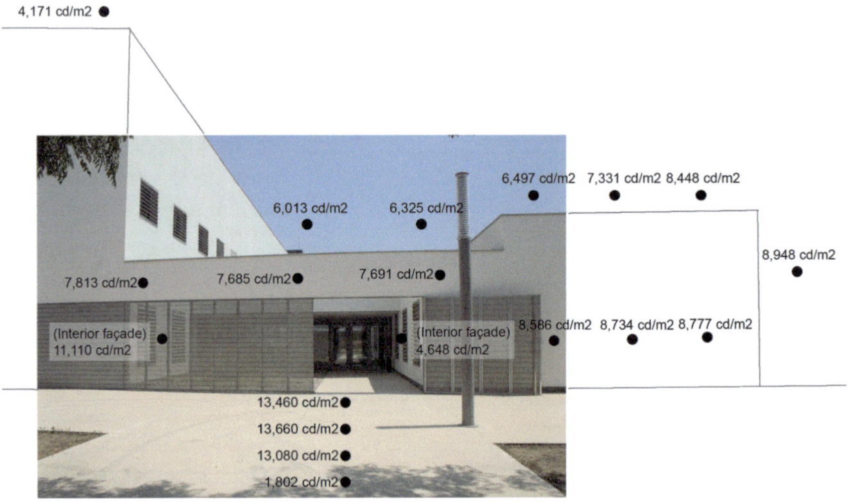

Fig. 7 Luminance measurements taken in situ

4 Results and Discussion

The analysis of the visual scene in the case study shows that 37–48% is occupied by the pavement, 25–49% by the façade, and 14–27% by the sky dome (Fig. 4). Obviously, the façade is the most important zone of the scene when approaching to the building, but the pavement also plays an important role in the appearance of the scene, covering a high percentage of the visual scene. In proportion, the vision of the sky is limited. The weight of each zone on the basis of its luminance can be deduced comparing the results obtained by the three explained procedures (Table 1). As regards to the average luminance, the table shows that the pavement (Zone 1) is the lightest surface in comparison to the others. According to the standard picture analysis, it has an average luminance of 6047 cd/m^2, due to the absence of shadows and the light grey concrete used as finish in the pavement. The minimum average value corresponds to Zone 2, the façade. Although the walls are bright white, there is a shaded porch in the entrance area which contributes to reducing the average value in this zone. The results also show that the sky has a lower luminance than the pavement (Fig. 8).

With reference to the luminance range, the measurements at the points indicated in Fig. 7 show that the highest and lowest values of luminance were measured on the pavement (13,660 cd/m^2 in the sun and 1802 cd/m^2 in the shadow), at the bottom of the visual scene (Zone 1). The measured luminance values of the sky next to the building were around 6000 cd/m^2, increasing towards the position of sun. Zone 2 showed values for the façade in the sun of around 11,000 cd/m^2, and 4600 cd/m^2 in the shade, but these points were behind the wall, in the entrance court. The surfaces outside were around 8700 cd/m^2. The standard picture analysis

Table 1 Luminance data obtained from different procedures

	Procedure	L average (cd/m^2)	L range (cd/m^2)	Standard deviation (cd/m^2)
Zone 3	Stand. Pict. An.	5657	8411–132	1343
	HDR image		10,300–1100	
	Measurements[a]		8998–4171[a]	
Zone 2	Stand. Pict. An.	3926	7703–298	1357
	HDR image		5600–1100	
	Measurements[a]		8777–4648[a]	
Zone 1	Stand. Pict. An.	6047	7692–408	1318
	HDR image		10,400–3600	
	Measurements[a]		13,660–1802[a]	
Total image	Stand. Pict. An.	5078	8411–132	1645
	HDR image		9792–515	
	Measurements[a]		13,660–1802[a]	

[a]The measurements correspond to the main surfaces, in the sun (first value) and in the shade (second value)

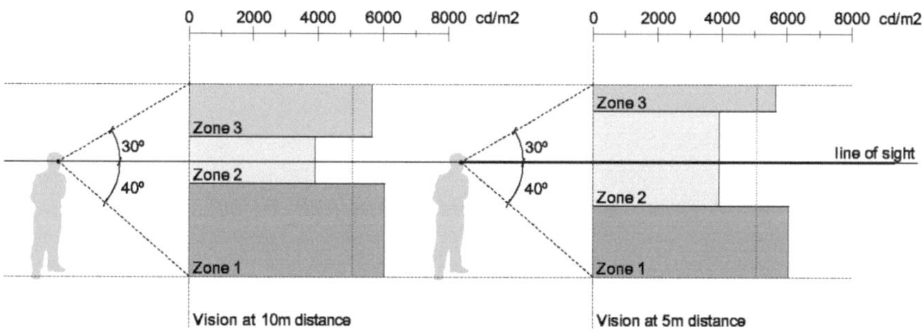

Fig. 8 Visual scene zones in the case study with its average luminance

does not entirely agree with the luminance measurements. If we look at the range of values in the standard picture analysis, the highest and lowest luminance is located in Zone 3. It may be caused because there is a little part of the building into this area which corresponds to a bright lit façade as well as small dark areas from the vegetation and windows. In addition, the standard picture analysis shows that the brightest points are not in Zone 1, as expected from measurements. In general terms, with this procedure, the range of values in all zones is much more similar because all pixels have been contemplated in the process. On the other hand, the luminance map easily shows the distribution of light within the scene with a scale of values, where the pavement has higher values than the other zones, and Zone 2 is clearly the darkest.

With respect to the methodology and the procedures employed, it can be said that:

- With the standard picture analysis the values obtained are very accurate, but the accuracy and its correspondence with a real environment are subject to the framing of the picture and the characteristics of the environment. However, is it a useful tool to determine the average value and homogeneity of a zone within the scene.
- The luminance maps obtained by HDR images show a useful depiction with false colour patches which permit an intuitive identification of the brightest and darkest zones of a scene, more intuitive than a normal picture. It is useful to estimate the distribution of light in a scene.
- Lastly, the measurements carried out in a case of study check the validity of the luminance values obtained from the maps, and at the same time can be used as a reference to convert the light values from the standard picture analysis into cd/m^2. Their reliability in terms of magnitude is linked to the accuracy of the measurement device, even though the measured values are a valuable source data to get a global description of the luminosity of the surfaces in a visual scene.

5 Conclusion

Determining the luminance of a daylight urban scene in a Mediterranean context is a complex task which led to an analysis of the visual scene as well as the determination of a series of luminance values with several procedures. The complexity of the scene and its elements imply that the results must be considered as a rough estimate of luminance values illustrative of a Mediterranean environment, which are intended to serve as a reference. In the case study, the pavement was the zone with higher values of luminance, though the sky dome was also bright and more uniform. Finally, the zone corresponding to the vertical surfaces has the lowest average luminance and the largest standard deviation. In the case study, if we had an interior with a luminance under 100 cd/m^2 on the main surfaces, the contrast with the exterior would produce that the visual system would have to cope with a 100:1 ratio of luminance between both situations (provided that values over 10,000 cd/m^2 were measured). In fact, the interior in a Mediterranean environment show luminance values around this value (Lopez-Besora et al. 2016), which confirms the existing contrast between inside and outside. The use of shaded zones and other materials, just before the entrance, would lead to reduce the luminance in the exterior scene, contributing to improve visual adaptation between the interior and exterior of the building. As a consequence, it would not be necessary an overuse of energy for artificial lighting through the increase of luminous intensity in order to improve vision inside. In addition, if lighting in a transition space is focused on increasing the luminance of the pavement, since this is the exterior zone with higher brightness (compared with sky and façades), it seems possible to improve the visual quality of the users with less energy demand. The sustainability and efficiency of an illumination system, in this case, can be translated in the adjustment of illuminance levels into the space, even its elimination, during the hours in which daylight outdoors is more than enough.

In conclusion, high and low values of luminance are present simultaneously in a Mediterranean environment, configuring non uniform but a light contrasted composition. Special attention has to be paid to pavements, which are not secondary surfaces, but important in the visual scene.

Acknowledgements This paper is supported by the Spanish Ministry of Economy and Competitiveness under projects BIA2013-45597-R and BIA2016-77675-R.

References

Araji MT, Boubekri M, Chalfoun NV (2007) An examination of visual comfort in transitional spaces. Architectural Sci Rev 50(4):349–356

Baker NV, Fanchiotti A, Steemers KB (1993) Daylighting in architecture: a European reference book. James & James, London

Hernández-Andrés J, Romero J, Nieves JL, Lee RL Jr (2001) Color and spectral analysis of daylight in southern Europe. J Opt Soc Am A: Opt Image Sci Vis 18(6):1325–1335

ISO, 15469:2004 (E)/CIES011/E: 2003 (2004) Spatial distribution of daylight—CIE Standard General Sky. ISO, Geneva

Lasagno CM, Pattini AE, Rodríguez RG, Colombo EM (2011) Developing a modelling factor index for transition spaces: a case study approach. Architectural Sci Rev 24(3):215–224

Lopez-Besora J (2015) La llum Mediterrània i els espais d'accés a l'arquitectura. Doctoral thesis. Universitat Politècnica de Catalunya. http://www.tdx.cat/handle/10803/334410

Lopez-Besora J, Serra-Coch G, Coch H, Isalgue A (2016) Daylight management in Mediterranean cities: when shortage is not the issue. Energies 9(9):753. doi:10.3390/en9090753

Panero J, Zelnik M (2007) Las dimensiones humanas en los espacios interiores. Gustavo Gili, Barcelona

Ruggiero F, Serra R, Dimundo A (2009) Re-interpretation of traditional architecture for visual comfort. Build Environ 44:1886–1891

Web HDR http://www.jaloxa.eu/webhdr/index.shtml

Part II
Architecture and Society

Heritage and Community Space as Contemporary Housing Project Matters. Neighbors Courtyards

María del Carmen Martínez-Quesada

1 Introduction

To address the issues raised in this research, it has been necessary to previously approach a series of matters such as: *what the contemporary housing is, what the act of inhabiting is, what the room is, what the community space is, what the project materials are and why neighbors "corral", or courtyards, can be a way to review the housing.*

It has also been proved the feasibility of the contemporary housing interpretation from common spaces, both public and private ones, analyzing: ***common spaces related to the design*** *of contemporary housing, social and familiar* ***community spaces*** *as result of* **architectural mechanisms**, *how* **architectural mechanisms help to design the domestic space**, ***neighbors courtyards allow to review spaces of encounter of the contemporary housing*** *and* ***neighbors courtyards*** *as a new* ***space for the community***, that have helped to define the scope of this study and bring new concepts in order to perform a diagnosis of the situation and to establish the theoretical reference frame of the object of this research: the contemporary housing and its review from neighbors courtyards and the concepts related to both areas.

The hypothesis established is that there are traditional architecture models that are current and reusable for the review of the contemporary and future domestic architecture. Therefore, from the methodology, as a way of rapprochement, it has been established a series of concepts and knowledge elements that are used in domestic architecture project as it has been verified by using these in specific proposals as study cases, performed by this research.

M.d.C. Martínez-Quesada (✉)
Department of Architectonical Projects, Higher Technical School of Architecture, University of Seville, 41012 Seville, Spain
e-mail: mquesada@us.es

P. Mercader-Moyano (ed.), *Sustainable Development and Renovation in Architecture, Urbanism and Engineering*, DOI 10.1007/978-3-319-51442-0_8

In them, it is verified the possibilities that these specific tools provide, in which the reference framework is placed, the *project descriptors*, for the reconfiguration of the structures of the selected rooms. Conclusions are made according to their values and aims reached in the performed actions in accordance with the descriptors to which they are subjected.

They become apparent by the experimental projects that are performed, since, although they depart from certain intentions for their development, the objectives initially exposed are explicit in both graphical verification of the different project tools used, as in their own conclusions made on various issues related to the projects.

Among the reached conclusions, there are the articulation strategies between public and private spaces, development of a methodology to study and verify the results of this research and the confirmation of the possibilities that traditional architecture has to provide by analyzing its characteristics and qualities for the reconfiguration of the contemporary housing.

2 About the Contemporary Housing

At this time, in which the heritage of the housing bubble has halted the construction of dwellings and has hampered the access to a fundamental right as is the housing and that contemporary housing, as a design problem, turns around a series of common issues, as shown in the different study cases performed by this research, such as that new compositions, the ways of living and the familiar necessities are constantly changing, reddening its members roles, it has been detected that current housing structures, based on the sum of more or less specialized spaces and directed to the traditional family, are no longer valid: it is not a result of the sum of functionally stable spaces (Kronenburg 2007).

Nowadays, housing extracts services, shifting the gravity center from the home to the community; the family concept has changed; there are dwellings that are empty or underused; the limits to which it is subjected difficult the socialization process of inhabitants and their weight in decisions and the adaptation to the changing needs of these are increasingly its importance. It leads to a second matter, such as the need to adjust the housing to these new situations from new points of view that will contribute more appropriate answers to each of the existing micro-realities, away from neutral and universal solutions and from positions that will allow to understand housing as an open process that answer to social evolution, professional and personal development of the individual and his family, by the interpretation of variables and unpredictable ways of changing life and allowing people to reflect their identity and privacy (Eleb and Simon 2013; Monteys and Fuertes 2001; Monteys 2014).

Therefore it is necessary to provide a new way of research able to review the housing concept and to establish useful tools for the construction of the domestic project.

3 The Construction of the Architectural Space from Common Spaces

3.1 Heritage and Memory, Popular Architecture for the Construction of Dwellings

Exposed the initial situation, and among possible values from which to address the matter, finding a private space with the need to establish relationships with others constitutes a possible research line for the reconfiguration of the livable space from the construction of spaces for community use, for both household and community (Chermayeff and Alexander 1968), that makes domestic architecture a way of thinking the coexistence, for sharing resources, reorganizing public and urban spaces, and enabling the search of new agreements and commitments with society (Fig. 1).

Given this possibility, it turns out that there are examples that bring this relational value in their spaces (Fig. 2). In these, tradition is their fundamental principle, and the relationship that constantly establish with the past justifies their future representation. There coexist individual and collective projects, intermediate mechanisms, such as galleries, "patios", or covers, where non-regulated negotiation processes are developed, which allow to understand a social structure that shares a space.

This is a traditional architecture whose value is provided by its ability, learned in time, in order to build the domestic space from human relations that accept adaptation as a way of living and that explain the space (Morgado 2003).

Therefore it is proposed that knowledge and qualities of traditional architecture spaces serve to rethink contemporary and future housing. The hypothesis is the following: if certain vernacular architecture already contemplated favor changes over time, established special relationships between private and public,

Fig. 1 Sarugaku, Hirata/Sohigaya house, Be-Fun (Hildner 2014)

Fig. 2 Living in a neighbor courtyard, 1961 (ICAS-SAHP, Fototeca municipal Sevilla. ge18_e-v1_61_002_01-3 BDI)

and expressed design criteria shared by the society that inhabits its. Why not use them as support for the review of contemporary housing? Why not make them a possibility of experimentation of neighbors contact spaces that will make them places for new ways of living and adaptable to our needs?

After this main matter, the first issue is that the typology of neighbor courtyards still has not been reviewed from the reconfiguration of social relations and space/inhabitant that made them a living building (Fernández 2003).

And the second one is that from the project it can be established a research methodology of the architectural discipline that expands knowledge through innovative proposals that are halfway between the recovery and review of an architectural type.

3.2 A Contemporary Domestic Project

How do we proceed to prove this hypothesis and its derivatives? First, and to answer the main matter, it is proposed a new field of experimentation with multiple and open contributions that allow to register events that happen in a model of traditional architecture from other indicators that provide new materials and data, in order to be able to approach the issue of community space and its involvement with the reality of housing and city (Anderson 2008).

As for the reviewing of neighbor courtyards, they are adapted to new ways of living, social habits, linked to information activities, and new technologies in space management from a different attitude and interests, not influenced by the conditioning model topics and from the consideration of the potential of the project of these buildings as a result of its historical experience (Alexander 1969).

And regarding the proposed interventions, these are used to exemplify the spatial consequences of the suggested tools (Moneo 2004), issues raised by this research and are involved in the design process, comparing them with each other.

3.2.1 From New Tools

Tools used were three: the neighbor's courtyards of Triana, the architectural design and the domestic descriptors.

First, neighbors courtyards have been used as a tool to consider that they are models based on the construction of supports for the community social activity (Habraken 1975; Habraken et al. 1979), allowing the development of the person in community and that have managed to incorporate the time dimension, so its use allows to think from a cultural reality and the memory (Alexander et al. 1980).

They also are not conditioned by principles of serialization of a standard cell for a standard family that, on the other hand, are rarely established as a community, nor by policy issues, which are difficult to be considered in these examples. They are also less influenced by market needs than other models of collective housing.

Within this typology, this research is focused on the examples of Seville because it allows to study the presented matter from within, not in the distance. Thus, it was possible to meddle in the reality of neighbors courtyards, which has been a first-hand source by visiting, measuring and observing how people live.

And from those examples in Seville (Fernández 2003), it has been chosen courtyards from the Triana district because it is a physical framework in which existing units still allow to establish a map, a network between them (Fig. 3). The formal characteristics achieved by this typology in this urban context, fueled by the shape and size of plots, and the precise formal and social definition of the district, which has only come to break with the growth of the second periphery of Triana, have remained until today (Trillo and Martínez-Quesada 1995).

Furthermore, in spite of the number of missing examples, this district is still has very significant presence of these, which makes that the field of work does not depend on morphological variations of the city but of the type itself (Martí 1993). This homogeneity in the localization and planning regulations also facilitates the comparison of the obtained results and the proposed research.

The second tool, the architectural project, is a result of the lack of models with features and spatial mechanisms for the study and verification of the hypothesis; therefore it is necessary to generate own examples to experiment with other tools and proposed concepts (Fig. 4) (Alexander 1969; Moneo 2004).

Fig. 3 Map of existing neighbor courtyards, 2015 (Martínez-Quesada 2016)

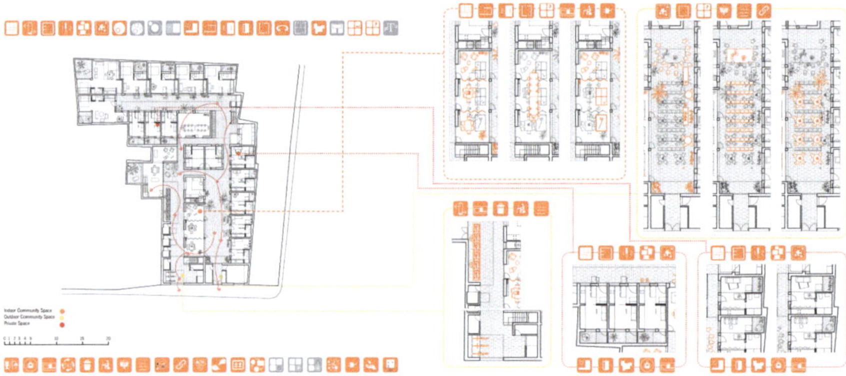

Fig. 4 Intervention verification, Bernardo Guerra 5 (Martínez-Quesada 2016)

By the action of the architectural project, it is studied what happens and it allows to display the potential research that it has because of its scanning capability, analysis and verification of specific items from which to determine conditions of practical type and its possible use.

Finally, and within the used tools, it has been detected, through different approaches and texts that are being made on the subject (Eleb et al. 1988; Sabater 2009; Miranda et al. 2009; Montaner et al. 2011; Hernández Pezzi et al. 2014), concepts interrelated with users experiences of the dwellings that are repeated, which have been studied and are introduced in this research as reviewing tools of the domestic space. These are the thick wall, the active band, the satellite habitat, the extra space, the unused room, the threshold, the permeability of the facade, the dysfunction, the un-partitioning, the residential hybrid, participatory processes…, mechanisms of diverse nature than are offered to the inhabitant as instruments with which can interact in the reconfiguration of the housing, establishing other different relationships from the traditional ones, more appropriate to the changes that are taking place, or as strategies that attempt to articulate processes as a means of control of the space which would no longer be established through shape but through tactics rules of game (Gausa 1996).

Therefore it is performed an approach to a number of basic concepts of domestic spaces and a review of the way of understanding the domestic project from other concepts that are specified in the project descriptors, which are contrasted with spaces from neighbors courtyards that are selected as an experimentation basis, and analyzed on developments of particular solutions in these supports through project processes, which are not intended to constitute normative models but a way of checking the proposed concepts and their effect on the way it is lived and imposed conditions.

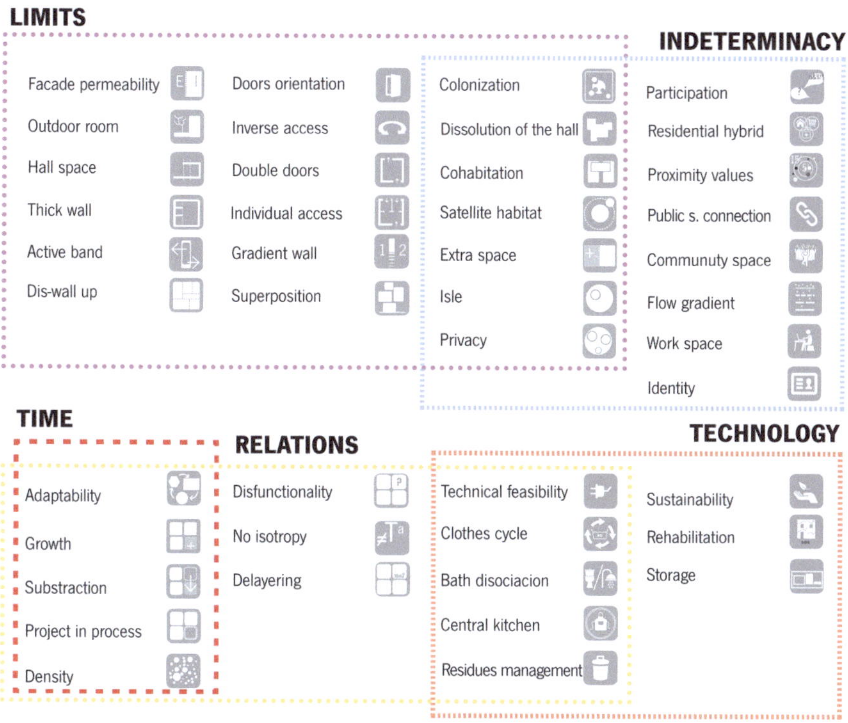

Fig. 5 Aggrupation of descriptors according to proposed categories (Martínez-Quesada 2016)

Descriptors are concepts with possible links to each other, but can be recognized as autonomous domestic resources. This double reading has established categories and descriptors (Fig. 5). Each concept—category—identifies a collection of tools—descriptors—which with its own meaning, help to define the first one.

The inclusion of the descriptors into categories involves some difficulty since, although they have been analyzed in this research, they are concepts of ambiguous limits. The *limit*, *relationships*, *time*, *indeterminacy* and *technology* have been the chosen categories in which descriptors are registered and identified, but it is recognized that their inclusion is not definitive and unique, but only one that is possible, as there are so many interested glances that can occur.

The adjustment of the descriptors and therefore their suitability for the purpose that characterizes it (Alexander 1969), necessarily happens because it is sufficiently specific and is well defined in terms of its objectives, for which there has been a catalog, and each one has been exemplified according to the existing projects and proposed study cases.

The problem of quantifying these in complete projects, although it has been influenced by the way they have been evaluated, has been expressed by two tables: (Fig. 6) one of presence, in which it is included the data of the approach to the descriptor concept, and another of the intensity perception of the descriptor presence, both in housing and community space.

The presence and frequency of these descriptors represent the potential of common areas in a particular model, the neighbor courtyard, which verifies the feasibility of the requalification of these from the proposed contemporary standards.

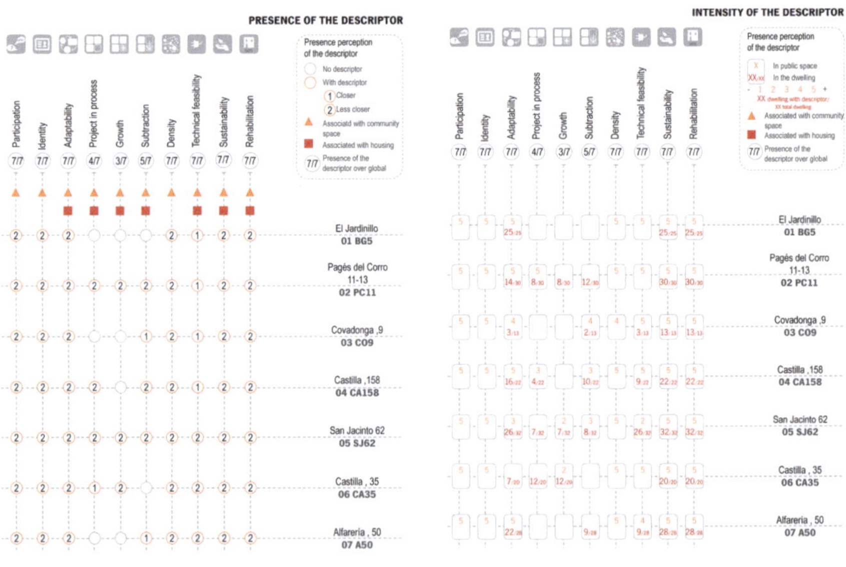

Fig. 6 Extraction of the quantification (Martínez-Quesada 2016)

However, those data have the value to establish other spatial and formal relations from other different approaches for future research relations.

3.2.2 Experimental Dwelling Projects

Neighbor's and descriptors are reviewed by performing five plus two own intervention proposals (Martínez-Quesada 2016) that serve to show that used elements, tools and models, are valid and usable in performing contemporary housing. They are original proposals that produce specific architectural situations, but, because of the used method, may represent a way of usable work elsewhere or with other techniques.

It has been chosen five examples (Fig. 7) of the 51 that are inventoried and two unoccupied plots, in order to establish a heterogeneous field of experimentation that serves to verify the validity of the results: two of the locations are adjusted to the original plot—01 BG5 and 09 CO9—in other two, it is annexed a small residual plot from the urban fabric, related to access spaces to block interiors, which is difficult to currently use—01 PC11 and 05 SJ62—and the last one—04 C158—annexes two adjacent neighbor courtyards, prioritizing the one which still retains its use.

In the unoccupied plots—06 C35 and 07 A50—it has been valued their shape, size and representativeness of the characteristic of the historic suburb, as a place to test the feasibility of the type reviewing.

Interventions have been managed by templates that in the first part serve typological characteristic and the second correspond to the graphical verification of descriptors and the interpretation of their implication in the definition of the domestic and collective space of neighbors courtyards, attending to the articulation of public and private domains and the relations between spaces, which eventually materialized in own conclusions.

This research has incorporated a new way of perceiving the world that provides information (Anderson 2008), where data are the means of explaining complicated processes of visualizing abstract phenomena, mapping the territory and its interpretation or building systems to generate an own one and it is from this approach how the analysis of the resultant data of the proposals made is understood.

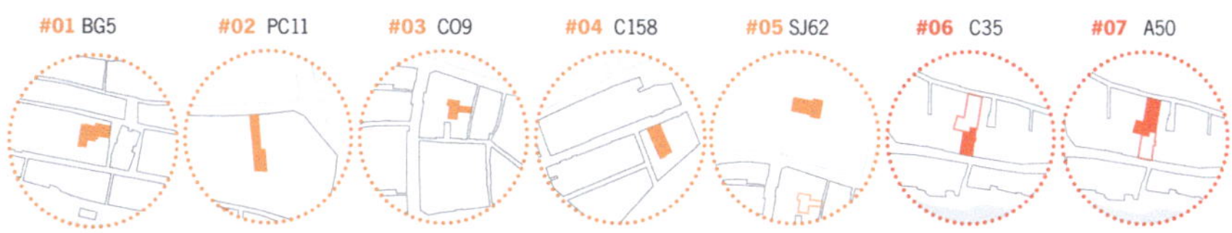

Fig. 7 5 + 2 Proposals (Martínez-Quesada 2016)

3.3 Results of Intervention Proposals

Conclusions are presented in a set of diagrams obtained from the seven approximations performed, which are understood as a set of models that, once analyzed, have allowed to collect the results relative to a number of typology characteristics in a set of graphics/summaries that possibilities the establishment of comparisons attending to them (Figs. 8, 9 and 10), gaining true value those of general character that allow to detect the potential of neighbors courtyards as the new community housing.

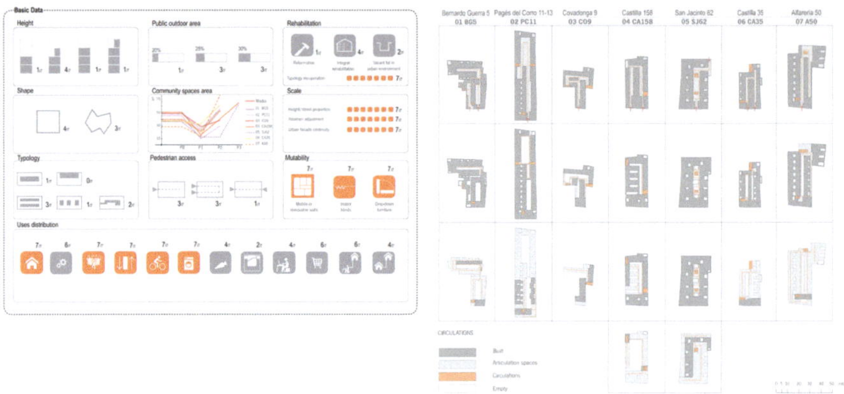

Fig. 8 Basic data and relation of built density versus free articulation spaces (Martínez-Quesada 2016)

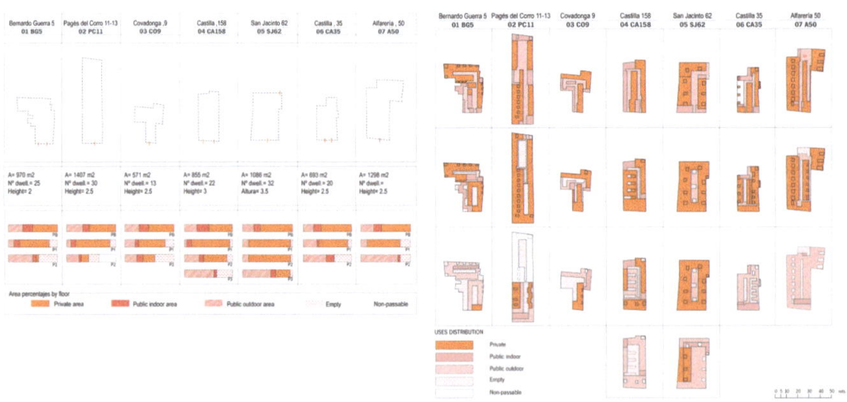

Fig. 9 Articulation of indoor public space (Martínez-Quesada 2016)

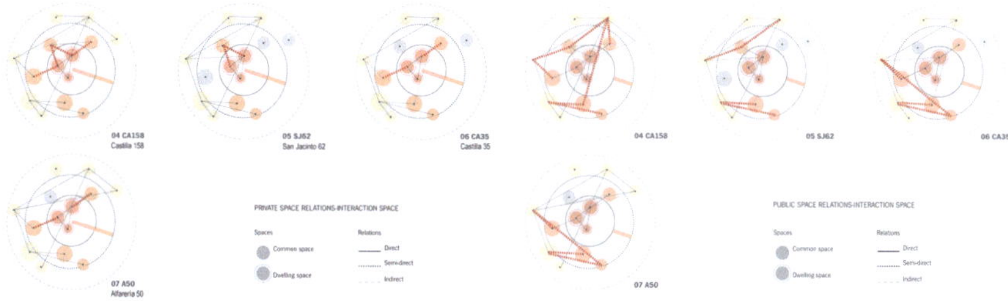

Fig. 10 Incidence of spaces and uses in the use of the interspace and its influence in private and public space (Martínez-Quesada 2016)

4 Conclusions

It is established mechanisms and non-regulated negotiation process that allow to understand the social structure, improve the coexistence relations and redefine a new sociability that makes sense to the urban space when it exceeds the proper limits of the chosen supports and creates an area of influence in their environment, which allows not only to continue to the constant extension of the city with the construction of new elements, but from mechanisms that allow to internally improve the city and to rehabilitate the community life through the review of neighbors courtyard.

The revision of the domestic project has been approached with the aim of developing strategies from which to review it and to propose elements that can be used for its contemporary construction.

It has been developed a methodology in which the studied object—housing—the place—neighbors courtyard—and the tools—domestic descriptors and project—are used to build this research and also to verify the achieved results by an innovative architectural project, raised from current budgets, which sets a new position in reference to the contemporary housing and a new study and analysis framework, and also allows to build a vision of the future based on tradition.

Neighbors courtyard are proposed to establish a temporal continuity that, from what already exists, beyond the social, urban or symbolic aspects, allows to think the future of these spaces and housing itself for the city and community as a representation and encounter places, using the architecture as a vehicle.

References

Alexander C (1969) Ensayo sobre la síntesis de la forma. Ediciones Infinito, Buenos Aires

Alexander C, Ishikawa S, Silverstei, M (1980) A pattern language/Un lenguaje de patrones. Ciudades. Edificios. Construcciones. Gustavo Gili S.A., Barcelona

Anderson C (2008) The end of theory: the data deluge makes the scientific method obsolete. In: Wired Magazine, issue 16.07, July 2008 http://archive.wired.com/science/discoveries/magazine/16-07/pb_theory/. Accessed 24 June 2015

Chermayeff S, Alexander C (1968) Comunidad y privacidad. Hacia una nueva arquitectura humanista. Ediciones Nueva Visión, Buenos Aires
Eleb M, Simon P (2013) Le logement contemporain. 1995–2012. Ed. Mardaga, Bruxelles
Eleb M, Châtelet AM, Mandoul T (1988) Penser L´Habité. Le logement en questions, PAN 14. Ed. Pierre Mardaga, Lieja
Fernández V (2003) Vivienda modesta y patrimonio cultural: los Corrales y patios de vecindad en el conjunto histórico de Sevilla. Scripta Nova, 146(070)
Gausa M (1996) Sistemas abiertos: nuevas interpretaciones del Hábitat. In: Nuevos modos de habitar. COACV, Valencia, pp 271–288
Habraken J (1975) Soportes: una alternativa al alojamiento de masas. Alberto Corazón, Madrid
Habraken J, Boekholt JT et al (1979) El diseño de soportes. Gustavo Gili, S.A., Barcelona
Hernández Pezzi E et al (2014) Vivienda contemporánea. Apuntes para una teoría. In: Montaner et al (ed) I Congreso Internacional de Vivienda Colectiva Sostenible. Barcelona, pp 350–353
Hildner C (2014) Future living. Colective Housing in Japan, Birkhäuser, Basel
Kronenburg R (2007) Flexible. Arquitectura que integra el cambio. Editorial Blume, Barcelona
Martí C (1993) Las variaciones de la identidad. CoAC y Ediciones del Serbal, Barcelona
Martínez-Quesada MC (2016) Los espacios de comunidad como materia de proyecto en la vivienda contemporánea. El caso de los corrales de Triana. Tesis Doctoral. Univ. de Sevilla
Miranda A, Mauri N, Amann B (eds) (2009) Instrumentos de proyecto. Concurso Viva. 5.688 Paisajes Domésticos vol. 6. SEPES Sociedad Estatal del Suelo. ARKRIT, Madrid
Moneo R (2004): Sobre la noción del tipo. El Croquis, 20 + 64 + 98:584–606
Montaner JM, Muxí Z, Falagán D (2011) Herramientas para habitar el presente. La vivienda del siglo XXI. Máster Laboratorio de la vivienda del siglo XXI, Barcelona
Monteys X (2014) La habitación. Más allá de la sala de estar. Gustavo Gili, S.L., Barcelona
Monteys X, Fuertes P (2001) Casa Collage. Gustavo Gili, S.L., Barcelona
Morgado R (2003) Uso del espacio y sociabilidad en los corrales de vecinos trianeros. Universidad de Sevilla, Tesis Doctoral
Sabater T (2009) Nuevos descriptores, nuevos operadores proyectuales. UPC, Universidad de Sevilla, Barcelona, Escuela de Arquitectura del Vallés
Trillo JL, Martínez-Quesada MC (1995) Los corrales de Triana. Proyecto y Patrimonio. Junta Municipal de Triana, Ayuntamiento de Sevilla, Sevilla

Restoration of Pier-Dock of Clevedon. An Example of Involvement of Society in Defense of the Cultural Heritage

Concepción González-García de Velasco, Andrés Agudo Martínez and Miguel González Vilchez

1 Introduction

Clevedon is an English city, placed on the western coast, over Wales, in the Channel of Bristol. Still nowadays it is kept relatively in altered, preserving the Victorian character of its better epoch. Its development was never spectacular, and the tourism of leisure did not change excessively, probably for the competition of other near ports of importance as Bristol, Penarth, Newport and Weston, all of them to less than 30 km. Its name has stayed ligature forever to its pier, for the lovers of the engineering, the architecture or the industrial archaeology, undoubtedly the most elegant of all the British wharves, happily preserved in its integrity.

The author of the wharf was the British engineer John William Grover, who projected the pier in 1866 for a private company, being constructed during 1867 and 1868. Lucky, Grover presented before the I.C.E. of London a work on the construction of this wharf, which means today we can know the technical, structural and constructive information of the project itself.

Grover projected for Clevedon a wharf of 298 m of length, the first 55 executed ones on work of stone whereas the 243 remaining ones were resting on eight vains of 30 m each one, formed by porticoes of four props of wrought iron, on which paths were resting, on both sides of the wharf, girders of double profile T, of wrought iron. These big lights, the height of the wharf on the sheet of water and the sensitivity of design of the porticoes, like lattices finished off by thin series of arches, give the pier a singular, slender and fragile image, which constitutes its sign of identity and its beauty. Though Grover had projected a simple metallic kiosk as

C. González-García de Velasco (✉) · M. González Vilchez
Department of Building Constructions I, Higher Technical School of Architecture, University of Seville, Av. Reina Mercedes, 2, 41012 Seville, Spain
e-mail: conchitaggv@us.es

A. Agudo Martínez
Gabinete de Proyectos, Universidad de Sevilla, Seville, Spain

P. Mercader-Moyano (ed.), *Sustainable Development and Renovation in Architecture, Urbanism and Engineering*, DOI 10.1007/978-3-319-51442-0_9

Fig. 1 The castle at the entry of the wharf (González-García de Velasco, Agudo Martínez y González Vilchez 2016)

cabin of control for the entry of the wharf, the property entrusted the architect Hans Price the project of a building with appearance of small castle that sheltered the dependences of control and that was contrasting its solid and medieval appearance with the delicate design and levity of the metallic structure. Today this contrast has remained assumed and constitutes one of the singular and landscape characteristics of Clevedon's wharf (Fig. 1).

2 Constructive Characteristics of the Wharf

The superstructure, resting on these porticoes, was composed of both metallic parallel already mentioned girders, and of other transverse girders, likewise metallic, placed in the same plane of the portico. On the principal girders they were resting other minor girders, made out of wood, on which planks of wood were fixed parallel to the wharf main direction. The banks were simple and compact, executed also in wood. The set was armed with San Andrés's crosses, formed by bars that were crossing in a central hoop. There also existed platens placed under the horizontal plane of the cover (Fig. 2).

The supports were formed each of them by four pillars of wrought iron, constituted by Barlow rails, clinched at the back, forming a tubular hollow profile. The arches were formed by a curled Barlow rail, clinched on a sheet or flat platen, and its tympanum was dividing for bars of the same type, like lattice. This typology of supports had been already used by Isambard Brunel in constructions for railroad bridges, and Grover found a sufficient stock of these elements, so the structure

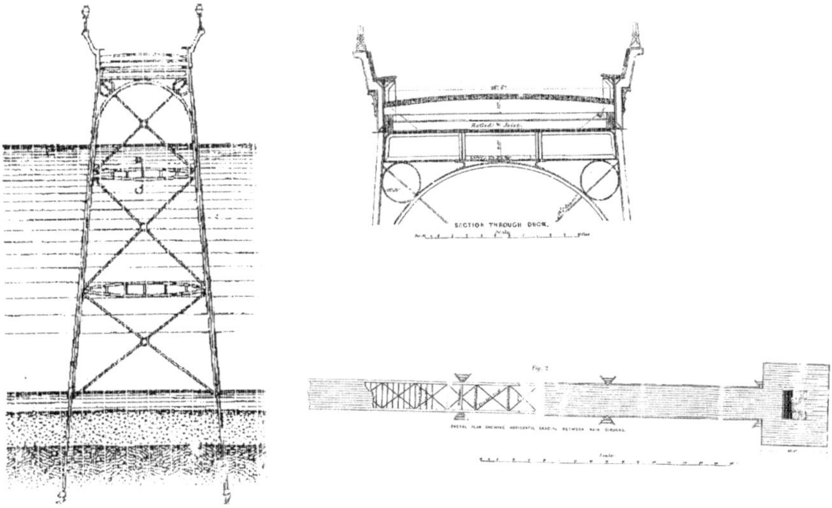

Fig. 2 Floor and section of the wharf [Grover (o.c. pág. 135)]

could be mostly prefabricated, from these materials. Each of four supports, or legs of support of the portico, was fixed to a stem of wrought iron of 12.7 cm of diameter, fixed in its turn to a helicoide Mitchell type, of cast iron, of 5.08 cm. The piles had been screwed before by means of a capstan of six arms, always placed at the end of the last elevated section, round which there was a passing rope on which several men were pulling, making the capstan turn (Fig. 3).

The original headpiece of the wharf, a rectangle of 15 × 14 m, excessively small for budgetary reasons, was executed by supports of wrought iron of the same characteristics, and had five platforms of different heights, to embark in the different tide levels. No construction was projected originally in the end of the wharf, though years later, ruined the structure of shipment due to a storm, got up a new headpiece on piles of cast iron and on it there got up an oriental style pavilion of (influenced by Nash's work in Brighton), that has remained to the present day (Fig. 4).

The description of the structure and the process of construction were narrated by Grover in his work already mentioned. Once screwed and placed in position the piles of wrought iron, the porticoes were mounted on the beach, one to one, and were transported in barges up to their place of anchorage, being hung up to the end of a pulley fixed to a tower of 27.43 m of height, which one was anchoring to four before placed timber piles. The pulley was placing the portico in its definitive position, riveting this one on the cylindrical piles of forged iron previously turned. Afterwards the girders launched in the same way and were raised to their position on the props by means of the mentioned pulley and a derrick installed on the previous already finished section. The pulley and the derrick were moved to the adequate position for executing the following section, up to finishing the complete wharf. The construction was quite rapid and the works - an approximate total of 350

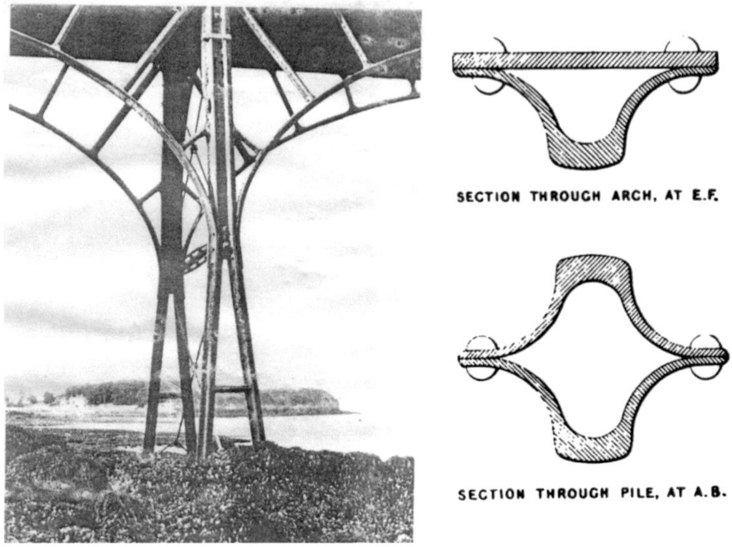

Fig. 3 Barlow rails forming curled pipes [Mallory (o.c. pág 13)]

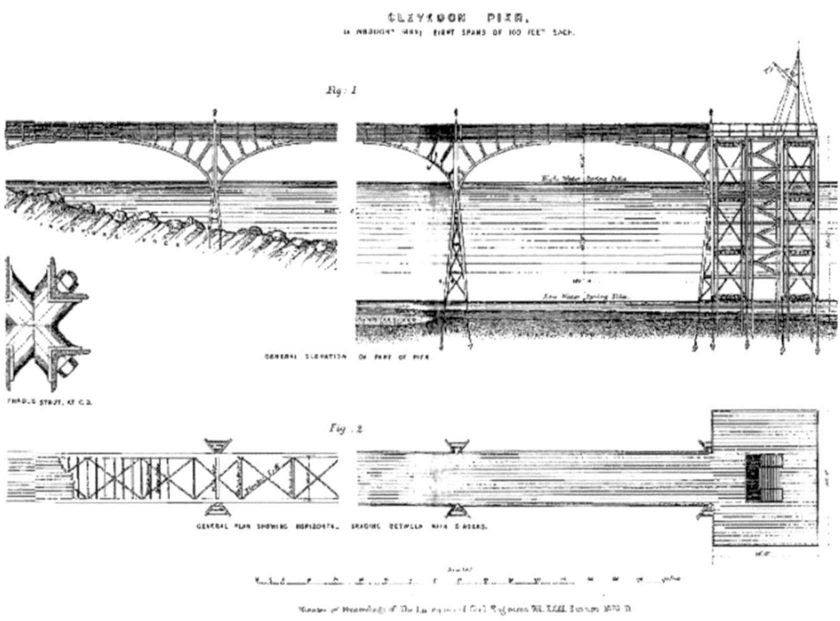

Fig. 4 Floor and gathering of the wharf with the first headpiece [Grover (o.c. pág. 137)]

tons of wrought iron—were executed in 18 months, with an average workforce of 60 men, a number, then, comparatively small for a project of that size. The second part of the report that we got in the I.C.E. of London on Clevedon's wharf, contains the discussion of the work presented before the presidency and members of the institution. Before them, Grover justified the placement of massive piles of wrought iron, instead of cast iron, basing his decision on his own experience consisting on its good performance, on which elements of wrought iron had been placed ten years ago in South Wales's wharves, which had not suffered scarcely any corrosion in all that time. But probably the most interesting point of this discussion was the theory of calculation that Grover presented, considering both hypotheses that he had taken into account in the posing of the structure: firstly considering each central vain as formed by two symmetrical arches, like brackets fixed in the supports that they had a punctual load in their end and, secondly the theory of continuous girder, with supports in the heads of the supports, contributing for it several pages of equations and considerations of calculation large developed, as well as the graphs of bending moments and deformations of the second hypothesis (Fig. 5).

Clevedon's wharf was never an economically prosperous attraction, given the competition of other close wharves. In addition, Grover had projected it in a direction not perpendicular to the coast, for what the platform of the pier was not

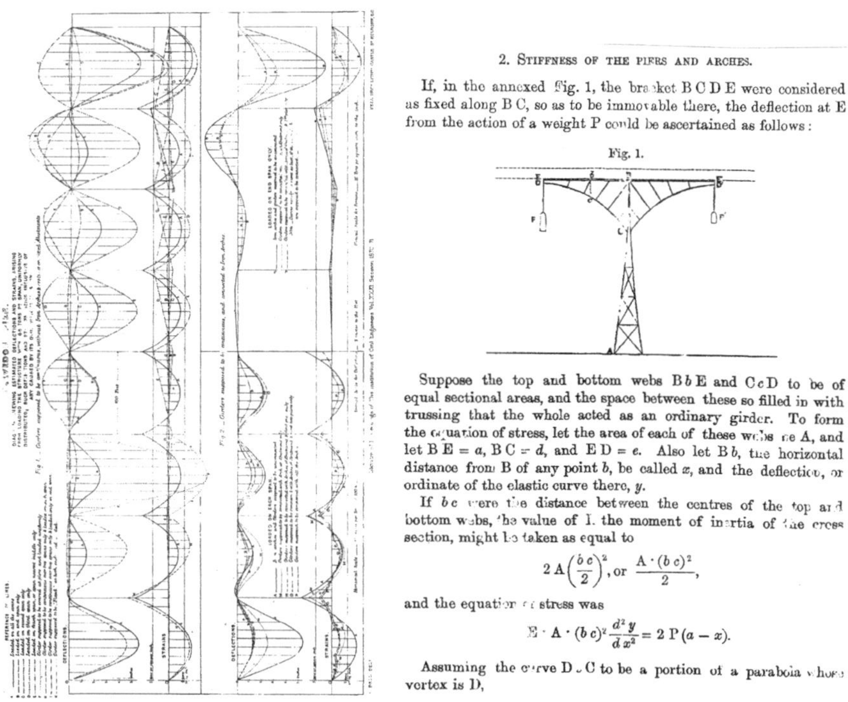

2. Stiffness of the piers and arches.

If, in the annexed Fig. 1, the bracket B C D E were considered as fixed along B C, so as to be immovable there, the deflection at E from the action of a weight P could be ascertained as follows:

Fig. 1.

Suppose the top and bottom webs $B b E$ and $C c D$ to be of equal sectional areas, and the space between these so filled in with trussing that the whole acted as an ordinary girder. To form the equation of stress, let the area of each of these webs be A, and let B E = a, B C = d, and E D = e. Also let B b, the horizontal distance from B of any point b, be called x, and the deflection, or ordinate of the elastic curve there, y.

If $b c$ were the distance between the centres of the top and bottom webs, the value of I. the moment of inertia of the cross section, might be taken as equal to

$$2\,A\left(\frac{b\,c}{2}\right)^2, \text{ or } \frac{A \cdot (b\,c)^2}{2},$$

and the equation of stress was

$$E \cdot A \cdot (b\,c)^2 \frac{d^2 y}{d\,x^2} = 2\,P\,(a - x).$$

Assuming the curve D c C to be a portion of a parabola whose vertex is D,

Fig. 5 Grover's calculations [Grover (o.c. pág. 185)]

Fig. 6 New central pavilion in the end of the wharf (González-García de Velasco, Agudo Martínez y González Vilchez 2016)

remaining parallel to the current direction of its estuary. This caused not few difficulties to the ships and the consequent blows of these against the structure of the headpiece, which in the year 1891 was sensitively damaged. After diverse negotiations, the shareholders achieved that the wharf was transferred to property of the local authority, for a reasonable price for both parts. The town hall repaired the general damages of the wharf and arranged the demolition of its headpiece, entrusting the engineer McDowell the execution of a new headpiece for the wharf, more robust and of larger dimensions, raised on piles of cast iron. This headpiece, which was finished in 1893, was designed this time orientated in sense of the running, and therefore lightly listed with regard to the directive of the wharf. On this platform were executed a few constructions that, in the shape of a central pavilion, both with of oriental style covers, had been projected for the new headpiece by municipal order. The wharf was re-opened in 1894, with the appearance that it has preserved to the present days (Fig. 6).

3 The Decadence of the Wharf

As all the piers, it suffered important damages along its life. In 1910 it was closed after being destroyed part of the cover of wood by a gale, reopening in 1913. In 1953 its economic results were bad, decreasing year after year, in spite of the efforts of the local town hall to promote its use. Its maintenance was not the suitable one and, before the danger of collapse in any sections of the beginning of the wharf,

Fig. 7 Collapse in the end of the wharf (http://clevedonpier.co.uk. 2016)

cylindrical piles of steel were placed, as prop shores of the wrought iron arches, spoiling its image. In an inadequate test of load that took place on October 17, 1970, filling water rafts placed on the cover, one of them was deformed, displacing the seventh portico, which was ruined dragging with itself both vains that were resting on it.

The wharf remained closed for many years, and the town hall, faced with the costly of its repair, decided to knock down it. But a massive popular response in favour of its conservation and the longed official cataloguing like protected building in “Degree II”: buildings that are of special interest, warranting every effort to preserve them, avoided its end, appearing in 1980 a project of reconstruction for an amount of 870.900 Eur. Thanks to the funds of the English Heritage the works began in 1985, of whose development we will deal later, considering the works finished in 1998.

4 The Restoration of the Wharf

We have reflected that, already in the year 1953, the economic results of the pier were poor, decreasing year after year, in spite of the efforts of the local town hall to promote its use. In the year 1970, the local town hall thought about the need to know the structural state of the wharf to establish if its repair and its possible impulse was profitable as local attraction. Since we have commented, in an inadequate test of load realized on October 17 of the mentioned year, filling water rafts placed on the cover, one of them was deformed, displacing to the seventh portico of the pier, which was ruined dragging with itself both vain that were resting on it (Fig. 7). The wharf remained closed for many years and, finally, the town hall, before the uncertainty of its state and the future of its repair, decided to knock down it.

But in the village of Clevedon a massive popular response took place in favour of its conservation, which came out of the regional and national press, which originated a great drift in favour of its rehabilitation stimulated by industrial archeologists, cultural societies (The National Piers Society) and the own British administration for heritage (The English Heritage). The wharf was catalogued officially as protected building "Degree II": buildings that are of special interest, warranting every effort to preserve, what prohibited its demolition and set the town hall of the city to study a project of future that allowed its definitive conservation.

In the year 1980 a project of reconstruction of the pier was presented for an amount of 750.000 GBP, part of which would be supported by funds of English Heritage. This project was modified later beginning the works in 1985 and finishing them in 1998. Today Clevedon's shines its brilliance and constitutes a model of action in defense of the architectural industrial heritage, which has preserved this monument for the posterity.

The principal constructive characteristics of Clevedon's wharf might be summarized in the following points:

Foundation of massive piles of wrought iron, screwed to the subsoil with coils of cast iron Mitchell type. Hollow props formed with Barlow rails of wrought iron, like ridge of four supports, from which they emerge in the shape of arch other rails held up on structural bars placed on the tympanums of the arches. The wharf is composed for 8 vains of 30.5 m each one. Any of them is formed for two metallic girders of wrought iron, of full core, parallel to each other, of 1.07 m of end and 0.45 of width, tied transversely by other similar in every portico. The wrought is constituted by iron girders of minor end, supported on the low wings of the principal girders, and on there were a few beams of wood and a parquet floor of strips in disposition "open".

The condition of the wharf was studied by engineers Roughton and Fenton who undertook a very detailed inspection of it, valuing the condition of corrosion to which its elements had come and proposing a project of consolidation of its structure consistent in keeping the Barlow supporting, refilling them with concrete of epoxy resins to increase its strength and stiffness (replacing the ruined props by others executed also with Barlow rails), and in reinforcing the metallic girders which join firmly for its interior a metallic plank mould, with a wing of polystyrene, to concrete inside it a few big girders of concrete post-tightened, that would get rigid the horizontal structure, relieving of load the metallic girders without losing its physiognomy.

The subsidy obtained from the English Heritage was determined to certain modifications in the project, the most important of which was of not allowing the reinforcement with concrete girders after which the project was reformed, re-calculating the structure with thicknesses and real loads. According to Fenton, several girders and numerous joists were replaced with others executed in soft steel, and were repaired those which allowed it. Also, according to Fenton, some supports were replaced using new Barlow rails, identical to the original ones acquired to the British Rail that still preserved some bars in its stores, realizing soldered joint tests between the wrought iron and the soft steel.

After our visit to Clevedon's city and to its wharf, of which we have to say that with its one contemplation deserves the efforts to have studied and investigated it in depth from a city as distant as Seville. We observed that the last three ridge of the structure, contiguous to the platform of the pier and correspondents to the ruined structure, show off to be executed by normalized profiles, in any case, perfectly integrated in the global solution.

The system used to execute the work of repair was curious, which consisted of installing a great platform of work fixed to half tide, at a height of the half of pier, with a great derrick. Once dismantled the board, the joists and girders were coming untied one by one, and with the derrick they were loaded in barges and were removed to the beach. There, in dry, in a built-in workshop at the foot of the beach to this end, they were examined thoroughly and it was decided if they were repaired or they were replaced with similar others. Likewise there were dismantled the props that were considered in poor condition and replaced with others mounted on the beach and moved to its position in the pier, whereas those which were allowed to be repaired (Fig. 8).

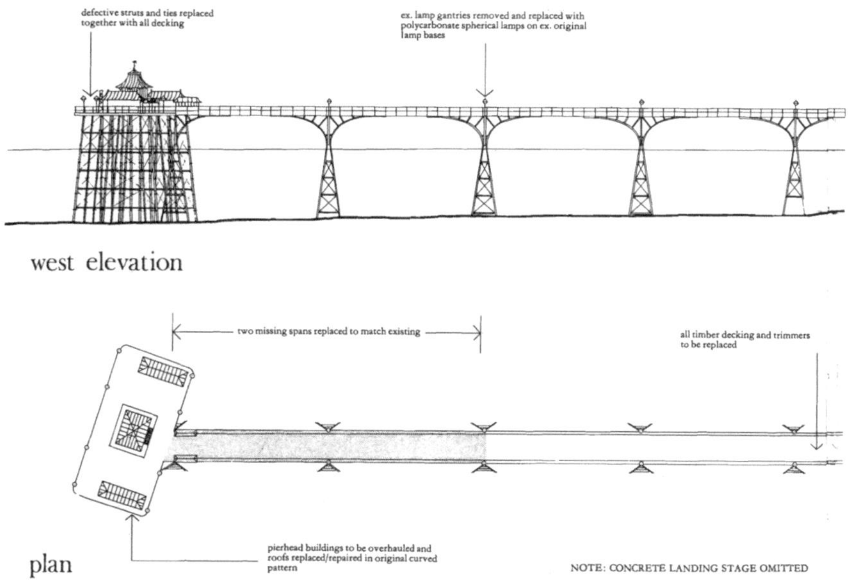

Fig. 8 Floor and gathering of the wharf with the new headpiece [Mallory (o.c. pág 53)]

Once repaired or replaced the big girders, they were moved again to its place of origin, the suitable works of fixing were executed. Later the joist, the beams and a new pavement of wood were installed. In the end of the pier three pavilions were placed, now specially glazed and dedicated to the stay of the public, in protection of the rain and the wind, whereas the ground floor of the principal pavilion was use for entering by stairs to the low space, from which people could embark in different levels.

The works of the bridge ended in 1989, date when it was opened to the public with the exception of the final end, which continued a few years more in repair, and were carried out by the British company Sheppard Engineering Ltd, from Gloucestershire, who were specialists in restoration of metallic structures. Today this company can show the work of restoration of the wharf of Clevedon as the best and the most international of its accomplishments.

Later, in 1998, the extreme platform was re-opened, with its rehabilitated pavilions, which was celebrated by the population of Clevedon with a great festival, in which the persons travelled up to the wharf in adorned coaches of epoch and dressed in Victorian suits (Figs. 9 and 10).

A recording that sells in the small castle of entry to the wharf turned into shop of souvenirs, place this inauguration on record, which supposed the victory of the constancy and of the effort of the people for recovering a jewel of their past, about to disappear forever. Meanwhile the Brighton Pier, the pre-eminent figure of the

Fig. 9 Repair of trestle or support of the wharf (https://www.ethex.org.uk. 2016)

Fig. 10 Party in the reinauguration of the wharf in 1998 (https://www.ethex.org.uk. 2016)

British piers of leisure, has succumbed a few years ago, and has disappeared of the beautiful beach on the one that was located, passing Clevedon's pier to promote Grade I: Buildings of Exceptional Interest, as the best sample of the British wharves of the 19th century.

The principal use of Clevedon's wharf at present is, purely and simply, a place of walk and of fishing with cane, though also it is in used as port of mooring place for pleasure boats that lead passengers of this point to other near coastal cities (Fig. 11). Its owner is a municipal company, which charges a small quantity from the visitors who walk on it, and does contracts with the local fishermen who use it as privileged place from which going fishing. It also rents the wharf for outdoors celebrations, like weddings, meetings and parties, which are organized in the extreme pavilions and in the central walk covered by a marquee. It has been chosen numerous times as setting of musical attractions (for example, One Direction's musical recent recording), presentation of records and events, etc. But the most important thing is that the name of Clevedon's city continues ligature to its pier, which it could have restored exemplary and to preserve as example of all and that has turned into the best tourist and cultural claim of the population.

Fig. 11 Current image of the wharf (García de Velasco, Agudo Martínez y González Vilchez 2016)

5 Conclusions

Luckily, the village of Clevedon is a sample of how a awared society of supporting its architectural industrial heritage, imposed its will to the political decisions that were claiming its disappearance. In this way, they obtain the protection, rehabilitation and the maximum recognition of its soft pier by the cataloguing as Monument of Exceptional Interest (Grade I), which guarantees its survival. May serve as reference of similar implications of the society: San Pelayo's Chapel (19th century, Asturias, Spain), which was in total condition of abandonment, and with the neighbours's effort was rehabilitated and reopened for the people in a celebrated inauguration; or the Cinema Odeón (neighbourhood of the Candelaria, Bogotá, Colombia), closed and totally abandoned which was restored by the people's initiative, reopened again to the whole people. Both are clear examples of rescue and conservation of the historical heritage of the peoples to instances of themselves.

It is tried with our work, to establish a model to be based on such future lines of investigation as: the conversion of works of Architecture in society heritage, or the definitive inclusion of the Industrial Archaeology to the Architectural and cultural scene.

References

Adamson S (1977) Seaside Piers. London
Anónimo (1988) Clever restoration of Clevedon Pier. Architects J Lond
Betjeman J (2016) The most beautiful pier in England. http://clevedonpier.co.uk
Clevedon Pier & Heritage Trust Ltd. (2016). https://www.ethex.org.uk
Fenton R (1984) Clevedon Pier. Special problems relating to its restoration. Papers presented at the first symposium of the Piers. Department of the Environment, London
Fenton R (1991) Clevedon Pier. Special problems relating to its restoration. Papers presented at the second symposium of the Piers. Department of the Environment, London
González-García C (2008) Caracterización de los sistemas constructivos de los embarcaderos británicos del siglo XIX. Influencia de su tecnología en la ingeniería civil española. Tesis, Universidad de Sevilla
Grover JW (1871) Description of a wrought iron pier at Clevedon, Somerset, Institution of Civil Engineers. Minutes of Proceedings, London
Mallory K (1981) Clevedon Pier, London

The Empathic City. Towards a New Model of Urban Sociability

Isabel Jiménez-López, Angela Barrios-Padura and Marta Molina-Huelva

1 Introduction

We understand public space as the place where urban life develops, a space of relationship in which contacts and connections between people occur and enables the realization of collective activities. Jacobs says that: "The streets and sidewalks are the main public places of a city, most vital organs" [...] "They are a means of communication and contact, an authentic social institution of the city" (Jacobs 1961).

In this sense, the architect Allan Jacobs identifies that "The streets are places of meeting and exchange from socially and commercially, are the medium where people meet each other and that is, the end of the day, the main raison to be of the cities" (Jacobs 1996). He identifies the streets and places full of meetings and feelings that accommodate many chance encounters, understanding them as a meeting place and urban sociability.

"Empathy" reflects the feeling of identification with something or someone (RAE 2001). Within the context of urban space, Leadbeter described it as the capability to connect with different oneself people, to find common ground, to exchange and share, "it is the dark matter of urban life: invisible but always present" Leadbeter (2014). In this regard it suggests that the success of a city is mainly due to the combination of two components that are systems and empathy, i.e., a support, the public space, and a relational structure and cooperation between people and spaces, the empathy.

I. Jiménez-López (✉) · A. Barrios-Padura
Department of Architectural Constructions I., Higher Technical School of Architecture, University of Sevilla, Seville, Spain
e-mail: isabeljimenez@us.es

M. Molina-Huelva
Department of Structures, University of Sevilla, Seville, Spain

P. Mercader-Moyano (ed.), *Sustainable Development and Renovation in Architecture, Urbanism and Engineering*, DOI 10.1007/978-3-319-51442-0_10

The main objective of this research is to study the relationships happening between people, actions and spaces, establishing so a new measurement scale that we will denominate Degree of Empathy, and generate a research methodology that gets to join in the assessment of the public space the systems and their relationships, i.e., quantitative and qualitative methodology, with the aim of improving the quality of life and livability.

Thus you need to understand public space as a space which by its very nature facilitates collective life in the city, which enables the encounter turning the streets into the most important structure of the city where they can develop countless scenes of great diversity. Therefore the street becomes an important value today, if understood as a mere circulation space in the beginning, right now we understand it as the space of relationship and development of everyday life, and thus it becomes a relevant study space for a new livability.

2 Public Space, Paradigm of an Empathic City

The urban renewal processes are essential for the development of life in the city. The design in free space is presented as a key to improving the quality of life of the habitants component, not only through the prism of their quality it is spatial or physical, but also through its cultural and environmental component, with a common goal which it is to achieve higher levels of collective welfare on urban spaces, improving livability in public space.

Borja raises the need for a new approach to analysis of urban dynamics through the "study of public space and the relationship between the configuration and the exercise of citizenship, understood as the statute that allows exercising a set of rights and civic duties, political and social" (Borja 1998).

The need for study of free space is increasingly "evident as a space for cultural relationship, leaving aside or minimizing its administrative and legal concept linked to an area of the free city building for public use, and giving qualities as a place relationship and identification, contact between people, urban animation and sometimes community expression" (Borja and Muxi 2000).

Segovia and Jordán say the role of public space is essential as meeting place. On their own words: "It is a place of relationship and identification, (…) contact between people, urban animation, sometimes community expression. (…) Sets the scope for deployment of imagination and creativity, the place of the party (where communication of all with all recovers), the symbol (the ability to recognize ourselves), the game, monument, religion" (Segovia and Jordan 2005).

Linkages and relationships established between people, the actions and spaces or environments create a sense of identity to the extent that the appropriation of it is generated. According to Vidal and Pol (2005) "The place attachment, place identity, urban social identity or symbolic urban space are some of the main concepts used for addressing the processes that account for the interaction of people with the environments and their main effects."

Citizenship builds a place in the public space at the time in which it appropriates and generates a sense of belonging endorsing the space. Definitely, it allows an identity is generated and thus a relationship of empathy.

2.1 The New Urban Livability

The Spanish Royal Academy defines the concept Habitability as: "Quality of living, and in particular that under certain legal standards, has a local or a house."

This definition, however, may be insufficient since it refers only, as noted Paola Jirón, to the objective aspects of housing, regardless of the subjective capacity that links to wellness and establishes a relationship of match between man and his environment. Similarly, Toni Solanas describes the need to define a new concept of habitability: "the new habitability must be redefined to adapt to new environmental and social constraints. A habitability can not be stated in the immediate future regardless of the resources needed to produce and maintain it over time. A habitability which, in order to be efficient, must be adapted to the demands of people, ways of living today, surpassing housing models linked to standards, increasingly less majority, of conventional life and whose generalization only serves to facilitate expression of value housing change over its use value" (Solanas 2010).

These considerations make us raise a new paradigm of habitability must, somehow, to contemplate the demands and needs of users in order to satisfy or ensure their comfort conditions, a new model that defines the quality of life of the inhabitants of the city beyond the concepts traditionally associated with housing, for habitability establishing a new scale of development that incorporates, likewise, these urban implications.

A redefinition of the concept also covers social and environmental issues, because "the number of needs that must respond habitability, instead of sticking to domestic as before, must expand the actual set is therefore imposes needs of people, i.e., to the characteristics of urban life" (Arcas et al. 2011). That is, we should not restrict ourselves only to the basic needs assigned to the interior of the houses but we must also incorporate other aspects which are essential in urban life, for example, the accessibility the various services offered by the city.

It is necessary, therefore, an analysis that allows us to recognize the different situations that develop in the city, in order to respond to the different needs of people in terms of parameters such as age, sex or culture. Thus, habitability recognize its quality of diversity, a prism through which we can provide a higher degree of satisfaction and quality of life.

On the one hand, basic needs can be the same for everyone, but how they are met will be different depending on each person or group. This leads to that we must contemplate a degree of flexibility and variety in uses the urban environment, to build public spaces and living urban fabric that meet the needs of most users. Any proposal seeking to regenerate the social life of public spaces must necessarily consider the people who inhabit it.

2.2 *Habitability, Life Quality, Social Sustainability, an Empathetic Relationship*

The Program of the United Nations about Human Settlements (UN-HABITAT) contemplates the relationship between habitability, quality of life and sustainability. It recognizes the need to improve the quality of life in the settlements, which affect the daily life and the welfare of the inhabitants of the city. It includes, in addition, that the quality of life of all people depends, among other economic, social, ecological and cultural elements, on physical conditions and spatial characteristics of our villages, towns and cities to contribute to the personal and collective wellness. Habitability directly is related to the qualities and characteristics of space, the social context and the environment, which are the aspects that contribute to achieve satisfaction. Therefore optimizing the livability of cities leads to an improvement of the quality of life of citizens, in a certain context and time.

Similarly, Alguacil (2000), understands the quality of life as the optimum level to meet human needs through the territorial scale, welfare and cultural identity.

This concept expands the definition of quality of life to a field in which social and emotional needs take center stage. In this regard, Salvador Rueda thus affects the quality of life is an assessment of the experience of the subjects, indicating that this assessment is a feeling rather than an act of reason. In his own words, "what better designates the "quality of life" is the "quality of the living which the subjects experience" (Rueda 1996), and so analyzing the quality of life of a society involves the analysis of subjective experiences that integrate it, to know how they live, what expectations they have and assessing the degree of satisfaction.

Closely related to the concept of habitability appears an aspect that has been studied in recent years, it is the urban environment. Aspects such as are land use, loss of value of green spaces, or the gradual increase in the use of private car triggers become an increasingly degraded and less livable urban space.

And we must recognize that this confluence of conditions on the urban environment has fueled a perception of the city, and open spaces, that constitute it as a hostile physical environment for the development of the most basic living conditions, coexistence and community relationship.

Thus, the efforts in building a model city and a system of open spaces alternative to the current model (increasingly degraded), in an attempt to provide a solution for improving the quality of life of the inhabitants and users city, can not ignore the resolution of environmental problems that have been developed in recent times. The planning involves incorporating sustainability issues not only as ecological aspects but with a broader approach, which aims to combine equilibrium aspects of ecosystems, social participation and balanced economic development, through proposals aimed at the rehabilitation of consolidated urban land and rehabilitation of neighborhoods and as a major factor in the recovery of open spaces, understanding that the new habitability involves the study of the urban fabric of the city as a result of interaction or environmental and social processes, and linking relationships that may arise between the inhabitants and users in public spaces and activities that can develop, since this relationship generates the life of the city and the improving of the quality of life and the social sustainability depends on it.

David Godschalk argues that in order to conceiving "sustainable development is fundamental and essential to proper urban planning" (Godschalk 2004). He proposes to restore the art of building and making community through citizens participatory planning and through urban design. He related sustainability-livability concepts understanding them as an instrument to make the most friendly and liveable cities. "The idea is to achieve a city that can live in a much more enjoyable, more pedestrian way, much more human scale, i.e., identity and habitability return to our cities, making them more sustainable" (Biondi 2007).

3 New Qualitative Tools for Urban Sociability

It is difficult to raise the complexity of urban reality based solely on quantification of parameters and standards that, although necessary, do not address the quality of life in the city, and just this must be the objective of urbanism.

In 1961 Jacobs argues that field urbanism as is stalled, because while urban planners hold onto to problem solving understood as a physical science, the progress will not be possible in the city. Today, half a century later, the processes of public participation in the general planning are virtually non-existent or reduced to a period of allegations or suggestions, which may suggest the concept of "participation", but it is just a process of consultation to the procedure already started. Thus, it is necessary to recognize the participation process as a collective fact prior to the beginnings of urban development. It involves, at first, to be aware of who are the subjects participating in the city, who are composing the inhabited city. Recognize that city are subjects of different genders, ages, ethnicities, and it is essential to detect and understand the diversity in order to avoid inequality.

Given the asymmetry between the different groups (differentiated by sex, age, etc. …) occupying and using the different spaces, it is pretended to obtain a global vision that collect the different transversalities of each population group from their own experience and that, besides representing a subjective assessment of the users themselves against the most objective quantification by the technical, it can be regarded as a basis for future actions to be taken for the regeneration of spaces, incorporating urban empathy in construction and humanization of public space.

We start from a conception of quality of complex life; on the one hand, it is possible to analyze and diagnose through indicators that quantify, but, as indicated by most authors, we must incorporate as an important element of the user experience. To this end, through proper selection and characterization of the elements that determine the public space, a diagnostic tool that allows evaluation of space from two different but complementary perspectives will be developed. These perspectives are: the valuation that technician himself can perform based on their own locations within urban contexts and the valuation users can give based on their own experiences. The participatory process is essential for research and reinforces its scientific, since the correct assessment of how an urban location responds to the needs of the people can only be described by them.

3.1 Methodologic Framework

The qualitative tool defines a series of urban indicators that allow us to know how spaces, the degree of social integration, the degree of satisfaction that the inhabitants and users have in a public space because of their spatial and environmental qualities are used. That is, identify which aspects define the livability in the city and public space to improve the quality of life of its inhabitants. Through analysis of the information the tool provides we can establish what we will call the degree of "urban empathy" that each location offers its citizens, and also allow us to establish bases to define possible strategies and lines of action designed to optimize this relationship environment/user.

The development of this research has been approached from a multiple methodology, using quantitative and qualitative tools and the combination of both.

The application of the methodology is mainly divided into three phases, and in each one of them there is a large component that intends to incorporate qualitative analysis through observation, events and actions that predominate in users to study. The three phases are: Urban Analysis (AU), assessment of the degree of citizen satisfaction (GSC) and obtaining the degree of Urban Empathy (GEU).

On the one hand data collection and development of urban space analysis based on the study variables it is proposed, which we will call Urban Analysis (AU). Research will address the issues raised in the proposed study through the analysis by the researcher as objective assessment. It uses a quantitative methodology that we find it essential to register conditions of urban boundary: the material qualities and the conservation status of the field of study, the accessibility and security, etc. Throughout the process we use a model of variables and indicators. The assessment proposed in this phase of implementation of the tool must be objective, it is necessary to obtain the same results whoever perform the analysis, and for this purpose a numerical scale of 1–5 is used, being one the less favorable value and 5 the optimum value.

For the application of some of the variables that influence the behaviour or actions of the users the observation as technique is used, "Observing is describing systematically events, behaviors and artifacts in the social setting chosen for study" (Marshall and Rossman 1989). This technique aims to incorporate data without involving people being studied, primarily it aims to collect data on habits and activities that take place in public space: what the social fabric of the area of study is, what the uses of space are …

In this initial analysis data on neighborhood characteristics and qualities of space will be obtained, how many there are and how the facilities and shops are distributed, how they use public spaces and what kind of people use them, de-glossing data by age and sex. For this we will develop descriptive maps of equipment, stores, uses of public space regarding mobility, accessibility and security.

Given the type of work proposed, it is considered necessary to use qualitative methods, as they can contribute and yield information which is not detected in Urban Analysis. This second phase of work it is call Citizen Satisfaction Rating (VSC). Through it we pretend to obtain first-hand information through participatory dynamics and direct interviews (Delgado and Gutierrez 1995) with citizens, in order to determine the degree of satisfaction or perception of users on different variables set for the field of study. Thus it may establish a direct link between the existing reality and the perception of citizens.

The scale measuring the degree of satisfaction is as follows: Excellent—very high = 5, Good—High = 4, Regular—medium = 3 Deficient—low = 2, Null—none = 1.

The last phase of the process refers to the methodological triangulation (Denzin 1978), which we call Measurement of Degree of Urban Empathy (GEU) and allows the linking of two different methodologies for the study of the same problem and which can extract new data obtained in the analysis of reality. In this research that triangulation will consist in comparing the evaluations obtained in Phase 1 and 2, in which different methods have been used (Patton 2002).

At this stage the tool aims to establish the comparative study between reality and satisfaction or public perception of it. To this end is proposed an evaluation matrix with two different inputs, on one hand data from urban analysis (AU), which described analytically what the state of the urban environment of study is, on the other component qualitative, which refers to the degree of citizen satisfaction (GSC). This matrix will return a result, the degree of urban empathy (GEU) that combines both aspects.

The final assessment will allow us to establish the relationship between the analyzed space, its equipment, its businesses and citizens who make use of them allowing us to establish improvement proposals for the neighborhood to increase the degree of citizen satisfaction and therefore the quality of life. To achieve a numerical result allows us to make comparisons and see what qualities of public space analyzed are those that generate greater degree of satisfaction, or why certain areas are more empathetic for children or more hostile for olders.

3.2 Definition of the Components

To define the qualities of habitability as components of an urban reality that is linked to users and establish links in the public space, and therefore can be estimated as empathetic qualities, six groups of urban variables encompass the basic principles of establishing new urbanism[1], which proposed a city where economic, social and environmental aspects are equally important, and needs of citizens are the starting point of urban design.

Six large groups of urban variables are set to parameterize and categorize the information, both in a qualitative and quantitative way, we will get from our tool when it is applied in each area of study: Activity and Use, Mobility, Accessibility, Safety, Complexity Urban and Environmental Perception. These variables will be analyzed through indicators that seek to respond to the problem of habitability and quality of life and, precisely for this reason, they must be qualitative in addition to quantitative because, "… in addition to its quantitative existence, it is not enough there is public transportation, this, in addition to satisfying requirements of a particular "number" of users, should be "qualitatively" satisfactory" (Luengo 1998).

Variables and indicators:

We propose six study variables, which in turn are broken down in (19) indicators, which will be formed in turn by a number of conditions to be evaluated (54) and which are measured by the determinings (221) for ensuring that the basic aspects of livability are analyzed.

Activity and Use: It will analyze the capacity of public space to allow different nature activities which can be developed on it: leisure, work, culture, learning, etc. It has five indicators: fitness for use, spatial quality, social vitality and identity and ownership.
Mobility: It will analyze design efficiency of the urban fabric, mobility and infrastructure of the fabric of the city. The variable is analyzed through two indicators to be adapting to mobility and spatial quality.
Accessibility: While accessibility is an obvious aspect that increasingly is taking more account, with the emergence of the new Decree and modification of the technical code to include in your security document of Use the mandatory parameter of accessibility, it remains a handicap in the consolidated city, and therefore it will be incorporated as variable study. We will consider the accessibility aspect in free space, access to buildings and public transportation. This variable has three indicators that are matching, spatial quality and vitality.
Security: It will analyze the social and spatial conditions that establish a degree of security of space itself against elements such as accidents, theft, etc. To study the safety we analyze three indicators that are adaptation, spatial quality, vitality.
Urban complexity: It will analyze the degree of diversity and complexity of urban space in terms of equipment and mixed uses it shows. The indicators proposed for the analysis of this variable are the study of the diversity of activities, the social diversity in the field and the possibility of participation.

[1]The concept New Urbanism was generated by Rober S. Davis in 1979, and later, in 1998, the Congress for the New Urbanism (CNU) rescued elaborating the ten principles of new urbanism: Pedestrian Urbanism, diverse city, mixed use, sustainability, accessibility, quality design, neighborhood public structure, compactness, friendly mobility, and participatory design.

Environmental perception: It pretends to reflect the degree of comfort of public space related to the presence of green areas, shadows, sitting areas… Understanding this variable covers three indicators that assess environmental conditions, green spaces, and human scale.

4 Application and Expected Conclusions to Develop an Emotional Atlas of the Urban Space

Once proposed indicators, with the objective that can diagnose the degree of empathy on public spaces, we proceeded to the application of the tool in a case study for evaluation. To do this, there has been a proper selection and characterization of the elements that determine the public space.

The participatory process is essential for research and reinforces its scientist aspect as the correct assessment of how an urban location responds to the needs of the people can only be described by them.

In order to know the perception of users regarding the two locations located in the northern town of Seville (Plaza José Luis Vila/San Luis and San Román/San Marcos), in the month of March this year 2016 a conference ("Pensar tu Barrio"/ "Think your Barrio") of citizen participation in public space and mobility was performed.

The development of this conference took place through six working groups plus a workshop held with children in the neighborhood. All the tables were based on graphic documentation about the urban environment in which the neighbors tipped their concerns and detected everyday problems and shortcomings on the public space.

While the workshops were conducted through spontaneous discussions and detailed discussions of various aspects, the workshop of children was carried out with a very different approach: it based on graphic material, worked with them in building the day map, expressing and describing how they move around the city, what kinds of spaces they are in their everyday tours, what activities can develop in them or what they need to develop these activities.

In parallel to the realization of the conference, in more detail, individual interviews were being carried out, in which, starting from the most widespread perception expressed in the conference, it was pretended to establish the degree of satisfaction with the indicators and variables of study. So far 60 have been conducted sample surveys equitable gender 30 men and 30 women. In turn are divided by age, 20 adolescents, 20 adults aged 18–64 and 20 adults 65 and older

The assessment tool aims to make a critical analysis of the use and benefits of it, always considering it is an open and flexible tool, as there may be parameters or indicators not initially considered and fieldwork with citizens could generate the need to incorporate along with other possible new areas of study. In this sense, the implementation of it, through critical analysis of the data obtained from both the assessment or suitability of the location, will allow us to draw many conclusions, including:

1. What qualities of a particular public space are those covering citizens ' needs, if those qualities satisfy only a collective or all of them, etc.
2. To establish what aspects of livability influence add more positive perception and user satisfaction, and therefore improving the quality of life.
3. Developing a cartography diagnosing the current state of studied urban environment.
4. Developing an emotional atlas that includes the results of urban spaces more and less empathetic.
5. Generating possible courses of action to achieve greater degree of comfort and satisfaction for each of the indicators.
6. Implementing measures or remedial actions, in such spaces, which rebalance situations considered dysfunctional or pathological.

References

Alguacil J (2000) Calidad de vida y praxis urbana. Nuevas iniciativas de gestión ciudadana en la periferia social de Madrid. Colección Monografías nº 179. Coedición del Centro de Investigaciones Sociológicas y Siglo XXI de España Editores. Madrid

Arcas J et al (2011) El futuro del hábitat: repensando la habitabilidad desde la sostenibilidad. El caso español. Revista invi N°72/agosto 2011/Volumen 26: 65–93. http://www.scielo.cl/scielo.php?pid=S0718-83582011000200003&script=sci_arttext-n1

Biondi S (2007) Cátedra de Nuevo Urbanismo en México. Investigación por el Tecnológico de Monterrey, Campus Querétaro

Borja J (1998) Ciudadanía y espacio público, en "Urbanitats" núm. 7: Ciutat real, ciutat ideal

Borja J, Muxi Z (2000) El espacio público, ciudad y ciudadanía, Barcelona. p 93. Disponible en: http://www.esdi-online.com/repositori/public/dossiers/DIDAC_wdw7ydy1.pdf

Delgado J, Gutierrez J (1995). Métodos y Técnicas Cualitativas de Investigación en Ciencias Sociales (pp 141–175). Madrid: Editorial Síntesis

Denzin N (1978) The research act. McGraw-Hill, New York

Godschalk D (eds) (2004) Land use planning challenges: coping with conflicts in visions of sustainable development and livable communities. J Am Plan Association 70(1):5–13

Jacobs J (1961, Trad. esp.1973) Muerte y vida de las grandes ciudades. Ediciones Península, Madrid

Jacobs A (1996) Grandes calles. Publicaciones de la universidad de Cantabria, Madrid

Leadbeter (2014) ¿Puede medirse la "empatía" de una ciudad? http://blogs.iadb.org/ciudadessostenibles/2015/11/25/empatia/

Luengo G (1998) Elementos para la definición y evaluación de la calidad ambiental urbana. Una propuesta teórico-metodológica. ponencia presentada en el IV Seminario Latinoamericano de Calidad de Vida Urbana, pp 8–11

Marshall C, Rossman B (1989) Designing qualitative research. Sage, Newbury Park, CA

Patton M (2002) Qualitative research and evaluation methods, 3rd edn. SAGE, Thousand Oaks

Real Academia Española (2001) Diccionario de la lengua española (22. aed.). Consultado en http://www.rae.es/rae.html

Rueda S (1996) Habitabilidad y calidad de vida. Cuadernos de Investigación Urbanística n 42:30
Segovia O, Jordan R (2005) Espacios públicos urbanos, pobreza y construcción social, Serie medio ambiente y desarrollo, No. 122 [en línea], Santiago, Chile: Naciones Unidas, CEPAL, División de desarrollo sostenible y asentamientos humanos. Disponible en Internet en: http://www.eclac.org
Solanas T (2010) Congreso Internacional Rehabilitación y Sostenibilidad. La necesidad de un nuevo concepto de habitabilidad. [En línea] septiembre de 2010
Vidal T, Pol E (2005) La apropiación del espacio: una propuesta teórica. Barcelona: Universitat de Barcelona.

Part III
Sustainable Renovation of Buildings and Neighbourhoods

(Re)Programa. Architectural Rehabilitation Incorporating Sustainability Criteria in an Andalusia Neighborhood

Begoña Blandón-González, Pedro Gómez de Terreros-Guardiola, Angela Barrios-Padura and Marta Molina-Huelva

1 Introduction

In 2015 the Research Group (PAIDI) TEP 954-InFact from the School of Architecture of the University of Seville, ended the Research Project "(Re) Programa: (Re)habitation + (Re)generation + (Re)programming", of the R&D Contract concerning the competence of the Department of Housing and Home Improvement of the Public Works Agency of the Government of Andalusia from 2012, funded by FEDER of the European Union (Barrios et al. 2015). It concludes that a study of an urban model must evolve in compliance with the requirements of current standards, but always covering the needs of groups that form it and, in particular, direct its transformation to the suitability of environments for active aging.

The project fulfils the objective of Law 8/2013 (Ministry of Public Works) on Urban Renovation, Regeneration and Renewal of "regulating the basic conditions that will ensure sustainable, competitive and efficient development of the built urban environment, by means of driving and promoting actions that will lead to the renovation of old buildings and the regeneration and renewal of the existing urban fabric, to ensure a suitable quality of life for citizens and the effectiveness of their right to enjoy decent and adequate housing".

The study allowed a critical review of the inspection protocol and diagnosis contained in Royal Decree of 7/2015 (Ministry of Public Works 2015), approving the revised text of the Land Law and Urban Rehabilitation, Title III dedicated to the "Building Evaluation Report".

B. Blandón-González (✉) · P. Gómez de Terreros-Guardiola · A. Barrios-Padura
Department of Building Construction I, University of Seville, Seville, Spain
e-mail: bblandon@us.es

M. Molina-Huelva
University Institute of Architecture and Building Sciences,
University of Seville, Seville, Spain

P. Mercader-Moyano (ed.), *Sustainable Development and Renovation in Architecture, Urbanism and Engineering*, DOI 10.1007/978-3-319-51442-0_11

In order to achieve the proposed (Re)Programa objective, a case study on a representative neighbourhood of European built park not responding to the needs of its population was developed. Dwellings and public areas including defects and detected pathologies were studied and characterization constructively.

After obtaining the data and diagnosis of the state of conservation of buildings, Building Evaluation Report (IEE) was developed for each of the locations. Fulfil the requirements of the Spanish Technical Building Code (CTE) (Ministry of Housing 2010) was also revised as a basis to value in detail the rehabilitation solutions and maintenance activities.

1.1 Case Study Description

The case study of the project (Re)Programa consisted of consolidated neighborhoods, with residential buildings built between the 40s and the 90s, public and private ownership, with housing programs of small, medium and large size, in the cities of Seville, Cordoba and Jerez de la Frontera, all in Andalusia in Spain. The socioeconomic status of the population is medium, usually pensioners.

1.2 Inspection and Diagnosis. Degree of Standard Compliance

Inspection and diagnosis were made applying the methodology and tools of Building Evaluation Report (IEE), according to the Royal Legislative Decree 7/2015 (Ministry of Public Works 2015).

The results (Barrios et al. 2015) from the constructive and social point of view, accessibility and security in the urbanization of neighborhoods and inside the building, originate the thought about the facilities they offer their users. Not only with regard to their type and suitability to the diversity of family structures, but the material deterioration, lack of maintenance and the need for technical upgrading, recycling and reprogramming.

Significant differences in the study of standards requirements in the current Basic Documents of the Spanish Building Technical Code (CTE) (Ministry of Housing 2010) on structural safety, fire safety, health, safety in use and accessibility, noise protection and energy saving were detected.

The results obtained regarding the compliance of the standard point out that the stability conditions; health and ornament are generally admissible. However, clear deficiencies in the thermal behavior of the envelope and in the accessibility of buildings, make us reflect about the conclusions. Regarding these aspects, the proposals for improvement in future works is necessary to carry out.

In any case, even assuming the important progress that the rehabilitation standards have had, comparing the results with the examples studied, it is observed that the inspection using the Building Evaluation Report impact only on some minor issues, pending most the requirements demanded in the CTE. Therefore, it is a partial diagnostic tool that requires under our point of view a critical review.

1.3 New Objectives and Work Methodology

Analyzing the current IEE, it can be said that the requirements of stability, health and ornament that have been required in Technical Inspections of Buildings (ITE) correspond with the basic requirements of Structural Safety (SE) and Healthy Conditions (HS) from CTE (Ministry of Housing 2010). They are the required in tools of analysis for energy certification with the basic requirements of Energy Saving (HE) and those considered in the study of accessibility (paragraph 9) of Document of Security of Use and Accessibility (SUA).

The Basic Document of Safety in Use (SU), Safety in case of Fire (SI), and Protection in case of fire (HR), are completely outside the content of the IEE. They are aspects of great importance, which should be included in the methodology of diagnosis and, above all, in its proposals for improvement.

In this regard, review and criticism in the research project, has led to the development of new methodologies for analysis, inspection, diagnosis and intervention focused on the current demands of comfort, energy savings and environmental sustainability and management environments from criteria of livability.

In this new stage of the research, we have developed a new tool, which provides to technicians, users and owners, a comprehensive diagnosis to make rehabilitation and recycling projects, technically and economically viable. It has been developed in the following phases:

1. Compliance with objectives and analysis of conclusions from the research project (Re)Programa.
2. New objectives and methodology
3. Deficiencies identified in the IEE
4. Extension of the study and a proposal of new Inspection Sheets of the CTE DB-SUA
5. Application and implementation of the new Inspection Sheets
6. Analysis of results and conclusions

In response to the deficiencies in the Building Evaluation Report (IEE) about Safety in Use Requirements in residential buildings, in this article the new Inspection and Diagnosis Sheets is presented.

2 New Diagnosis Protocol. Inspection Sheets

The analysis based on the Building Evaluation Report (IEE) focuses shallow and poorly in CTE standards requirements related to Basic Requirements of Safety in Use in residential buildings. Following this line, the research is focus in the development of new Inspection Sheets that include relevant aspects needed for intervention and give response to the population demands, joining the aim that ensures safety, healthy conditions and ornament in buildings.

2.1 Scope of Proposals Inspection Sheets

Once the study on Security of Use and Accessibility requirements and systematizing data collection started, the difficulty of its implementation was checked due to the singularity of the different models of neighborhood and residential typologies. Thus, testing each of the sections proposed in the IEE, the complexity of application on existing buildings was evident. The LOE and the CTE are written from the perspective of the new architectural project, complicating its application in specific interventions or smaller ones, although with important relevance in the people quality of life.

Therefore, in order to improve its implementations it was decided to redrafting the Inspection Sheet, considering separately each sector that can be distinguished in residential areas as Hallways and Distribution Spaces, Stairways Areas, Exterior Spaces and Courtyards, Interior spaces of Housings, Roofs and Lightning rods, Traffic circulating areas and Swimming pools. Thus, it is justified:

SECTOR 01. HALLWAYS AND DISTRIBUTION SPACES. It covers communal interior spaces, from the hallway to restricted spaces, the door of the dwelling or, the nucleus of vertical communication between floors. Steps and interior ramps of inner circulation road and whose laid down does not reach the character of vertical communication between floors are also included.

SECTOR 02. STAIRWAYS. It covers the vertical connection elements located in interior communal areas of the building and due to its use, they enable communication between floors. Also it is included in this sector staircase landings and distribution spaces corresponding to these elements (entrance and exit from the staircase) and the corridors that provide access to dwellings and other spaces with restricted access. This singularity is accepted after verifying the uniformity of each floor on the design and use of space.

SECTOR 03. OUTDOOR SPACES AND COURTYARDS. It covers the exterior spaces (including areas of common access to the building) that are part of the urbanization/community, to the interior of the building. Stairs, ramps and exterior galleries (without glass) that belong to the escape route to a safe area are included. Similarly, communal areas and road access to maintenance or parking if they invade sporadically these outdoor spaces are included.

SECTOR 04. INTERIOR HOUSING. It considers the spaces developed from the gateway to homes. Terraces, stairways and private ramps being reserved an exclusive sector to include them. Cantilever facade, communal external galleries, courtyards and communal roofs, etc. access is discarded. However, as is found in the monitoring carried out in some buildings, these elements can be currently found, limited and/or maintained temporarily for some neighbour.

Sector 05. ROOFS AND LIGHTNING RODS. It covers roofs, flat or inclined, and their access roads. Also the elements integrated in the hipper ends and those ornamental elements, complementary (clotheslines, antennas, etc.) or building systems are included. Stairs, ramps and galleries located outside roof space shall be governed by the same sector and have the corresponding section on the Inspection Sheet. Finally, as indicated by the CTE DB-SUA, paragraph 8 (Ministry of Housing 2010), safety in case of the risk related to the action of lightning, electrocution and fire caused by this action, through appropriate protection building systems is limited.

Sector 06. ZONE TRAFFIC FLOW. It covers those a access areas, exit and routes planned for neighborhood cars. Paths and connection elements (stairs and ramps) for pedestrians developed as a part of this application are included.

Sector 07. POOLS. This sector covers the bathing area and adjoining areas for the use and enjoyment of it as long as the area is limited and with a restricted access.

2.2 *Model and Content of Proposal Sheet*

In the model of Inspection Sheet developed, each of the sectors includes independently, analysis and monitoring of compliance with each section of the CTE, including aspects related to Safety in Use and Accessibility (security against the risk of falls, safety against the risk of impact or entrapment, security against the risk caused by inadequate lighting, security against the risk from moving vehicles, etc.).

It is possible to supplement it with new sections related to the type and location in the building that could be of interest in each specific sector. All this, along with the aim of reducing to acceptable limits, the risk of users suffering immediate damage to the intended use because of their design, construction, use and maintenance. The limits defined in order to ensure the safe use of a residential building are included.

3 Application and Implementation of Proposal Sheets

The neighborhoods case study are four new locations in addition to those defined in (Re)Programa, executed between the 40s and the 70s, with traditional structural systems and constructive solutions from its period of construction (Table 1). They are buildings from public or private ownership, with owners with different socioeconomics levels; small medium or high. They are in different consolidated

Table 1 List of cases of study and main characteristics

No	Reference	Construction year	No of floors	No of door way	No of dwellings	Patio	Garden	Lift	Parking
L01 ReP	SE-L.Remedios "Manzanas Cuadradas"	1938–1943	B + 3	54	324		x		
L03 ReP	SE-L. Remedios "Bloques Espiau"	1955	B + 6	8	98	x		x	
L04 ReP	SE-L.Remedios "B. lineales"[a]	1960	B + 4	60	600	x			
L08 ReP	SE-Triana "V. Esperanza"	1969	7	18	513	x	x	x	
L12 NP	SE-L. Remedios "Coop. Químicos"	1971	7	2	42	x		x	x
L12 NP	SE-L. Remedios "Manzanas A"	1950–1955	B + 4	37	575	x		x	
L12 NP	SE-L. Remedios "Manzanas B"	1950–1956	B + 5	36	540	x		x	
L12 NP	SE-Center "Sta. Catalina"	1966–1968	4	4	32	x	x	x[a]	

[a]Lift incorporated in 2005

áreas in Seville as the neighborhood of Los Remedios, Triana and historical center, resulting representative as a social, economic, technical and environmental challenge preserving the original interest in urban issues defined in the Project (Re) Programa.

On each of these locations and, once defined the scope for different sectors, we proceed to visit the buildings, tracking and review each section described in the corresponding Inspection Sheet. The implementation and monitoring of the limits and peculiarities found in each of the buildings are included as part of the document and allows exhaustively evaluate the results obtained in different standards and social points of interest.

The inspections revealed deficiencies and irregularities in each sector as described below.

As indicated in the CTE DB SUA, in Sect. 1 concerning safety against the risk of falls, the risk of users from falling at the same or different level will be limited, for which the floors will be adequate to encourage people slipping, stumbling or producing mobility difficult.

Thus, in the sheets drawn up for this purpose, matters relating to the slipperiness of floors, discontinuities in the pavement and all the requirements applicable to the slopes, with or without steps, are included.

Therefore, after the monitoring carried out at different locations, it is contemplated that, in the sectors 01 and 02 corresponding to hallways, distribution areas and stairways, most of the buildings studied show the original pavement at interior communal areas, smooth texture and good quality (marble, granite and terrazzo). However, due to the type of surface treatment applied or worn, 41% of these do not comply with the minimum degree of slipperiness set by the regulations. In other cases, the lack of maintenance gets a texture that can be found within the limits of class C1. As a general data for any of the locations (87% of cases), the steps, stairways and ramps in the buildings have a lower degree of slipperiness of the indicated and, in any case, 98% of access to hallway steps is of the same type tan interior ones, and therefore it does not fulfill the limit imposed.

Moreover, due to lack of maintenance and regular deterioration in buildings of this age, the pavement of communal areas has major damage in corners and edges around the building. However, the original quality of these floors has allowed an aging and an exemplary durability.

Inside the dwellings, the cases are very varied, being renovated in recent years according to owner preferences in many cases. In our context, interest in polishing and buffing treatments, leads to not fulfil the defined limits. In this regard, the original pavement, in those dwellings that maintenance it, is in a very good condition. Only in bathrooms and kitchens that it have not been renovated or replaced, major breakages and chipped originates sometimes cutting surfaces are presented.

Regarding the possible discontinuities that can be found, there are small rugs that can cause slips in the access corridors to housings, in staircase landings and in distribution corridors. The 36% of these corridors in different floors and in different buildings are invaded by their presence. In some cases (5%) the small rugs are cut not to cause landslides and even they are fixed to the floor with the added risk of causing setbacks in its path. In the remaining 59% there are no small rugs or they do not interrupt the corridor because they are more than a meter wide.

Similarly, 45% of the hallways present in the building entrance, the registration of caskets with impaired cap. The perimeter metal profile is warped, highlighting on the pavement and can cause tripping.

Inside the dwellings, the projections on the bumpers, latches and joinery fixing step, in dining rooms and lounges (at 5% of cases) are remarkable. 41% of dwellings can present these discontinuities, given the type of each case. This value is not absolute, since access to all dwellings in the case study has not been possible. It is obtained from the cases of the same type and location that have actually been visited.

The cantilevers floors and accessible terraces are adequately protected, although some have ridges due to exit carpentry.

Overall, the interior patios and gardens inside the estates (although due to its bad situation they are less trodden in 70% of cases) have significant slopes and damage/breakage in their coatings, which precludes a safe use. Similarly, the exterior communal areas are lack of maintenance and have got important deterioration of its surface.

Regarding changes in elevation, in the proposals sheets slopes (difference in height between two adjacent surface along the same route) solved or not with steps, in stairs half-pace or ramps (considered as vertical nucleo of relationship) are distinguished.

It can be said studying this, that 48% of the hallways has got in the interior path a gap grater than 5 cm, solved with steps (there are 2 steps in 92% of cases) whose height, in some cases are more than 28 cm. Only in three of the locations defined, there are no ramps or mechanical systems, platforms or chairs, usually paid by some of the neighbors who were interested on.

Regarding the different levels lower than 5 cm, the existing ramps exceed the recommended slope and, like indoor steps, they are not sufficiently illuminated or highlighted. They have got no protective barriers in most cases, although they have handrails on one of the sides.

Regarding the different levels higher than 55 cm in communal areas, 98% of cases are solved with steps and have got guardrails although not always appropriate according to regulations.

In the types studied, vertical communication areas are with two half pace stairs in 43% of cases and with three pace stairs in 57% of cases. There are no ramps or lift in 53% of the hallways. In any case, the design steps and stairs compliance to established dimensions, near the set limit in some locations and, 10% show a reduced riser so the climb is performed in a milder form. 97% of the steps in the hallways and stairs have embossment and therefore do not meet the specifications defined.

Design conditions and safety of protective barriers on slopes, stairs, ramps, balconies and communal galleries, are included in Sect. 1 of the sheets proposals to facilitate data collection.

In relation to the protection barriers, it highlights that 40% of existing ones, are designed from a brick screed on a railing on vertical supports fixed. This solution allows climbing on the element with the risk of falling to a lower level. In all other cases the protective barriers are solved with balustrades, also originals that although scalable, do not present serious risk of falling. Regarding the height of the handrail, in most cases, it is less than 90 cm and is located on the inside of the footpath.

Inside homes and terraces, although compliance is not required, the corresponding section is also included, allowing inspection of the barriers of protection in these areas. Moreover, as indicated, maintenance and renewal of the materials has been developed in recent years according to need and economic conditions of their owners. Thus, except the need to adapt housing to improve conditions for people with reduced mobility, maintaining its building systems and interior cladding, it is better than in the communal areas.

Section 2 is on security against the risk of impact. As indicated in the CTE DB-SUA, the risk that users may suffer impact or entrapment with fixed or openable elements of the building will be limited. In the proposal sheets, the risks arising from the presence of elements along the way, including fragile or imperceptible (glazed surfaces) and those related to possibilities entrapment (sliding doors or windows in communal areas) are included.

The access doors to buildings are made of glass in all cases, with outer metal grille for protection, facilitating their identification avoiding impacts. However, the weight of the door is very large and it is difficult to open it. It causes in its closure the risk of entrapment due to closing speed. Also, in no case we have found opening devices from the outside with internal locking mechanism. It is especially important to study hallways and corridors due to the presence of large pots and planters in 70% of the buildings. This reduces, in the case of one-meter wide corridors, the useful width of the path interfering the route.

Also, in 57% of the hallways of the building, there are fixed elements that stand in communal areas of circulation, such as mailboxes, counters or electrical boxes.

Similarly, the existing lateral luminaire in corridors can cause impact in 70% of cases. Outside, windows, planters and fencing, protruding from the facade can cause the risk of impact to the neighbors in the area ($h < 1.5$ m).

Section 3 on safety against the risk of entrapment in enclosures, is reflected, as indicated by the CTE DB-SUA, in the sheets trying to limit the risk of users accidentally getting caught in these spaces.

Section 4 includes aspects with regard to security against risk caused by inadequate lighting in communal areas and footpaths, as well as emergency lighting. As indicated in the CTE DB-SUA, the risk of damage to persons as a result of inadequate lighting in circulation areas of the buildings will be limited, both indoors and outdoors, even in case of emergency or failure of the normal lighting. In this respect, the lighting of interior areas, in hallways and distributors and in interior corridors is insufficient in 93% of the hallways. The existence of external windows, such as vertical communication areas in 70% of cases, allows to compliance this requirement during the daylight hours.

In nighttime hours, artificial lighting is insufficient in all cases (40 lx on average). The presence of mirrors on the side walls of the hallway at some locations tripling the lighting of these spaces. Regarding the emergency lighting in the communals areas, there are in 28% of the locations.

Finally, regarding the cleaning of the glass surfaces, existing openings in the stairs areas are not a problem for the interior cleaning. However, in 40% of cases the exterior windows can not be cleaned from the inside part because some of its breads is fixed. However cleaning doors from the hallways is easy, but the outside fences impede the work and only in some cases, cleaning is easy folding down the fence.

In Sect. 5, on safety against the risk of drowning in pool areas, as indicated by the CTE DB-SUA, the risk of falls that may result in drowning by restricting access elements is limited. In any case, the studied locations have not this area (Ministry of Housing 2010).

In Sect. 6, aspects of safety relation to the risk caused by vehicles are included. As stated in the CTE DB-SUA (Ministry of Housing 2010), the risk caused by moving vehicles according pavement types and the signaling and protection of the wheel tracks and people circulation, will be limited. In this regard, the elaborate Inspection Sheets distinguished communal areas that form the car roads (access to garages and car parks, etc.), and the necessary ones in the pedestrian path with appropriate protections and signaling. The percentage of residential buildings without a garage in the basement, surface parking integrated into the community and presented in this study, can not be considered extended to all of the buildings that are considered into this type of location (only in one case there is a garage basement). Although the lack of alternative path for pedestrians is observed, an inadequate lighting and lack of protective measures etc. are not representative results; therefore, the study must be referred to another selection of buildings.

4 General Conclusions

The need for intervention on the built park in order to respond current standards requirements and adapt dwellings to the evolution of the great diversity of family structures, without forgetting the peculiarities of the percentage of the population, is with the issue of energy efficiency, one of Europe's most important challenges in the XXI century. The difficulty of making a deep diagnosis in areas of consolidated city encourages researchers and professionals in the field of architecture to develop new approaches. The aim is to find tools to improve the state of the building and the public space, its shortcomings and requirements, from technical, environmental and social criteria.

Based in prioritizing rehabilitation and maintenance activities, the Inspection Sheets proposals raised in advance of the existing Building Evaluation Report (IEE). It impacts are not only on basic aspects of safety, health, accessibility and ornament, but also on those aspects related to the requirements of functionality, livability, comfort and environmental sustainability, in order to realize the existing needs, and provide guidance about the most successful project in each case.

The research carried out has allowed the implementation and review of the methodology of diagnosis on about 50 buildings from different locations and its findings reflect highly relevant results regarding the use, accessibility and livability. The need to prioritize interventions that facilitate the use of domestic and urban spaces is enhanced.

Supporting the creation of a Basic Stamp that reflects "the quality, achievements and potential of the building" (CSCAE 2012), as an instrument certifying the performance improvement of the housing stock. The Inspection Sheets proposed in this work determine a model that is integrated into the IEE, defining possible actions and improvements to achieve the desired benefit.

In this study the need and interest in continuing this project with new standards sections is shared. The advance in this topic will provide users with more complete information, including all current requirements. The possibilities of future intervention will offer doubtless, significant benefits in many areas of society, and represents an improving in the life quality of citizens, being necessary to launch public programs with subsidisation.

Acknowledgements The authors would like to thank the ERDF of European Union for financial support via project "(Re)Programa: (Re)habilitación + (Re)generación + (Re)programación" of the "Programa Operativo FEDER de Andalucía 2007–2013". We also thank to Public Works Agency and Regional Ministry of Public Works and Housing of the Regional Government of Andalusia.

References

Barrios Á, González E, Mariñas JC et al (2015) {Re}Programa: (Re)habitación +(Re)generación +(Re)programación. El reciclaje y la gestión sostenible del parque edificado andaluz. Gestión de entornos habitables desde criterios de envejecimiento activo, género y habitabilidad urbana. Universidad de Sevilla, Sevilla. https://idus.us.es/xmlui/handle/11441/28314. Accessed 25 Ago 2016

CSCAE (2012) El sello básico del edificio como propuesta para la rehabilitación arquitectónica, integral y sostenible al servicio de la sociedad. Paper presented at the 11° Congreso Nacional de Medio Ambiente, Fundación Conama, Madrid, 26–30 Nov 2012. http://www.conama11.vsf.es/conama10/download/files/conama11//GTs%202010/23_final.pdf. Accessed 29 Ago 2016

Ministry of Housing (2010) Real Decreto 173/2010, de 19 de febrero, por el que se modifica el Código Técnico de la Edificación, aprobado por el RD 314/2006, de 17 de marzo, en materia de accesibilidad y no discriminación de las personas con discapacidad, BOE de 11 de Marzo. Orden VIV/561/2010, de 1 de febrero, por la que se desarrolla el documento técnico de condiciones básicas de accesibilidad y no discriminación para el acceso y utilización de los espacios públicos urbanizados. Diario Oficial, Madrid

Ministry of Public Works (2013) Ley 8/2013, de 26 de junio, de rehabilitación, regeneración y renovación urbanas. Diario Oficial, Madrid

Ministry of Public Works (2015) Real Decreto Legislativo 7/2015, de 30 de octubre, por el que se aprueba el texto refundido de la Ley de Suelo y Rehabilitación Urbana, BOE 261 de 31 de Octubre Título III. Diario Oficial, Madrid

Steps Towards the Integration of Regeneration Processes Obsolete Buildings Envelope Spanish in the Paradigm of Sustainable Development

Manuel Ramos Martín and Pilar Mercader-Moyano

1 Climate Change, Greenhouse Gases and Sustainable Development

Countless research results on climate change is impacting shown as present at such essential aspects as the quality of air, water, food production, etc. However international agreements, achieved during the second half of the twentieth century and early twenty-first directed to "urgently" address the problems of climate change, have not yet been able to reverse the trend. Given the need to reduce greenhouse gases-such provocateurs greenhouse gas (GHG) as were already identified in the Montreal Protocol of 1987 comes the most important line of political action on environmental matters. In this way and through these agreements the states involved are committed to reducing their GHG emissions, which not is paradoxical if we consider the fact that limit a development model and traditional growth without in any way affecting their processes and social and economic implications (González-Díaz 2013).

Everything suggests that the sustainability paradigm illustrated by the famous Brundtland Report, filed with the World Commission on Environment and Development in the late eighties, has not been played yet adequately, may in part be due to the extent ambiguity its core definition:

> Meet the needs of the present without compromising the needs of future generations.

This definition includes intelligently concept of necessity, which inevitably sets a number of limitations, but also demands a more equitable distribution of resources.

M. Ramos Martín (✉)
University of Seville, Seville, Spain
e-mail: manrammar3@alum.us.es

P. Mercader-Moyano
Departament of Building Construction I, University of Seville, Seville, Spain
e-mail: pmm@us.es

P. Mercader-Moyano (ed.), *Sustainable Development and Renovation in Architecture, Urbanism and Engineering*, DOI 10.1007/978-3-319-51442-0_12

This equity demand in turn support political systems to ensure the participation of citizens, and to show a political will to change that makes it possible to achieve a level of sustainable development that addresses three interrelated areas: environmental, social and economic (European Conference Sustainable Cities 1994).

The European Council welcomes these measures through the creation of a mechanism for monitoring emissions of greenhouse gases and assessment of progress in meeting the commitments regarding those emissions (Decision 93/76/EEC), which allows Member States to determine the total quota of allowances to be allocated, considering its potential reduction. This supposed the starting gun of a series of measures based on the limitation of climate change, but in any case a necessary propulsive transformation based on sustainable development.

1.1 European Buildings and Emissions

Due to the significant environmental impact of the building sector in Europe they took a series of measures on the limitation of emissions of carbon dioxide by improving energy efficiency (Directive 93/76/EEC). The Directive required Member States to develop and implement energy efficiency programs in the building sector and report on its implementation.

It is not until nine years later that in Europe start to talk of energy efficiency of buildings (Directive 2002/91/EC), taking into account outdoor climatic conditions and local conditions, as well as indoor climate requirements and the relation cost-effectiveness.

The European Council held in Brussels in March 2007 laid the foundation on a European short and medium term Energy Policy (EEP). the need to increase energy efficiency in the Union to achieve the objective of reducing energy consumption by 20% and increase by the same percentage the use of energy from renewable sources by 2020. Thus defining the long-term commitment to maintain the global temperature increase below 2 °C and 2020 reduce the total greenhouse gas emissions by 20% or more compared to 1990 levels. Those commitments were set out by the Directive 2010/31/EU on the energy efficiency of buildings, which merged the previous Directive 2002/91/EC and set new targets for the period 2010–2020 in relation to the minimum requirements for energy efficiency, energy certification and periodic inspection of thermal installations of buildings. This Directive determines and completes aspects already addressed in the 2002 Directive introduces modifications and new features such as the methodology for calculating the minimum energy efficiency, which should be set in accordance with a common "comparative methodology framework" and established based on an optimal level of profitability where investment costs, maintenance, operation, energy is taken into account, etc. Calculated for the period of life of the building (European Union 2010).

In March 2011 the Commission confirmed that the Union was not, despite progress in national policies, track to reach its goal of energy efficiency. In the Council conclusions on the Energy Efficiency Plan 2011, it highlighted that existing buildings of the Union account for 40% of final energy consumption in the Union, which seemed to indicate that Member States should create a strategy to long term beyond 2020 aimed at mobilizing investment in the renovation of residential and commercial buildings to improve energy efficiency of the housing stock. That is why approved in October 2012 Directive 2012/27/EU, establishing a common framework of measures for the promotion of energy efficiency within the Union in order to ensure the achievement of the main objective of energy efficiency of the Union 20% savings by 2020 and pave the way for further energy efficiency improvements beyond that year (EU 2012).

1.2 Building Energy Efficiency in Spain

Currently in Spain the basic procedure for certifying the energy efficiency of buildings is regulated by Royal Decree 235/2013, in line with what was stated in the aforementioned European directive of 2010, establishes a basic procedure for the certification of both energy efficiency of new buildings, as those buildings or parts of buildings which are sold or leased to a new tenant, provided they do not have a valid certificate (Spain 2013). Thus the obligation to have by the buyers or users of such certificate must include objective information on the energy efficiency of a building and reference values such as minimum energy performance requirements in order that the owners or set tenants of the building or unit to compare and assess its energy efficiency. These minimum requirements are regulated by the Energy Saving Basic Document HE of the Technical Building Code.

The purpose of the approval of the procedure is the "objective information" for buyers and/or users, without expressly provides a prescription of solutions for improving energy efficiency, and timelines for adopting them. However the need to present a document of recommendations for improving optimum or profitable levels of of a building is identified; consisting of active measures (by acting on the thermal and/or light) or passive (intervening on the envelope of buildings).

1.3 The Calculation of Energy Efficiency of Buildings in Spain

The calculation of energy efficiency regulated by the RD 235/2013 is supported by two main indicators: annual CO_2 emissions equate ($kgCO_2e.m^2$) and annual consumption of non-renewable primary energy (kWh m^2). In addition to these indicators considering the demand for both heating, for cooling, consumption of non-renewable primary energy disaggregated services and annual emissions disaggregated services, power consumption and other fuels (Ministry of Industry, Energy and Tourism 2015a, p. 4).

Thus the final energy rating of our building is established according to the annual consumption of it and emissions, these indicators may be determined by a thermal model of the building or equivalent simplified methods to the technician responsible for conducting the evaluation being both methods regulated by the state.

1.4 Recognized Tools for Building Energy Certification and Assessment

The state recognizes, through the Ministry of Energy and Tourism four software implementations that can generate and computationally evaluate the energy efficiency of buildings: CERMA, CE3, CE3X, based on simplified methods and tools Unified Lider Calener (hereinafter HULC, acronym in Spanish) which provides more accurate data through the use of a thermal model of the building.

They are all equally valid, however they use different parameters and methods of calculation which consequently makes the results are different in the determination of emissions, such as calculating the energy consumption of the building. A comparative analysis by the University of Granada in 2015 contrasted the different results than the three tools based on the simplified method offered for the same project with those obtained by CALENER PVY software, the previous version of HULC (Carpio et al. 2015, p. 100).

The comparative study showed that the tool CE3 increased by 2.21% calculate CO_2 emissions. By contrast documents as CEX (previous version to CE3X) and CERMA had lower values, with reductions of 26.29 and 24.49% respectively. Although these differences do not involve significant errors in obtaining corresponding to the energy rating of the building letter, if it is true that the use of a tool against other can lead to errors in the choice of specific solutions for enclosures of buildings.

In the case of interventions in existing buildings it is not required, except in the case of an extension thereof, which power consumption limitations indicated in section HE0 and only aspects contemplated energy demand is met. However the simplified methods through the standard proposed for the most used existing buildings (CE3, CE3X and CALENER) do not allow separately determine the energy demand for heating and cooling, which is seen as extremely necessary if we look at the major temperature fluctuations between summer and winter regime of temperate climates. This undoubtedly introduces distortions in the results does not allow to adequately project-based solutions that passive strategies, as seen from the perspective of LCA are those that have the greatest impact on the set of measures that could be taken.

1.5 Complementary Tools for Building Energy Assessment

Alternatively to the tools provided by the administration may be other, as long as the technical conditions concerning procedures for evaluating the energy efficiency of buildings in Spain established by the ministry (Ministry of Industry, Energy and Tourism are verified 2015b), either with the aim of using the results obtained for the energy certification of buildings or for verification of regulatory requirements of the CTE-DB-HE. This will provide, among other conditions, the general characteristics of the calculation procedures and the accuracy of the same. According to this document, the capacity calculation procedures to obtain results with sufficient accuracy for use in the assessment of the energy efficiency of buildings can be credited either by making use of a calculation engine reference [DOE2, BLAST, ESP, SRES/SUN (SEIRES/SUNCODE), SERIRES, S3PAS (LIDER/CALENER), TAS, TRNSYS, EnergyPlus] whose accuracy is considered proven, or by specific justification for their accuracy in a report of results for based procedures UNE-EN ISO 13790. Serve to illustrate the feasibility of raising the alternative assessment using BIM tools, the use of Revit DOE2 calculation engine for evaluating the energy performance of the building both in the design phase and use, something already demonstrated work of the UPC (North Sosa 2014). With this points to the possibility that the legislation itself provides for grounded adoption of alternative methods, with the advent of computer-based models of building programs, would allow also consider related to sustainable building measures improvements in energy efficiency parameters of buildings.

2 Sustainability Assessment of Buildings

Evaluation systems and tools that will see consensually envisage the definition, scope and objectives set by the international standard 15392, which set no political agendas or provides priorities for specific concerns internationally established, although their requirements and objectives if they are related to those goals policies aimed at identifying the general principles for sustainable construction (ISO 15392 2008, p. vi). As recognized by the international standard, the evaluation systems of buildings should consider the three main aspects of sustainability: environmental, social and economic; understood from a perspective of life-cycle assessment (Andrade and Bragança 2016); without their having to be confused with the social aspects of usability conditions, durability, safety and comfort, intrinsic to any architectural project of new plant.

2.1 Tools for the Evaluation and Certification of Sustainability

We start talking about evaluation systems that, in the case of our study, apply to existing buildings, which is not the only possible context in which to consider them, as we shall see. These systems, in addition to the assessment of sustainability, propose a classification within a range of scales, can be incorporated later certification through an accredited institution or agency. They are those entities responsible for evaluating, classifying, certifying sustainability and develop specific tools for each method makes their particular considerations and establishes a system of scales around the concept of sustainability on working.

It should be noted important differences between the methodology and areas that provide different tools to demonstrate the sustainability of the projects evaluated, qualified and/or certificates. While some tools understand sustainability through energy efficiency operational phase of the project, as for example the PassivHaus standard, there are others that progress on the consideration of other environmental burdens linked to the use phase of the building and directly related with the construction or deconstruction and demolition of the building (BEAT 2002, DGNB, LEnSE and other) processes. In the end of this section we indicate a table with the different phases that the systems and tools collected includes or not; likewise the geographical context in which they arose and which mainly have been applied is identified, however many of the tools have been applied in various countries (DGNB, BREEAM, LEED and SBTool), which would support a comparative analysis of sustainability through its structure.

In the synthesis chart review of these tools (Fig. 1) you can comparatively observe the main advantages and disadvantages of the various tools for sustainability assessment and certification studied.

2.2 Assessment Tools Based on Sustainability LCA

As we have seen at the beginning of this section, to contemplate the life cycle of our buildings and their relation with the various dimensions that comprise sustainability becomes a general vocation of all existing systems, due in part to the importance this has in making initial decisions. Using the technique of life-cycle assessment we can understand and address the environmental impacts associated with a product or service; through the collection of inputs, outputs and potential burdens on the environment, throughout the different stages of the life of the product or service (UNE-EN ISO 14040 2006, pp. 8–10). By this method you can: identify improvements in the environmental performance of the products used; provide information to those involved in the design process and decision making; select measurement techniques and indicators for the identification of impacts and

		PRINCIPIOS GENERALES DE LA CONSTRUCCIÓN SOSTENIBLE (15392 2008) / INFORMACIÓN DE LA EVALUACIÓN DEL EDIFICIO (SG. EN 15804 2012 + A1 2013) / INFORMACIÓN DEL CICLO DE VIDA DEL EDIFICIO / ETAPA ANTERIOR AL USO / ETAPA DE PRODUCTO			ETAPA DE CONSTRUCCIÓN		ETAPA DE USO / ESTRUCTURA DEL EDIFICIO			FUNCIONAMIENTO DEL EDIFICIO		ETAPA FIN DE VIDA				INFORMACIÓN ADICIONAL	DETALLES DE HERRAMIENTAS / TIPOLOGÍA / USO							FASE DE APLICACIÓN					
		A1 EXTRACCIÓN Y PROCESADO DE MATERIAS PRIMAS	A2 TRANSPORTE AL FABRICANTE	A3 FABRICACIÓN	A4 TRANSPORTE A LA OBRA	A5 INSTALACIÓN EN EL EDIFICIO	B1 USO O APLICACIÓN DEL PRODUCTO INSTALADO	B2 MANTENIMIENTO	B3-B5 REPARACIÓN, SUSTITUCIÓN Y REHABILITACIÓN	B6 USO DE ENERGÍA EN SERVICIO	B7 USO DEL AGUA EN SERVICIO	C1 DECONSTRUCCIÓN, DEMOLICIÓN	C2 TRANSPORTE HASTA VERTEDERO	C3 TRATAMIENTO DE RESIDUOS	C4 ELIMINACIÓN	D POTENCIAL REUTILIZACIÓN RECUPERACIÓN Y/O RECICLAJE	RESIDENCIAL	PEQUEÑO TERCIARIO	GRAN TERCIARIO	NIVELES	CÓDIGO PAÍS DE ORIGEN (SG. ISO 3166/2)	AÑO DE APARICIÓN	DIMENSIÓN DE LA SOSTENIBILIDAD CONSIDERADA [AMBIENTAL: A | SOCIAL: S | ECONÓMICA: E]	PRE-DISEÑO	DISEÑO	EDIFICIOS EXISTENTES	NUEVOS EDIFICIOS	REHABILITACIÓN	SOPORTE [S:SOFTWARE | T: TABLA DE CÁLCULO]
SISTEMAS DE EVALUACIÓN, CALIFICACIÓN Y CERTIFICACIÓN DE LA SOSTENIBILIDAD	BEAT 2002	•	•	•	•	•	•	•	•			•	•	•	•		•	•	•	9	DNK	2000	A+E	•					S
	BREEAM			•	•	•		•		•	•				•	•	•	•	•	6	GBR	2000	A+S		•	•	•	•	T
	CASBEE			•	•	•				•	•					•	•	•	•	5	JPN	2001	A+S			•	•	•	T
	DGNB	•	•	•	•	•	•	•	•	•	•					•	•	•	•	3	DEU	2008	A+S+E		•	•	•		S
	GBTool	•	•	•	•	•	•	•	•	•	•	•	•	•	•	•	•	•	•	7	CAN	1998	A+S+E		•		•		T
	LEED	•	•	•	•	•				•	•	•	•	•	•	•	•	•	•	4	USA	1998	A+S		•	•	•	•	T
	Minergie	•	•	•						•							•			3	CHE	1994	A+S+E		•	•	•	•	T
	PdC	•	•	•						•							•			1	ESP	2008	A+S		•		•	•	T
	PassivHaus									•							•			1	DEU	1988	A		•	•	•	•	S
	HQE			•				•		•	•	•	•				•			5	FRA	1974	A+S+E		•	•	•	•	T
	SBTooL	•	•	•	•	•	•	•	•	•	•	•					•	•		7	INT	1998	A+S+E	•		•			T
	VERDE	•	•	•	•	•	•	•	•	•	•	•					•	•	•	6	ESP	2009	A+S+E		•	•	•	•	T
	LEnSE	•	•	•	•	•	•	•	•	•	•	•	•	•	•	•	•	•	•	7	EUR	2008	A+S+E	•	•	•	•	•	S
	ZCB									•							•			1	GBR	2008	A	•					T
HERRAMIENTAS DE EVALUACIÓN BASADAS EN EL ACV	ATHENA	•	•	•	•	•	•	•	•	•		•	•	•	•	•	•	•	•		CAN	1989	A	•	•				S
	BEES	•	•	•	•	•	•	•	•			•	•	•	•		•	•	•		USA	1994	A+E	•	•				S
	CEV						•	•	•								•	•	•		INT	2011	A+S+E		•	•	•	•	T
	EcoEffect	•	•	•	•	•	•	•	•			•	•	•	•		•	•	•		CHE	2008	A+S	•	•				S
	EcoProP				•	•	•	•	•								•	•	•		FIN	1999	A	•	•				S
	Eco-Quantum	•	•	•	•	•	•	•	•			•	•	•	•		•	•	•		NLD	1999	A	•	•				S
	ECOSOFT	•	•	•	•	•						•	•	•	•	•	•	•	•		AUT	2008	A	•	•				S
	EPIQR						•	•	•								•	•	•		EUR	1998	A	•	•	•		•	S
	EQUER	•	•	•	•	•	•	•	•			•	•	•	•		•	•	•		FRA	1995	A	•	•				S
	ESCALE	•	•	•	•	•	•	•	•	•	•	•	•	•	•		•	•	•		FRA	1999	A+S	•	•				S
	GaBi	•	•	•	•	•						•	•	•	•	•	•	•	•		GER	2011	A	•	•				S
	HQE/PIMWAQ						•	•	•								•	•	•		FIN	1998	A+S	•	•				S
	IMPACT	•	•	•	•	•						•	•	•	•	•	•	•	•		GBR	2012	A+E	•	•				S
	LCA-House	•	•	•	•	•	•	•	•			•	•	•	•		•	•	•		FIN	2000	A	•	•				T
	LCAid	•	•	•	•	•	•	•	•			•	•	•	•		•	•	•		AUS	2008	A+E	•	•				S
	LEGEP	•	•	•	•	•	•	•	•	•		•	•	•	•	•	•	•	•		GER	2004	A+E	•	•	•	•	•	S
	LEGOE	•	•	•	•	•	•	•	•			•	•	•	•		•	•	•		DEU	1995	A	•	•				S
	LEnSE	•	•	•	•	•	•	•	•	•	•	•	•	•	•	•	•	•	•	7	EUR	2008	A+S+E	•	•	•	•	•	S
	Open House	•	•	•	•	•	•	•	•	•	•	•	•	•	•	•	•	•	•		EUR	2013	A+S+E		•	•	•	•	S
	SimaPro	•	•	•	•	•	•	•	•			•	•	•	•	•	•	•	•		GBR	1990	A+E	•	•				S
	TCQ-2000	•	•	•	•	•						•	•	•	•	•	•	•	•		ESP	1996	A+E	•	•	•	•	•	S
HERRAMIENTAS DE EVALUACIÓN DEL COMPORTAMIENTO ENERGÉTICO	BDA				•	•	•	•	•	•							•	•	•		USA	1997	A+E	•	•	•	•	•	S
	BSLCA	•	•	•	•	•	•	•	•	•		•	•	•	•		•	•	•		FIN	2000	A+E	•	•	•	•	•	S
	CYPECAD MEP	•	•	•	•	•	•	•	•	•							•	•	•		ESP	2011	A	•	•	•	•	•	S
	DesignBuilder									•							•	•	•		GBR	2011	A	•	•	•	•	•	S
	EnergyPLUS									•							•	•	•		USA	1996	A	•	•	•	•	•	S
	eQUEST									•							•	•	•		USA	2013	A+E	•	•	•	•	•	S
	HULC									•							•	•	•	7	ESP	ESP	A	•	•	•	•	•	S
	OpenStudio									•							•	•	•		USA	2011	A+E	•	•	•	•	•	S
	Revit MEP									•							•	•	•		USA	2015	A	•	•	•	•	•	S
	SBI-LCA	•	•	•	•	•	•	•	•	•							•	•	•		DNK	1996	A+E	•	•	•	•	•	S
	TRNSYS									•							•	•	•		USA	2015	A	•	•	•	•	•	S

Fig. 1 Comparative analysis of the various tools. *Source* Own figure

certification of results through a classification process based on that assessment. However, an analysis of this kind covers all stages of the life cycle, which may require a large number of data and application of a considerable amount of resources for its implementation. That is why simplification of this analysis was normalized in order to reduce the complexity of the study and hence the cost, time and effort to do it (UNE 150041 EX 1998, p. 4). Something that points directly to the suitability of LCA simplified models for use in construction.

The study of the various tools investigated throughout the state of art (indicated in Fig. 1) has shown the interoperability of these systems with assessment and above certification (such as LEED and BREEAM), who find these tools the means to assess the environmental impacts of construction, use and end of lifecycle of the physical structure of the building. It has also studied the possibilities these tools proposed for integration into new models of work, verifying that some of them (TCQ2000 and IMPACT) have begun to operate in BIM environments or contemplate sharing files in IFC format (BRE Group 2012; ITEC 2016).

2.3 Environmental Impact Assessment of Building Envelopes

Any self-respecting construction process involves a series of environmental impacts that could be evaluated through many of the above tools. However, as we have identified, not always based on the LCA of a building systems will achieve results sufficiently transparent, since they contain an important complexity. It is therefore understandable that both regulatory level (AENOR Standards), as recent research bet on a model of sustainability assessment of buildings based on simplified LCA techniques. There are now investigations that use such techniques, developing tools for selection of constructive solutions envelopes by using models that study the environmental impacts of the most common solutions (Huedo et al. 2016).

3 Energy Efficiency and Sustainable Construction in Building Rehabilitation

As we have noted, internationally limiting GHG emissions established is reflected in building through the concept of energy efficiency. However the certification process that we not go beyond marketing and characterization of the problem and it is found through the limited interventions on energy rehabilitation in Spain; something that is completely contradictory if we consider that our country has today with 5.48 million obsolete residential buildings built before 1980, according to INE 2012. The reality, according to Euro-construct (2013) it is that the rehabilitation of buildings is not widespread and fifteen points lags behind the European average, around 41% of the construction sector (De Val 2015).

Addressing the urban rehabilitation process from the perspective of sustainable development illustrated above, necessarily involves a number of implications that go beyond environmental considerations and shall consider, among other issues, economic responsibility between neighbors and management; something that will require finding new ways of management and financing, for example, by capitalizing on energy savings, support from private entities and other strategies for the activity to develop sustainably (Cervero-Sánchez and Agustín-Hernández 2015). The complexity of its implementation and the high level of demand to agents involved in the process It thus appears. However, the crisis of the traditional model and policy requirements at European level on energy rehabilitation, generate a land of opportunity for the emergence of new models of management of rehabilitation processes.

Likewise specialization and participation of different professionals throughout the construction process management needs this information for planning, quantification of costs/economic, environmental and social benefits. In this complex context, the definition of an integrated and collaborative design methodology based on performance considering the use of BIM and simulations throughout the entire process (Mediavilla et al. 2015) represent an important aid in management process and evaluation of actions in rehabilitation. In addition the BIM environments provide a collaborative workspace promises to be the usual means by which different professionals in the building sector to collaborate simultaneously on different projects. Some BIM tools allow and to integrate environmental impact assessment of the buildings (Motawa and Carter 2013); the implementation of certifications like LEED are now a reality in constant development (Jalaei and Jrade 2015). In this sense the BuildingSMART association works in the development of guidelines and the promotion of BIM methodology through open standards (BuildingSMART Spanish Chapter 2016), being a national reference and clarifying the real opportunities that the implementation of these models work represent.

As a final thought, we can say that the evaluation of the sustainability of each transformation process is vital in the light of the data mentioned above. That is why the definition of an evaluation methodology developed for interventions regeneration envelopes of buildings and their integration into new work platforms allow a gradual professionalization of trades related to architecture, enabling the optimization of processes and the easiest transmission information.

References

Andrade J, Bragança L (2016) Sustainability assessment of dwellings—a comparison of methodologies Sustainability assessment of dwellings—a comparison of methodologies. Civil Eng Environ Syst 33(2):125–146. doi:10.1080/10286608.2016.1145676

BRE Group (2012) IMPACT. Recuperado 27 de abril de 2016, a partir de http://www.impactwba.com/

BuildingSMART Spanish Chapter (2016) BuildingSMART, Spanish Home of openBIM. Recuperado 17 de noviembre de 2015, a partir de http://www.buildingsmart.es/

Carpio M, Martín-Morales M, Zamorano M (2015) Comparative study by an expert panel of documents recognized for energy efficiency certification of buildings in Spain. Energy Buildings 99:98–103. doi:10.1016/j.enbuild.2015.04.022

Cervero-Sánchez N, Agustín-Hernández L (2015) Remodelación, Transformación y Rehabilitación. Tres formas de intervenir en la Vivienda Social del siglo XX, 67:1–11

De Val JR (2015) Potencial del nuevo marco normativo para el impulso de la rehabilitación y la regeneración urbana en los ámbitos autonómico y local 67:1–17

European Conference on Sustainable Cities (1994) Carta de las Ciudades Europeas hacia Sostenibilidad, Aalborg

European Union (2010) Directiva 2010/31/UE del Parlamento Europeo y del Consejo, de 19 de mayo de 2010, relativa a la eficiencia energética de los edificios (refundición), Diario Oficial de la Unión Europea, de 18 de junio de 2010, núm. 153, pp 13–35

European Union (2012) Directiva 2012/27/UE del parlamento Europeo y del Consejo de 25 de octubre de 2012, Diario Oficial de la Unión Europea, 14 de noviembre de 2012, núm. 315, pp 1–56

González-Díaz MJ (2013) Sostenibilidad y construcción. In: García-Navarro J (ed) Sostenibilidad en la construcción. Instituto de Ciencias de la Construcción Eduardo Torroja, Madrid, pp 13–54

Huedo P, Mulet E, López-Mesa B (2016) A model for the sustainable selection of building envelope assemblies. Environ Impact Assess Rev 57:63–77. doi:10.1016/j.eiar.2015.11.005

ISO 15392 (2008) Sustainability in building construction—general principles

ITEC (2016) ITeC lanza una versión del software TCQ2000 para trabajar en un entorno BIM. Recuperado 15 de abril de 2016, a partir de http://www.coaatavila.es/doc_boletin/BIM_02_16.pdf

Jalaei F, Jrade A (2015) Integrating building information modeling (BIM) and LEED system at the conceptual design stage of sustainable buildings. Sustainable Cities Soc 18:95–107. doi:10.1016/j.scs.2015.06.007

Mediavilla A, Izkara JL, Prieto I (2015) HOLISTEEC—Plataforma colaborativa en la nube basada en BIM para el diseño de edificios energéticamente eficientes. Span J BIM 15(1):4–11

Ministry of Industry, Energy and Tourism (2015a). Calificación de la eficiencia energética de los edificios, Madrid

Ministry of Industry, Energy and Tourism (2015b) Condiciones técnicas de los procedimientos para la evaluación de la eficiencia energética de los edificios, Madrid

Motawa I, Carter K (2013) Sustainable BIM-based evaluation of buildings. Phys Procedia 74:419–428. doi:10.1016/j.sbspro.2013.03.015

Norte Sosa E (2014) Modelado de información para la edificación en BIM y su aplicación frente a las exigencias de la demanda energética con LIDER Y ECOTECT

Spain (2013) Real Decreto 235/2013, de 5 de abril, por el que se aprueba el procedimiento básico para la certificación de la eficiencia energética de los edificios, 89Boletin Oficial del Estado, 13 de abril de 2013, núm. 89:27548–27562

UNE 150041 EX (1998) Análisis del ciclo de vida simplificado

UNE-EN ISO 14040 (2006) Gestión ambiental. Análisis del ciclo de vida. Principios y marco de referencia

From Recovery Constructively Towards the Social Reactivation. The Integrated Knowledge of Traditional Architecture as a Sustainable Strategy

Victoria Domínguez Ruiz

1 Introduction

The theoretical reflection on the Protection of the Built Heritage comes to devise, since the beginning of the twentieth century, in parallel to the problem of knowledge. The concept of the Heritage has been expanded so that it does not allow to contemplate reality without the iteration of multiple factors (Servicio de Protección del Patrimonio Histórico de la DGBC 2003). As an example in Conservation, in the last decades in our country, we have frequently resorted to the figure of Historic Site[1] as collective typology, precisely because of this concept more inclusive and complete. In this willingness to link to each village with its history, it arises more recently the Intangible Heritage, which we could insert within systems of knowledge of philosophical and spiritual dimension, but that they are also involved in this integrated and comprehensive understanding of the object. From the moment in which we identify the architectural event; not only as construction material and built reality (container space), but as a cultural event (container of human activities), then, new patterns of significance are revealed.

This project seeks to contextualize the flour mills of the Guadiana river when it passes by Mértola (Portugal) and focuses on a particular stage of a more detailed study: water mills in the Ribeira do Vascão. There are some small constructions, barely documented graphically, forgotten and scattered on the surface, with a state of considerable deterioration or a large part of them lost. In the territory of Mértola

[1]"Places linked to events or memories of the past, to traditions, cultural of nature creations or and human works, that have a significant value from the point of view of historical, ethnological, archaeological, palaeontological or industrial" according to BOJA number 248 of 19/12/2007.

V. Domínguez Ruiz (✉)
Departamento de Construcciones Arquitectónicas 1, Escuela Técnica Superior de Arquitectura. Universidad de Sevilla, Seville, Spain
e-mail: v_dominguez@us.es

P. Mercader-Moyano (ed.), *Sustainable Development and Renovation in Architecture, Urbanism and Engineering*, DOI 10.1007/978-3-319-51442-0_13

specialists describe how the permanence of the settlements has been throughout history (Macias 2005; Torres 1995) since it was founded in time pre-roman. In particular, these hydraulic engines proliferating throughout Portugal from the s.X, but they probably existed from the roman era (Veiga de Oliveira et al. 1983). At present there is an effort on the part of municipal, cultural and environmental institutions, that recognize and maintain the presence, both physical and social, of the Guadiana river as a key element of Mértola. This policy of global activation (the river and its surroundings) has led recently the restoration of some of the existing water mills in its river bed. The recovery of its historic heritage achieved the sociocultural dynamization of the area. However, the situation for the mills of the Ribeira do Vascão is far away from this reality: it is only on one occasion that we can see their integration into a network of cultural, environmental or educational landmarks (the case of the Moinho do Alferes). In the line of the incipient regenerating activity observed in the area, the advance that we proposed lies in extending this initiative to other mills calling for a collaborative action research, i.e.; gathering methodologies present in the patrimonial management teams—that usually operate separately—and generating an interdisciplinary knowledge base that set an interesting starting point for this scenario of Vascão, smaller in scale than the Guadiana.

2 Objectives and Expected Results

We start from the hypothesis of the benefits of implementing new technologies in the area of Heritage Conservation. In this particular case, focuses on a few architectural pieces that enclose particularities, so they could not tolerate the generalization of the traditional information systems (that obviate the architectural scale). We therefore advocate an interdisciplinary methodological review with the aim of generating knowledge of the object to be studied. In this regard, the objectives and results to be achieved are:

- Disclose this methodological approach, which seeks a review of the knowledge and skills of the architect, directing them toward a larger conceptual horizon and participating in other disciplinary approaches. As a result, we will extend it to other examples of traditional architecture of the area in the project "Sustainable guardianship of Cultural Heritage through Digital BIM and GIS Models: contribution to Knowledge and Social Innovation" (ref. HAR2016-78113-R), with the provisional approval within the Call 2016 of R+D Projects, corresponding to the State Program of Research, Development and Innovation oriented to the Challenges of Society.
- To support this hypothesis, the characterization of the mills has been carried out in accordance with protocols of data load from the Information System for Architectural Heritage of Portugal (SIPA), deepening and expanding the aspects studied in it. As a result we intend to enter the information of the inventory data sheets in this system, being accessible to any user via the Internet. We are

awaiting the approval of the contents. We would be providing, through a digital information model, a greater knowledge of some of the aspects that characterize these parts, known to play a key role in the limited productive system of the cereal several decades ago, but often only treated since the imaginary visual.
- The data-sheets of 29 water mills in Ribeira do Vascão have been offered to the Campo Arqueológico de Mértola (CAM). This is intended to make a contribution for future research processes (inventories, case studies, theses and hypotheses, etc.), conservation (management plans, analysis of pathologies, draft of archaeological and architectural intervention or maintenance plans) and diffusion (queries on heritage information system, digital edition, map compositions, etc.), all of them inherent in the architectural and urban rehabilitation.

3 Justification and Methodology of the Integrated Knowledge: From the Territory to the Built Object

Only if we know our daily events (many authors call microhistory) we can establish ties with the scale of the territory that surrounds us and, after a global analysis, go toward a macro-scale (da Silva Rafael et al. 2010), where the social structure spreads out. By documenting the architecture of the past, people can attempt to reconstruct something that happened a long time ago to improve the understanding of the world, of present, and also to dream of a future.

According to this perspective—from built architecture towards the immaterial dimension—, this investigation aims to establish a basis for the debate on the insertion of these small parts in the cultural framework of Mértola, enabling a sustainable strategy in our society of knowledge. It motivates us to analyze the constructive event both in the physical sense and material sense as in its social and cultural perception; consider not only the historical value of the building, or the chronology of its construction, but all those values (landscape, artistic, geological, environmental, scientific) inherent in its frame. To avoid a fractional study, the methodology supplies an interdisciplinary approach, in which the immersion in other sciences dedicated also to the built heritage (such as geography, the archeology or ethnology), complements the capabilities of the architect.

The Archeology of Architecture, insofar as it is able to identify changes in constructive stages or interventions, applies the same method of analysis for archaeological sites and building in surface (it is justified because they share the same origin and development as a result of the continuous constructive activity from humans). It has allowed us to the consideration of the space and its architectural concept beyond its material essence, which is an aspect of great conceptual importance.

The studies of historical geography in Portugal constitute a relevant contribution to local development strategies. Professor Orlando Ribeiro—as maximum representative since around half of the twentieth century—defended an integrated

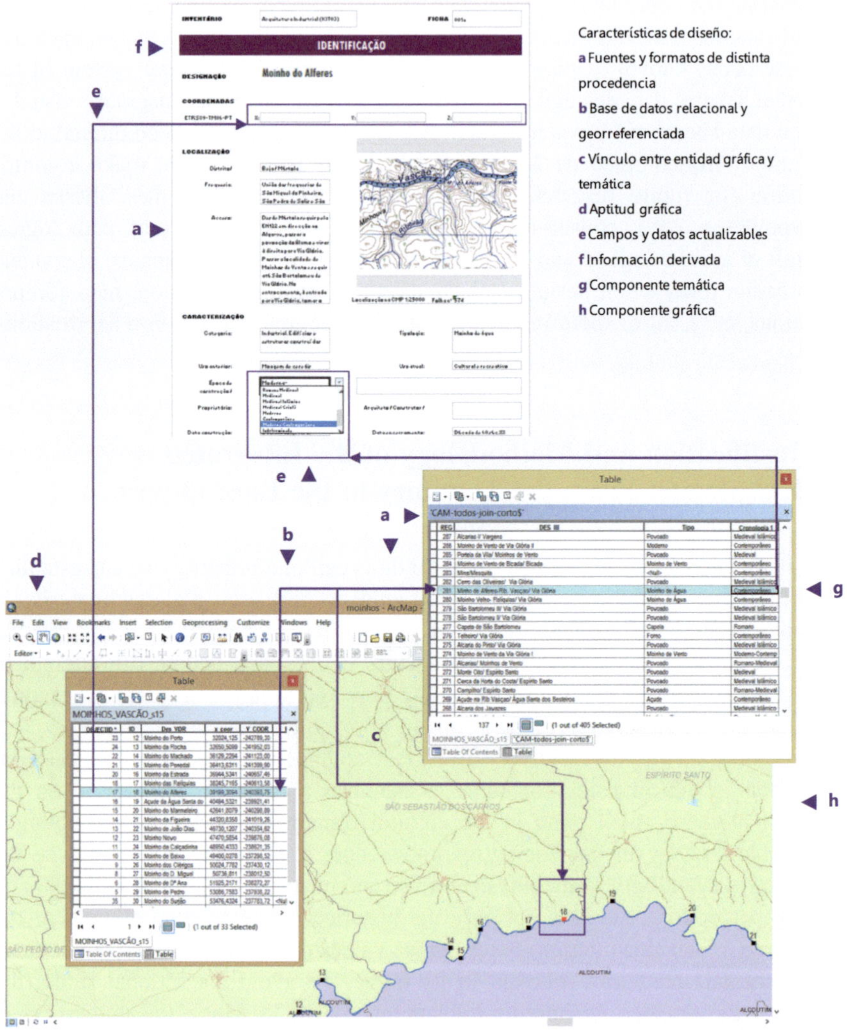

Fig. 1 Conceptualization and design features of the proposed information system. *Top* Tab containing the information created for a mill. *Bottom* Location of mills on different layers: hydrology, roads, settlements. Image by author

knowledge so as not to separate the partial aspects (physical, cultural and social) of the geographic entity object of study.

With these premises we create a geographical database (GDB), and a variety of graphical layers superimposed for the analysis, as shown in Fig. 1.

In relation to the procedures and materials used, we turn to the potentiality of new technologies applied to the study of the heritage and its territorial aspects. The GIS technology represents a qualitative leap regarding storing and treatment of data; it has allowed us the implementation of the GDB, in which will seat the digital model of information created. For this purpose we differentiate two processes: (i) consultation to cartographic and patrimonial official databases as digital support and (ii) fieldwork (graphic and thematic) relating to the mills studied.

For the first point (i), one specific source of reference was the SIPA, by ini-initiative of IHRU (Instituto da Habitação e da Reabilitação Urbana). This work was interesting because gives to know a part of the information structure in three volumes of the digital collection Kits-Património (Vieira and Lacerda 2010a, b). In this case SIPA registers five water mills in the territory of Mértola. However, only one of them belongs to the margins of the Vascão. The treatment of the contents is only made on a large scale, with lack of data loading or records updating.

We had raster and vector maps at territorial level, published and/or transferred by authorities (central and local). Were acquired for the study of the landscape aerial photos of the American Flight 1958 in the CiGeoE to an area of approximately 30 linear km around the Ribeira do Vascão. It is also obtained digital material compatible with GIS from the cartographic department of the municipality of Mértola.

For the second of the points (ii) the sources were found in several sites, such as publications of local studies (Campo Arqueológico de Mértola 2012; Guita 1999), unpublished material as tabs of archaeological sites and (Campo Arqueológico de Mértola 2006; Orlando 2006) or doctoral theses (Guita et al. 2006).

Equally important was the fieldwork and the subsequent production of data such as: metric and architectural lifting of the mills, cross sections of the ground in the enclave of each mill, complementary terrestrial and aerial photographs (with the support of a drone).Some necessary work for the storage, management and publishing of the information gathered and produced are: the capture and georeferencing of existing raster maps, import and conversion of files in several formats, establishment of an internal protocol of entry, management of tables, error corrections, management of map projections, etc.

4 Results of the Proposed Information Model

4.1 Constructive Recovery: Local Knowledge as a Sustainable Key

It provides for the identification of techniques and construction materials (Fig. 2), traditional hydraulic technologies, date of construction, conservation status, visible architectural fragments, functional elements (Fig. 3), etc. These attributes These attributes will form the GDB and with them we will be characterizing each item in the sample.

These hydraulic stations (Guita et al. 2006: p. 183) cannot be ignored in the functional study. Knowing the solutions of energy utilization, balance and respect for nature and controlled exploitation of the resources will contribute to the

Fig. 2 Constructive analysis. Comparison of constructive technical solutions (Alferes water mill) with typological studies in traditional architecture of the area (Mesquita, Mértola). Drawing by Agostini and Vannetiello (1999)

implementation of the value of the mills. In turn, the formal analysis of the mills evidence actions of minimal impact, however we admit that were categorically designed in order to contend with the force of the river. Figure 4 shows vaulted buildings near the influence of Guadiana; while, at greater distances, the ceramic roof consists of reeds on wooden logs.

4.2 Social Reactivation: Essential Function of Cultural Heritage

As a final reflection, any transformation or abandonment of a pre-existing architecture happens when users consider it inadequate to develop their usual activity in construction, function or form. According to this triple perspective—we suppose social recovery as a consequence of the action research and not an objective—we collect some nearby initiatives and suggest what role they could play these preexistences.

Since the scope of the praxis, each year the CAM recruits volunteers to work in archaeological excavations. In the archaeological map of Mértola (Campo Arqueológico de Mértola 2012) are represented, among other archaeological sites, 42 mills (water or wind) distributed by all the municipal term, of which only three belong to the Ribera del Vascão. The characterization that we have developed will do a great deal to lead in the not too distant future campaigns of excavations in other water mills that were not treated until now. The tabs proposals could be updated with new data collected.

Fig. 3 Functional analysis. *Top* Identification of elements of the Alferes water mill, Ribeira do Vascão. *Bottom left* Weir and dam. *Bottom right* House and oven. Image by author

In the field of research the CAM also organizes meetings such as the I International Congress on traditional architecture in the Western Mediterranean (CIATMO), where gathered architects, archaeologists, teachers, researchers, professionals and interested citizens. Pre-industrial structures of popular architecture were the subject of several oral interventions (AAVV 2015).

The traditional citizen participation in Mértola reaches also the practice constructive. The recovery of the Alferes Mill was carried out in 1992 by the Association for the Defense of the Heritage of Mértola (ADPM) with the criterion of giving continuity to the techniques and construction materials precedents, from conversations with knowledgeable of the mill (according to direct information supplied by Jorge Revez, president of the ADPM). The graphic documentation of this rural water structures, generated in our study after the architectural lifting—which includes the mill, the house, the oven and its settlement—, it might be useful for the ADPM. Equally for the owners of the Figueira Mill, Relíquias Mill (Fig. 5) and Calçadinha Mill, where the level of development of the study has been greater than in the rest.

The experience in recovery, archaeological intervention, musealization of small buildings—where the internal dynamics (Professional School ALSUD, Municipallity of Mértola, ADPM, CAM) are supported by national and international actors—are local development strategies for a balanced enjoyment of the cultural heritage of the villa on the part of all. And we are convinced that there is room for the water mills and other examples of architecture at risk.

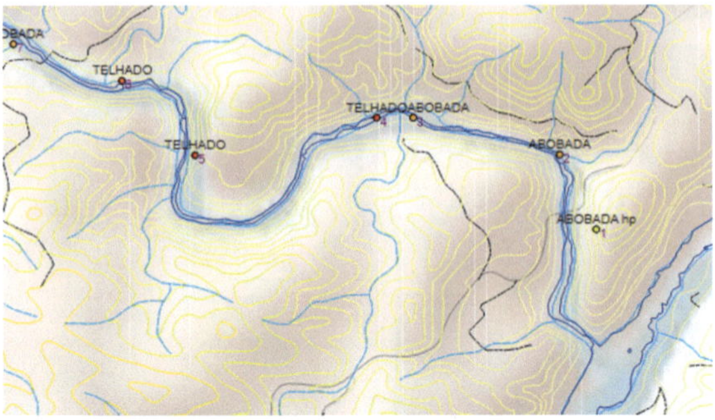

Fig. 4 Formal analysis. System coverage in the mills of the Ribeira de Oeiras. Localization on layers of hydrography, orography and roads. Image by author

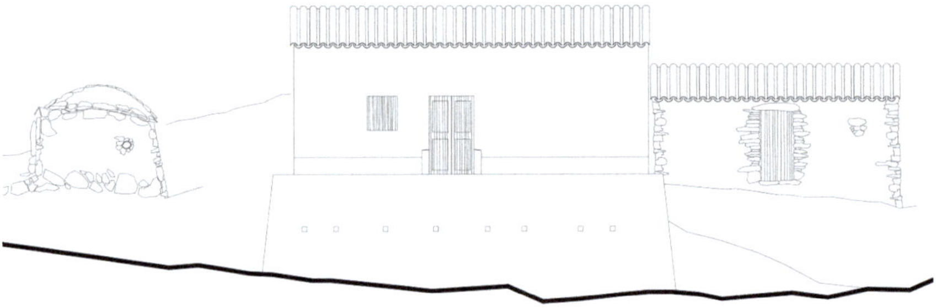

Fig. 5 Lifting the Reliquias Mill: elevations of oven and miller house. Image by author

The dissemination of the built heritage as a strategy of sustainable development have already been demonstrated in projects of lifting of traditional household typologies of Mértola, historical characterization, landscapes and study of its urban evolution (Costa 2016).

4.3 *Results: Sheet Inventory*

These tabs identifying each mill (Fig. 6) include a structured information, oriented to the different phases in the process of architectural rehabilitation (identification, description of elements, graphic definition, conservation or intervention).

INVENTÁRIO Arquitetura Industrial **FICHA** 17a

IDENTIFICAÇÃO

DESIGNAÇÃO Moinho das Relíquias

COORDENADAS (m)

ETRS89-TM06-PT	X: 38245,7165	Y: -240613,58540	Z: 89

LOCALIZAÇÃO

Distrito/ Município: Beja / Mértola

Freguesia: União das freguesias de São Miguel do Pinheiro, São Pedro de Solis e São Sebastião dos Carros

Acceso: A partir de Mértola seguir pela EN122 em direcção ao Algarve, passar a povoação do Álamo e virar à direita para Via Glória. Passar a localidade de Moinhos de Vento e seguir pela estrada que vai de Via Glória para Giões. Fica do lado esquerdo da ponte da Ribeira do Vascão.

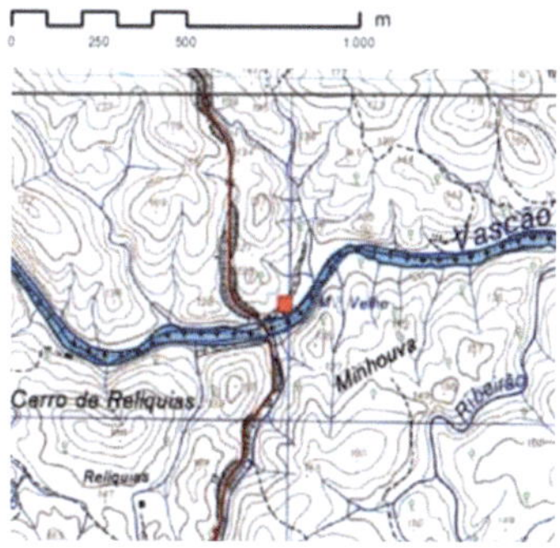

Localização na CMP 1:25000 Folha nº 574

CARACTERIZAÇÃO

Categoria:	Industrial. Edifícios e estruturas construídas	Tipologia:	Moinho de água
Uso anterior:	Moagem de cereáis	Uso atual:	Sem uso
Época de construção /	Moderna-Contemporânea		
Proprietário Arquiteto / Construtor / Autor: Moleiro	Moleiro: António Ildefonso Eugénio, S. Bartolomeu de Via Glória (Mértola).		
Data construção:		Data encerramento:	Década de 90 séc.XX

Fig. 6 Example of sheet inventory created for characterizing each mill. It integrates both of graphical and thematic information. Image by author

5 Conclusions

The following conclusions can be drawn from this research:

- The methodological developments presented in this work constitute an information base for the analysis, conservation and dissemination of a heritage at risk in its historical context as a vehicle to promote the critical development of the society. It is also extrapolated to other examples of traditional architecture in both the database and the structure of the model allows for variations.

- From a more implemental approach, an instrumental basis has been created, versatile and compatible with softwares that operate in the local management of the built heritage Mértola, which provides a high degree of applicability of the work.
- The synthetic organization in tables of information on the mills one by one (29 watermills) allows complete other information systems in heritage management. It is the case of the aforementioned SIPA; the existing record of the Alferes Mill (code on the system IPA.00033203) can be updated. The rest would be records of new creation by not having at the present time any reference in the system to the mills of the Vascão.
- We believe the sheet inventory as an action of putting in value of this popular architecture specific.
- The identification of potentialities in these places, in accordance with the characterization of landscapes, track accesses or topographic profiles, contribute to generating specific strategies for each mill in terms of local development, with different special conditions, in connection with a range of existing events (routes, Cycle of bread, architectural tourism, etc.).
- After carrying out the cartographic integration, we have been detected relationships of proximity between the mills and other traces of the colonization of the territory, as anticipated by the well-known historical activity of movement of the bread and the oil along the old tracks crossing the many rivers around the Guadiana. It aims as likely the physical presence of the mills in past decades, even centuries earlier, suggesting a more in-depth study.
- The study supports a contribution to improve the knowledge of these parts, in general very underanalysed and underreported, through an analysis of its presence in different types of documentary sources of different rigor and provenance.

Acknowledgements Archeological Site of Mértola (CAM, Portugal) and the University of Seville to provide logistical and financial support respectively during an international stay of in-research in this center (2013–14). Thanks to the Research Group TEP-130 by aid to the dissemination of research results. The work team "Un modelo digital de información para el conocimiento y gestión de bienes del patrimonio cultural" (HAR2012-34571) by the exchange of knowledge.

References

AAVV (2015) I Congresso Internacional de Arquitetura Tradicional no Mediterrâneo Ocidental. In: Costa MR, Gómez Martínez S, Ribeiro V (eds) 1º Congres. Argumentum, Campo Arqueológico de Mértola, Lisboa

Agostini I, Vannetiello D (1999) La casa rurale nel territorio di Mértola, studio tipologico. Arqueologia Medieval (6):267–275. misc

Campo Arqueológico de Mértola (2006) Fotografías y fichas de sitios del Levantamiento Arqueológico del término municipal de Mértola (2005/2006). Inédito

Campo Arqueológico de Mértola (2012) Carta arqueológica do Concelho de Mértola. Campo Arqueológico de Mértola, Mértola

Costa MR (2016) Mértola: a arquitetura da vila e do termo. (Campo Arqueológico de Mértola, Ed.). Mértola

da Silva Rafael LI, dir: Brigola JC, Torres C (2010) Os Trinta Anos do Projecto Mértola Vila Museu: Balanço e Perspectiva (trabajo fin de master). Universidade de Évora-V Curso de Mestrado em Museologia

Guita R (1999) Engenhos hidráulicos tradicionais. ICN-PNVG, BOOK, Mértola

Guita R, Moutinho M, Custódio J (2006) Museologia e ecomusealização global. Propostas para a musealização de engenhos, conjuntos e sistemas hidráulicos. Tesis doctoral inédita (THES). Universidade Lusófona de Humanidades e Tecnologias, Departamento de Museologia, Arquitectura, Urbanismo e Geografia, Lisboa. Retrieved from http://www.ceied.ulusofona.pt/pt/formacao-avancada/mestrados/141-msc-museologia-pt/497-msc-museologia-dissertacoes

Macias S (2005) Mértola: o último porto do Mediterrâneo. Catálogo de exposición Mértola: história e património (séculos V-XIII). (Campo Arqueológico de Mértola, Ed.) Mértola

Orlando J (2006) Fotografias do Ribeira do Vascão (inédito). Alcaria dos Javazes, Mértola

Servicio de Protección del Patrimonio Histórico de la DGBC (2003) Los bienes culturales en el territorio y en la historia. Boletín Del Instituto Andaluz Del Patrimonio Histórico (43):32–37

Torres C (1995) O espaço familiar e formas de habitar no Garb al Andaluz. In 1as jornadas de Cerâmica medieval e Pós-Medieval. Métodos e resultados para o seu estudo. misc, Tondela, pp 33–40

Veiga de Oliveira E, Galhano F, Pereira B (1983) Tecnologia Tradicional Portuguesa: Sistemas de Moagem. Instituto Nacional de Investigação Científica - Centro de Estudos de Etnologia

Vieira J, Lacerda M (2010a) KIT01 Património Arquitectónico – Geral | versão 2.0 | documento definitivo. Colecção KITS – Património. Retrieved from https://www.portaldahabitacao.pt/opencms/export/sites/ihru/pt/ihru/docs/publicacoes/KIT01nov.pdf

Vieira J, Lacerda M (2010b) KIT03 Património Industrial | versão 1.0 | documento definitivo. Colecção KITS – Património. Retrieved from https://www.portaldahabitacao.pt/opencms/export/sites/ihru/pt/ihru/docs/publicacoes/KIT03nov.pdf

Energy Retrofitting and Social Housing Instrumentation Attending Passive Criteria. Case Study in Winter

S.G. Melgar, J.M. Andújar and M.A. Bohórquez

1 Introduction

The building sector is key to reducing greenhouse gases, responsible for global warming phenomenon (Cellura et al. 2013; Mohareb and Mohareb 2014; Ramon and Burgos 2008; Stojiljković et al. 2015). There are multiple researches focused on how to design and build new buildings with a highly efficient energy performance as a research demonstrator (Aldegheri et al. 2014; Irulegi et al. 2014; Navarro et al. 2014; Ochoa and Capeluto 2008; Pataky et al. 2014; Rodriguez-Ubinas et al. 2014a, b; Serra Soriano et al. 2014; Terrados and Moreno 2014; Terrados-Cepeda et al. 2015). Other focuses the problem on the new prefabricated construction techniques, with high levels of industrialization processes (Avellaneda et al. 2009; Ruiz-Larrea et al. 2008), but almost always applied to new building, a marginal situation in the crisis of the construction sector we are involved. Far fewer studies focus the problem on how to improve energy efficiency in existing buildings, almost all of them based primarily on data obtained through energy simulations (Chow et al. 2013; Chuah et al. 2013; Kharseh and Al-Khawaja 2016; Murray et al. 2014), without measure and analyze the real behavior of renovated buildings with instrumentation systems and analysis of physical variables, using in situ measurements.

The current situation of the Spanish homes built park (Instituto Nacional de Estadísticas 2011) is 55.9% prior to the first regulation on thermal conditions in buildings (Urbanismo 1979), so their energy behavior could be described as ruinous, following the current regulatory standards. According to the same INE

S.G. Melgar (✉) · J.M. Andújar · M.A. Bohórquez
TEP 192 "Control y Robótica" Research Group, Higher Technical School of Engineering, University of Huelva, 21871 Palos de La Frontera, Huelva, Spain
e-mail: sergomel@uhu.es

S.G. Melgar
LAR Arquitectura, c/Puerto 5 entreplanta A, 21003 Huelva, Spain

P. Mercader-Moyano (ed.), *Sustainable Development and Renovation in Architecture, Urbanism and Engineering*, DOI 10.1007/978-3-319-51442-0_14

statistics, 29% are prior to 2002, and therefore long before the current regulation on energy saving (de Fomento 2006), which has recently been updated dated September 2013.

According to Suarez and Fragoso (2016), the fundamental step to achieve the target set by the Directive 2010/31/EU is limiting the consumption of non-renewable energy and primary energy demand in the building. European directive leaves undefined the concept of NZEB, relying to each country member the concretion of the concept.

The Passivhaus standard building (Asdrubali et al. 2008; Blunden 2009; Brew 2011; Cagna 2012) sets the limitation of primary energy consumption in 120 kWh/m^2 per year, and energy demand for heating and cooling 15 kWh/m^2 per year.

Moreover, the problem of energy poverty, defined as the need for family unit to invest in energy more than 10% of their monthly income (De Luxan Garcia De Diego et al. 2015) is increasing in our society, caused by the continued rise in energy prices and the economic crisis that impoverished most citizens of the eurozone. Under these conditions it is not reasonable to work in key demand reduction or repayment, but in improving the living conditions of users of these houses.

This paper wants to fill an important gap in the bibliography, focusing the study of thermal comfort inside rehabilitated social housing under energy poverty conditions. This is how to improve the living conditions of poor families to pay energy bills in their homes. And how to get from passive solutions in the thermal envelope of housing.

This paper is structured in the form that set out below. In the second section we describe the object studied housing, architectural and construction features. In the third section we present the activities of energy rehabilitation practiced in the house. In the fourth section we develop the material and method specifically developed for research that supports this work and that have allowed obtaining the results shown below. The fifth section discusses the main results obtenidos, which are expanded and discussed at the next. In the seventh and final section we reflect the main findings of the research.

2 Study Model

The property studied is located in the city of Huelva, with a mild climate in winter, hot in summer and a high level of annual insolation. Integrated in a multi-family building it is part of a representative type of residential growth in the 60s, 70s and 80s in Spain. It is located on the fourth floor of a building of five. It has two opposite facades: the exterior facade facing southeast occurs building access, and the interior facade facing northwest promotes cross ventilation. Housing under study is inhabited by a young family of five and has a built area of 104.44 m^2.

Constructive qualities of housing before rehabilitation are very poor, consisting in walls plastered and painted externally formed by 12 cm masonry, unventilated air chamber without interior insulation and a 4 cm masonry inside. The carpenters are aluminum sliding doors without thermal break, with monolithic glass 4 mm thick. The level of infiltration of outside air in the house is very high. It has no ventilation or air extraction in bathrooms. The interior comfort conditions of the pre-housing rehabilitation are very poor, both in summer and winter, according to their users report.

3 Retrofitting Works

The actions have been taken under the concept of passive housing or near zero energy building (NZEB) applying the passivhaus standard to housing rehabilitation, which means to improve the amount and continuity of thermal insulation and airtightness of the house energy envelope. These actions are part of the I+D+i research project EREBA2020 developed by the TEP192 Control y Robótica research group, funded from FEDER funds by Andalucia Government.

The energy performance envelope rehabilitation of housing involved the application of an external thermal insulation system from the outside (ETICS) based on 10 cm thick mineral wool (LM) (see Fig. 1), arranged in two layers 5 cm thick each. The particular system installed is called ISOFEX from Isover™.

The house has been installed a system of seven chambers PVC frames ($Uf = 1.1\ W/m^2\ K$) with 82 mm total thickness, harboring dual chamber low emissive glass, injected with 90% argon gas. The frame is 6/15/6/15/6 ($Ug = 0.6\ W/m^2\ K$) (see Fig. 3). To avoid compromising the functionality of space, in kitchens were installed five cameras sliding PVC frames ($Uf = 2.4\ W/m^2\ K$) with 72 mm total thickness, harboring single chamber low

Fig. 1 Image of the facade of the house during construction

emissive glass, injected with 90% argon gas. The frame is 6/16/6 (Ug = 1 W/m^2 K). The thermal bridge between the existing enclosure and the new frame (installed outside of facade, on the same plane as the insulation ETICS) is solved by SWS system, from Soudal trademark. It consists in a semipermeable film—continued expansion joint around the perimeter of the sealed framework—with flexible polyurethane foam and neutral silicone grouting with high durability (see Fig. 2). To mechanically fix the heavy carpentry sheet to the original outer wall and keep it safe (achieving a complete continuity of the thermal envelope, with absolute absence of thermal bridges in the encounter with building facade). It has chosen to install drawers complexes formed with phenolic panels in each of the windows.

Fig. 2 Construction details of windows

The vertical housing dividing walls have been treated inside (because they are inaccessible from the outside) using double-sided partition plate with continuous insulation (not between profiles), consisting in mineral wool 5 cm thick.

It has also been performed on the upper and lower slab of the house, as the energy poverty of most of the homes in the area could cause significant heat flows through the floors. In fact, the upper slab has been coated with continuous insulation panel of mineral wool 5 cm thick, fastened by galvanized low profile steel omegas (2.5 cm) and single-coated plasterboard. The floors deck of the house has been risen and reassembled, to place under it a low profile insulating sheet (8 mm), consisting in a expanded polyethylene, an aluminum sheet and air bubbles sheet.

The rehabilitation works have been completed with the installation of a new ventilation system that ensures indoor air quality for high sealing values obtained in the retrofitted house (h^{-1} = 1 on blower door test). The final solution consists of a forced ventilation system with double airflow with high efficiency heat recovery. The main machine (see Fig. 3) consists of a variable speed fan (100–350 m^3/h), with a total of four conduits: two for indoor air (round-trip) and two for outside air (taken and exhaust). The system recovers the 93% of the heat from the air inside the house before ejecting it outside in winter. The conduits through which the ventilation air is driven at very low speed run through the false ceiling of the corridor, with drives in dry rooms (living room and bedrooms) and returns in wet rooms (kitchen and bathrooms).

To operate in summer conditions, the machine is equipped with a temperature sensor which activates a bypass to operate the system in freecooling mode when a tendency to overheating occurs.

Figures 4 and 5 show the outline of the enclosure before and after rehabilitation, as well as its main physical parameters, analysis transmittances and interstitial condensation.

Fig. 3 Air heat recovery and duct system

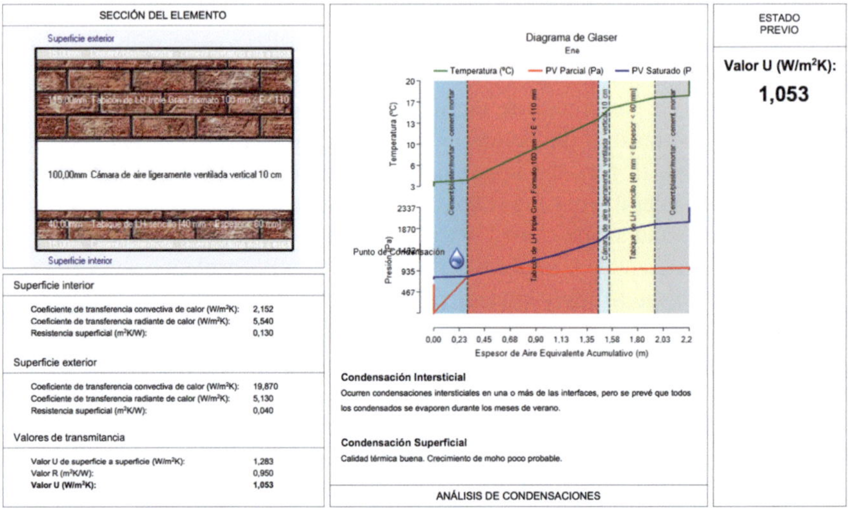

Fig. 4 Exterior walls of the house before rehabilitation

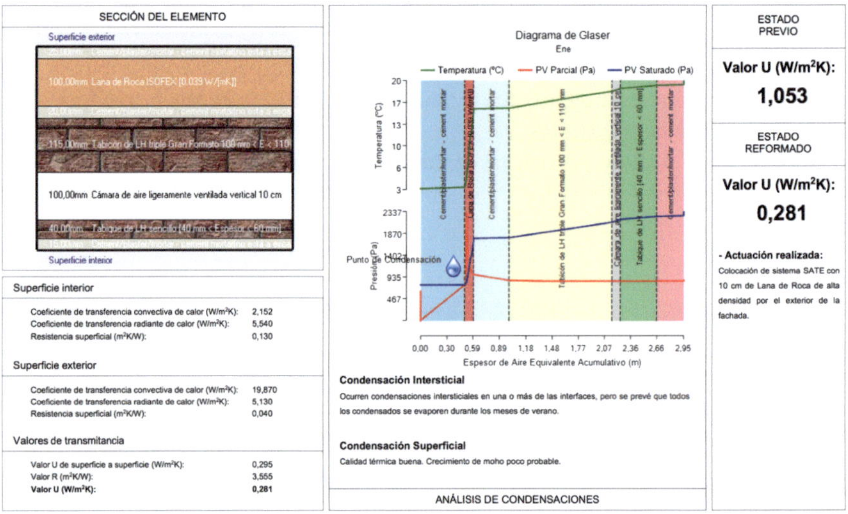

Fig. 5 Exterior walls of the house after rehabilitation

4 Material and Method

The methodology used in this research parts of a 3D modeling housing using the standard Building Information Modelling (BIM). This BIM model enables to optimize the decision-making phase thanks to it is a dynamic database that

Fig. 6 Interior sensor, exterior sensor and communication hub

integrates geometric, energetic and constructive information. Once the major decisions had been taken (geometry, implementation, guidance and protection gaps). For construction project, the energy simulation tool EnergyPlus was used because of its power and customizability. There are no temperature measurements before rehabilitation state, because although an instrumentation system was installed in some of the rooms, it could be measuring only for two months in the summer due to the work times; resulting not representative as compared to a study in cold weather.

It is considered that the thermal comfort in winter for the purposes of this research is determined from a qualitative point of view by the absence of significant fluctuations in temperature, and from the quantitative point of view for those conditions of temperature above 18 °C.

The data acquired in December 2015 included in this work have been carried out in occupied housing conditions, without any input from heating systems. In no case they have opened the windows, keeping indoor air quality thanks to the heat recovery ventilation installed.

For the data acquisition of temperature outside and inside the house, two sensors per room have been installed. For the purposes of this study we selected the ones located in the main living room and kitchen, because both are day use stays located in opposite facades the house. All sensing system installed is wired to signal level

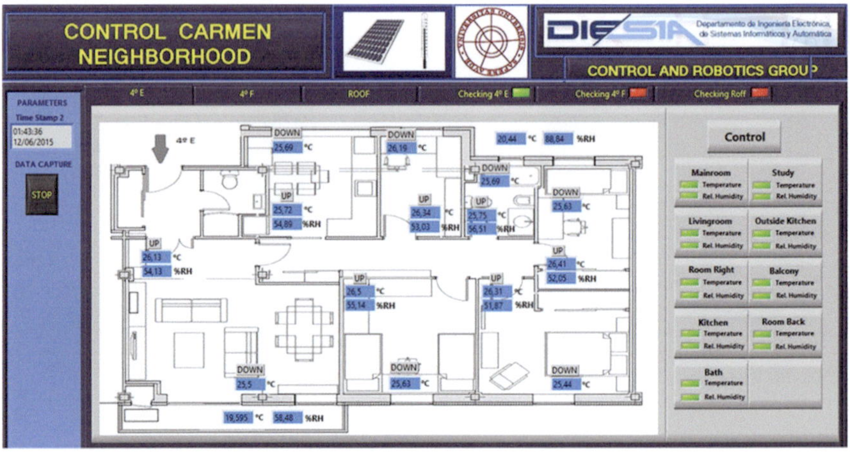

Fig. 7 Sensor location in the house

and governed by a principal master that centralizes all communications and sends them over the Internet to servers in the research group at the School of Engineering at the University of Huelva, where they are stored in real time. In Fig. 6 we can see two of the sensors installed inside and outside the home. Also we observe the centralization of communications located in an annex to the house electricity box next to the door.

For the acquisition and subsequent storage of all signals of housing under study it has developed a software application (virtual instrument) that takes the data, detects errors in communication, stores and allows the realization of graphs and reports. In Fig. 7, temperature measurements and locations of the sensors are shown.

5 Results

Figures 1 and 2 present the results of evolution of the daily temperature obtained in two representative days of cold weather (December 8 and December 12, 2015), from the full year of measurements already performed on the rehabilitated housing. Note the large temperature variability irradiated on the exterior walls of the house (outside facade in blue and outside yard facade in orange), as opposed to the stable temperature of the air inside the house (living room in green and kitchen in yellow).

In Graph 1, for the day December 8, 2015, We note that the outside temperature of the facades acquires a minimum value around 14 °C during the whole night, between 22:30 and 9:00 in the morning. The absolute minimum values are 11.26 °C at 8:30 for the outer facade and 12.81 °C at 8:54 for the interior facade.

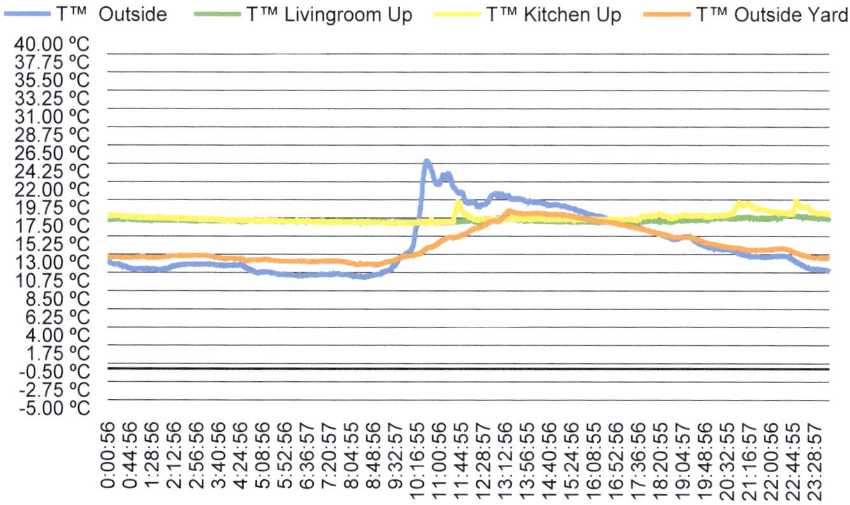

Graph 1 Evolution of temperature day December 8, 2015

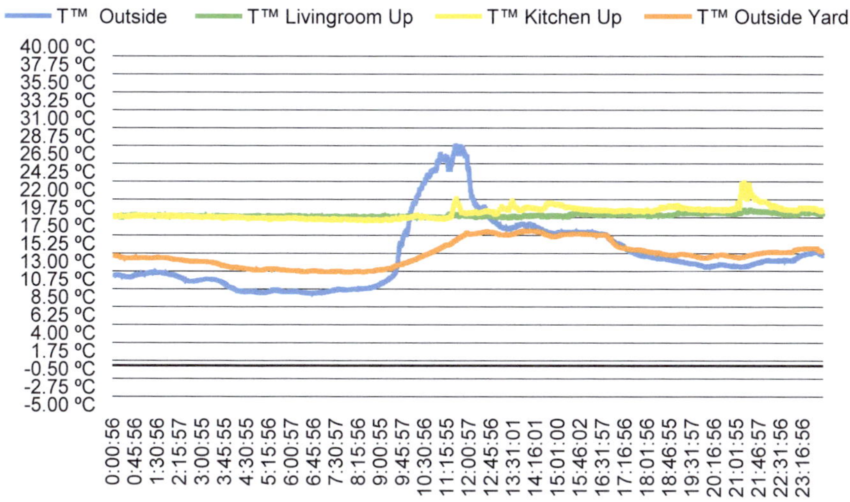

Graph 2 Evolution of temperature day December 12, 2015

On the outer facade facing southeast, the temperature rises rapidly to a maximum of 25.73 °C at 10:36, only to fall again this time much more slowly, as a result of the movement of the sun, to 11.26 °C. The maximum variation in surface temperature of the facade for that day is 14.47 °C.

In the outside yard facade—oriented northwest—increased temperature is more moderate and progressive because it does not receive direct sunlight, with up to 19.50 °C at 13:21. This temperature is maintained for about two hours closest to

19 °C, and the continue falling gradually to 12.81 °C with the sunset and overnight. The maximum variation of the facade surface temperature for that day is 6.69 °C.

For that day of December, the variation of temperature inside the house is only 1.16 °C in the sensor located in the living room (with a maximum of 19.13 °C, a minimum of 17.97 °C and a 18.32 °C average), and 2.87 °C in the sensor located in the kitchen (with a maximum of 20.78 °C, a minimum of 17.91 °C and an average of 18.63 °C). Note that we can see two visible peaks very localized of temperature variation in the kitchen sensor, which matches with periods of cooking, because the house is occupied. They don't produce significant variation in the daily average temperature in the room.

Graph 2 shows the results measured on December 12, 2015, for the purpose of comparing the overall pattern of behavior of the envelope in cold weather. It observed that the outside temperature of the facades acquires a minimum value around 11 °C overnight and much of the afternoon, in the interval between 18:30 and 9:00. The absolute minimum values are 9.01 °C at 6:43 for the exterior facade and 11.75 °C at 7:54 for the interior facade.

In the outer facade—oriented to southeast—the temperature rises rapidly to a maximum of 27.64 °C at 11:33, only to fall again abruptly at first and then more slowly as a result of the movement of the sun, up to 9.01 °C. The maximum variation of the facade surface temperature for that day is 18.63 °C.

In the outside yard facade—oriented northwest—increased temperature is more moderate and progressive because it doesn't receive direct sunlight, with a maximum of 16.94 °C at 13:53. This temperature is maintained for about three more hours around 16 °C to continue falling gradually to 11.75 °C with sunset and overnight. The maximum variation of the facade surface temperature for that day is 10.25 °C.

For that day of December, the variation of temperature inside the house is only 1.10 °C in the sensor located in the living room (with a maximum of 19.63 °C, a minimum of 18.53 °C and an 18.88 °C average), and 4.87 °C in the sensor located in the kitchen (with a maximum of 23.03 °C, a minimum of 18.16 °C and an average of 19.21 °C). Note again that two visible peaks of very localized temperature variation in the sensor of the kitchen, which significantly coincide with periods of cooking, because the house is occupied. As in the previous case, since the specific peaks, pose no significant variation in daily average temperature in that room.

6 Discussion

In winter conditions, the surface temperature varies with sun exposure, with increases of 14.47 and 18.63 °C in the outer facade and 6.69 and 10.25 °C in the interior, for the cases studied. These solar gains, are beneficial in winter for the interior comfort of the home.

The analysis of the variation of indoor air temperature in winter inside the house show values of 1.16 °C in the living room, and 2.87 °C in the kitchen. The average temperature measured during the day is 18.32 and 18.63 °C, respectively. Again almost identical for both rooms. Again the same pattern is repeated the next day of the study in winter conditions, with a variation of 1.10 °C in the living room and 4.87 °C in the kitchen; and a daily average of 18.88 and 19.21 °C respectively. Remember that the maximum variations in the kitchen are related to the heat generated for cooking, as a result of the house has been inhabited during the winter.

From the perspective of interior comfort in the winter, we check that it is possible to maintain the house indoor temperature in very stable values around 19 °C without heating, which is a very appropriate goal if we increase adaptive factors related to metabolic activity and clothing level of users.

7 Conclusions

The rehabilitation of the thermal envelope formed in the housing studied guarantees an appropriate indoor thermal comfort under energy poverty conditions in winter. It is an added value to the interior comfort of housing the thermal stability, which does not vary significantly for the user during the hours of the day or night.

Rehabilitation criteria set by the Passivhaus standard are appropriate to ensure proper thermal performance of near zero energy buildings in winter conditions in the area of study.

Acknowledgements The authors would like to thank the ERDF of European Union for financial support via project "EREBA2020" of the "Programa Operativo FEDER de Andalucía 2007–2013". We also thank all Public Works Agency and Regional Ministry of Public Works and Housing of the Regional Government of Andalusia staff and researchers for their dedication and professionalism. Similarly to Soudal and Weber, and especially LAR Architecture, Isover and Saban Construciones for their collaboration in the implementation phase of the work of energy rehabilitation of housing.

References

Aldegheri F, Baricordi S, Bernardoni P, Brocato M, Calabrese G, Guidi V, Vincenzi D (2014) Building integrated low concentration solar system for a self-sustainable Mediterranean villa: The Astonyshine house. Energy Build 77:355–363. doi:10.1016/j.enbuild.2014.03.058

Asdrubali F, Bonaut M, Battisti M, Venegas M (2008) Comparative study of energy regulations for buildings in Italy and Spain. Energy Build 40(10):1805–1815. doi:10.1016/j.enbuild.2008.03.007

Avellaneda J, González JM, Marques G, Vidal J (2009) Technological innovation in public housing developments: the INCASOL Technological Innovation Competition. Informes de la Construccion 61(513):87–100. doi:10.3989/ic.09.002

Blunden K (2009) Building research housing group. Build Eng 84(8):29

Brew JS (2011) Achieving Passivhaus standard in North America: lessons learned. Paper presented at the ASHRAE Transactions
Cagna J (2012) Montgomery primary school, exete a lesson to be learned. Build Eng 87(7):18–20
Cellura M, Guarino F, Longo S, Mistretta M, Orioli A (2013) The role of the building sector for reducing energy consumption and greenhouse gases: an Italian case study. Renew Energy 60:586–597. doi:10.1016/j.renene.2013.06.019
Chow DHC, Li Z, Darkwa J (2013) The effectiveness of retrofitting existing public buildings in face of future climate change in the hot summer cold winter region of China. Energy Build 57:176–186. doi:10.1016/j.enbuild.2012.11.012
Chuah JW, Raghunathan A, Jha NK (2013) ROBESim: a retrofit-oriented building energy simulator based on EnergyPlus. Energy Build 66:88–103. doi:10.1016/j.enbuild.2013.07.020
de Fomento M (2006) Código Técnico de la Edificación. CTE-DB-HE
De Luxan Garcia De Diego M, Gómez Muñoz G, Román López E (2015) Towards new energy accounting in residential building. Informes de la Construccion 67(Extra1). doi:10.3989/ic.14.059
Instituto Nacional de Estadísticas (2011) Censo de población y viviendas
Irulegi O, Torres L, Serra A, Mendizabal I, Hernández R (2014) The Ekihouse: an energy self-sufficient house based on passive design strategies. Energy Build 83:57–69. doi:10.1016/j.enbuild.2014.03.077
Kharseh M, Al-Khawaja M (2016) Retrofitting measures for reducing buildings cooling requirements in cooling-dominated environment: residential house. Appl Therm Eng 98:352–356. doi:10.1016/j.applthermaleng.2015.12.063
Ministerio de Obras Públicas y Urbanismo (1979) NBE-CT-79
Mohareb EA, Mohareb AK (2014) A comparison of greenhouse gas emissions in the residential sector of major Canadian cities. Can J Civ Eng 41(4):285–293. doi:10.1139/cjce-2013-0465
Murray SN, Walsh BP, Kelliher D, O'Sullivan DTJ (2014) Multi-variable optimization of thermal energy efficiency retrofitting of buildings using static modelling and genetic algorithms—a case study. Build Environ 75:98–107. doi:10.1016/j.buildenv.2014.01.011
Navarro I, Gutiérrez A, Montero C, Rodríguez-Ubiñas E, Matallanas E, Castillo-Cagigal M, Vega S (2014) Experiences and methodology in a multidisciplinary energy and architecture competition: Solar Decathlon Europe 2012. Energy Build 83:3–9. doi:10.1016/j.enbuild.2014.03.073
Ochoa CE, Capeluto IG (2008) Strategic decision-making for intelligent buildings: comparative impact of passive design strategies and active features in a hot climate. Build Environ 43(11):1829–1839. doi:10.1016/j.buildenv.2007.10.018
Pataky R, Áts Á, Áts-Leskó Z, Birtalan O (2014) Constructional considerations for the mobile Plus-Energy House. Energy Build 83:195–208. doi:10.1016/j.enbuild.2014.07.015
Ramon AP, Burgos AC (2008) Moving the entire building sector towards low CO_2 emissions. Paper presented at the PLEA 2008—Towards zero energy building: 25th PLEA international conference on passive and low energy architecture, conference proceedings
Rodriguez-Ubinas E, Montero C, Porteros M, Vega S, Navarro I, Castillo-Cagigal M, Gutiérrez A (2014a) Passive design strategies and performance of Net Energy Plus Houses. Energy Build 83:10–22. doi:10.1016/j.enbuild.2014.03.074
Rodriguez-Ubinas E, Rodriguez S, Voss K, Todorovic MS (2014b) Energy efficiency evaluation of zero energy houses. Energy Build 83:23–35. doi:10.1016/j.enbuild.2014.06.019
Ruiz-Larrea C, Prieto E, Gómez A (2008) Architecture, industry and sustainability. Informes de la Construccion 60(512):35–45. doi:10.3989/ic.08.037
Serra Soriano B, Verdejo Gimeno P, Díaz Segura A, Merí De La Maza R (2014) Assembling sustainable ideas: the construction process of the proposal SMLsystem at the Solar Decathlon Europe 2012. Energy Build 83:185–194. doi:10.1016/j.enbuild.2014.03.075
Stojiljković MM, Ignjatović MG, Vučković GD (2015) Greenhouse gases emission assessment in residential sector through buildings simulations and operation optimization. Energy. doi:10.1016/j.energy.2015.05.021

Suárez R, Fragoso J (2016) Estrategias pasivas de optimización energética de la vivienda social en clima mediterráneo. Informes de la Construcción 68(541). doi:10.3989

Terrados FJ, Moreno D (2014) "Patio" and "Botijo": energetic strategies' architectural integration in "Patio 2.12" prototype. Energy Build 83:70–88. doi:10.1016/j.enbuild.2014.03.081

Terrados-Cepeda FJ, Baco-Castro L, Moreno-Rangel D (2015) Patio 2.12: prefabricated, sustainable, self-sufficient and energy efficient house. Participation in the 2012 Solar Decathlon Competition. Informes de la Construccion 67(538). doi:10.3989/ic.13.138

Assessment Method of Urban Intervention in Social Housing Developments: The Rehabilitation of Caño Roto (Madrid) Case Study

Noelia Cervero Sánchez

1 Introduction

This study presents the results of the doctoral thesis *La vivienda protegida de promoción pública en España (1939–1976), estado e intervención: Metodología gráfica de análisis, (Government financed affordable housing in Spain (1939–1976), condition and intervention: a graphic analysis method)* applied to a case study: the urban rehabilitation of the Caño Roto social housing project (Madrid) . The research develops and validates a model that can be used to assess the condition of social housing developments and the quality of interventions that focus on their urban regeneration, and therefore represents a practical method for future action.

The subject of this study is one of the most vulnerable sectors in the urban weave of Madrid today, consisting of the affordable housing developments built between the Spanish Civil War and the Transition. As they have evolved, these estates have developed a series of shortcomings that mean they fail to comply with current acceptable living standards. The superposition of physical, conceptual and social problems means that any intervention model that focuses on rehabilitating these areas must aim to achieve integrated urban regeneration. This aim is closely linked to the right to the city concept promoted by the European Union, itself based on the line of thought developed by Lefebvre (1978, p. 111) in which citizens appropriate city spaces to generate conditions of diversity, wealth and social opportunity. This explains why any type of intervention that aims to achieve urban regeneration, whether through rehabilitation, renovation or a combination of approaches, faces the difficulty of it affecting a wide range of other factors i.e. urban planning,

N. Cervero Sánchez (✉)
Department of Architecture, School of Engineering and Architecture,
University of Zaragoza, Zaragoza, Spain
e-mail: ncervero@unizar.es

P. Mercader-Moyano (ed.), *Sustainable Development and Renovation in Architecture, Urbanism and Engineering*, DOI 10.1007/978-3-319-51442-0_15

environmental, construction, socio-economic, political and management (Tecnalia et al. 2011).

Since the 1990s, the European Union has linked urban regeneration to a model that achieves balanced development and competition between cities through processes that (European Commission 1990), as described in the 2007 Liepzig Charter, combine land-, sector- and time-related aspects. Its integrated urban development policy is an approach that encompasses key areas in urban policy, with intervention by economic agents and interested parties, primarily the government and inhabitants (Castrillo et al. 2014). Furthermore, *Spanish Law 8/2013 of 26 June on urban rehabilitation, regeneration and renovation* provides a comprehensive view of urban regeneration, defining it as any activity that is "not only open to meeting energy efficiency and economic recovery aims but also to actively contributing to environmental sustainability, social cohesion and to improving the quality of life of all inhabitants, in both housing, buildings and urban spaces".

2 Objectives

The recent nature of the legislation and the imprecision, in conceptual and methodological terms, that can be observed in interventions performed to date in Spain highlight the need for a method that can systematically assess conditions before an intervention starts and the effects that it will cause. The ultimate aim is to contribute to regulating these processes through a forward-looking instrument that provides an all-encompassing evaluation method addressing every decisive aspect. This would mean that, when developing an operation on a territorial level, it would be possible to establish its sustainability.

Having developed the integrated assessment model, this paper aims to verify how far it can be applied. To this end, it presents the results of an assessment of the urban rehabilitation of the Caño Roto social housing project (Madrid) (Figs. 1 and 2). Caño Roto, projected by architects J.L. Íñiguez de Onzoño and A. Vázquez de Castro,

Fig. 1 Caño Roto social housing project. Original (1994) and rehabilitated conditions (2015). *Sources* CREA Arquitectos file and author's photograph

Fig. 2 Caño Roto social housing project. Original (1994) and rehabilitated conditions (2015). *Sources* CREA Arquitectos file and author's photograph

was built between 1957 and 1963. The intervention was based on a project by E. Hernández Fernández, J.L. López Delgado and G. Ruiz Palomeque between 1994 and 2004. This case study is relevant for two reasons. It covers a pioneering process, whose review reveals how it has contributed to a policy of which it itself formed a part. And, it is now completed with time having passed, so it provides an opportunity to assess its evolution to the present day.

3 Methodology

The graphic model for urban intervention assessment condenses the following key points to assess:

- Firstly, the condition of the housing development before the intervention; this is known as the Original Condition. The result of the intervention is known as the Rehabilitated Condition. This allows for objective comparison between the two.
- Secondly, the Intervention Model i.e. the planning, management and participation processes that determine the action that is taken.

Analysis of the Original Condition leads to a diagnosis that clarifies the shortcomings that have evolved in the area; these may be considered threat factors. It also determines any favourable conditions or opportunity factors when implementing the regeneration. The Rehabilitated Condition analysis is based on the intervention process applied to the housing estate and its evolution up to the present day. This evaluation is combined with the Intervention Model assessment to determine the success and failure factors of the process and its effects.

Assessment of the Original and Rehabilitated Conditions consists of three subject blocks defined by concepts of size:

- City. Approaches the housing estate from a territorial perspective, understanding it as part of a cultural fabric that needs to function inclusively. It has just one subject area: 01. Connectivity.
- Housing estate. Considers the area as a unit and evaluates it in terms of composition, condition and quality of its inhabitants life. It has three subject areas: 02. Morphological and social profile; 03. Public spaces; 04. Urban complexity.
- Building. Studies its stability, accessibility, living conditions, functionality and energy efficiency. It has two subject areas: 05. Typological criteria; 06. Construction criteria.

Assessment of the Intervention Model has just one subject block:

- Intervention. Contains one subject area 07. Intervention Model, which studies the administrative and management system of the operation, urban planning tools and the involvement and emotional attachment of inhabitants during the process.

These subject blocks contain items or analysis points: 20 are used to determine the Original and Rehabilitated Conditions and 4 to determine the Intervention. These items are information units, defined by evaluating a series of quantitative or qualitative parameters that aim to provide an interpretative tool. Data are examined according to orientation criteria (these may be urban planning, social or building criteria) and documentation is covered by three different frameworks:

- Theoretical Framework. Establishes non-regulatory criteria related to the specific characteristics of these types of housing estate. Contains the bibliographic sources that establish the conceptual context for the model and the general documentary basis for the study.
- Methodological Framework. Provides non-regulatory criteria that act as a non-binding strategic framework. Based on principles, aims, directives and measures, devised to achieve greater urban and building sustainability without interfering in the responsibilities of government administration. These criteria are drawn from guides, manuals or previous research that certify and authorise the quality and sustainability of the urban environment and buildings (Agencia de Ecología Urbana de Barcelona 2012).
- Legislative Framework. Provides criteria established by legislation that regulate urban regeneration in Spain. It is analysed by differentiating between: instruments that specialise in urban planning legislation, attendant legislation, development policy and cross-level instruments (Rubio del Val 2015).

Bearing the above in mind, each item and the sphere to which it belongs are qualified by applying the following terms: superior to satisfactory, satisfactory, could be improved or inadequate. When analysing an intervention that took place in the past, in terms of those parameters that are affected by regulations, they are taken into account: legislation applicable at the time and current legislation, which functions as the threshold. When assessing a current, or planning a future, operation, applicable regulations now in force determine the standards governing new building work activity; any results above these values are considered to be ideal. In any event, as these are interventions in public spaces and existing buildings that are not governed by specific regulations, the evaluation considers whether or not the intervention is rational and coherent in its implementation. When it is not possible to reach the required standards, the method assesses whether or not the solution found is acceptable or whether it is performing below admissible levels; for example, in issues related to security and accessibility, there are certain minimum standards that cannot be ignored (Vega Catalán 2010).

The study of urban regeneration of the Caño Roto social housing development involved field work and documentary work, that continued throughout the whole research period. Field work involved the systematic gathering of information that could be used to contrast the present situation in the neighbourhood with theoretical details to understand its evolution to the present day. The documentary work consisted of accessing:

- Bibliographic sources: viewed in the ETSAM and *Colegio Oficial de Arquitectos de Madrid* (Professional Association of Madrid Architects) libraries and the documentation centre of the *Consejería de Medio Ambiente y Ordenación de Territoro* (CMAOT) (Regional Ministry for the Environment and Territorial Planning) in the Community of Madrid;
- Oral sources i.e. interviews with technicians who were directly and indirectly involved in the operation;
- Graphic sources: urban planning maps and photographic material, consulted through various services provided by the Town Hall, Community of Madrid and National Geographic Institute;
- Documentary sources found in various departments of the *Instituto de la Vivienda de Madrid* (IVIMA) (Housing Institute of Madrid), the personal files of *CREA arquitectos* and the National Institute of Statistics.

4 Case Study: The Rehabilitation of Caño Roto

The Caño Roto social housing development is a 19.46 ha estate containing 1606 dwellings located in the south-east of Madrid, on an irregular-shaped plot on a steep slope. Adapting to the topography and population requirements meant developing a large number of different residence types that coexist with additional buildings. The architecture is rationalist in style and the buildings are outward-facing. There are 680 dwellings in straight line blocks running from GF+3 to GF+5 high; 324 dwellings in tower GF+5 high and 602 single-family dwellings. The estate draws from a variety of reference points such as the Italian neorealism, the British brutalism and the southern Spanish tradition (Fernández Galiano et al. 1989). The variety of volumes found in the buildings is used to create free spaces that are well-orientated and on a human scale. The diagnosis that began at the start of the 1990s highlighted the deterioration of both, the public spaces and the buildings,

Fig. 3 Results index after applying the assessment method to the case study

with stability problems, a failure to adapt to current standards, accessibility issues and façades that were in disarray. Therefore, the intervention was mainly focused on adapting the estate to urban planning and construction standards. It included structural and functional works that would not affect living conditions in the private spaces.

By applying the graphic analysis method to this case, there were obtained the following results, presented in an index (Fig. 3) and detailed below. The index shows the results of the Original and Rehabilitated Conditions assessment for the items in areas 01–06 in two parallel columns, while area 07 has just one column that corresponds to the Intervention.

4.1 Assessment of Original and Rehabilitated Conditions

4.1.1 Connectivity

The setting of the housing estate remains the same before and after the intervention because, although it is becoming more central over time, it remains part of a broken environment. The exclusion factors present in its context, consisting of urban vacuums with self-built settlements that create an unsafe atmosphere, are slowly being transformed into general, broader equipments. However, the lack of nearby services means that housing estate inhabitants are obliged to commute to perform their everyday activities. They have good public transport connections, which is an opportunity factor, however, there has been no success in having these connections developed further.

In conclusion, an analysis of items 0.01 Urban location, 01.02 Mobility, 01.03 Cohesion and biodiversity, 01.04 Urban exclusion and 01.05 Access and services, considers connectivity at the housing estate as a threat while the intervention measures could be improved.

4.1.2 Morphological and Social Profile

The intensity of estate use remains virtually the same with a gross density of 82.52 dwellings per Ha in the Original Condition and 84.10 Ha in the Rehabilitated Condition. There is minimum variation in the use of the site; almost 70% of the surface is used for residential purposes. The only differences are as follows: a 10% reduction in non-residential use and a 30% increase in road and parking areas (after the intervention these represent almost 25% of the total). The negative effect these changes have had in terms of developing the area mean they represent a factor that could be improved.

An analysis of occupancy in the estate reveals a significant increase in empty dwellings, rising from 4.69 to 12.78% with a 26% reduction in inhabitants. The current number of inhabitants is 3640. Despite the ingrained roots of the inhabitants, the social profile represents a threat factor for the smooth functioning of the estate due to its socio-demographic and socio-economic vulnerability and continuing crime problems. The lack of social activities focused on introducing diversity into the population establishes this item as a failure factor.

Therefore, the results for items 02.06 Intensity of use, 02.07 Uses and occupancy of site and 02.08 Social profile, convert this area, which was originally considered a threat, into an area where the intervention failed.

4.1.3 Public Space

The reurbanisation process in the intervention involved updating the accessibility at the estate and the infrastructures. However, no solution was found to the original failure to adapt the housing development to circulating traffic, resulting in heavy vehicle use within the public space; this issue could be improved.

The free areas retain the dimensions and proportions they had in their Original Condition, with distinct spaces that are carefully thought out in terms of their construction, perception and use, and also the types of materials and plants that characterise them. Their precarious state of conservation led to intervention with a series of actions that failed to appreciate their previous peculiarities. The proportion of green space was increased by 160% and finishing was used that has been repeated across the outskirts of Madrid, eradicating the identity that once defined them. This failure to reflect on what they once were, combined with a lack of maintenance (now more necessary than before due to the increased green zones), mean that these measures were inadequate.

Analysis of items 03.09 Roads and parking and 03.10 Public spaces reveals that while this area is originally considered inadequate, the intervention measures could have been improved.

4.1.4 Urban Complexity

Turning to non-residential use in the estate, although over time the initial lack of equipment has adequately improved for the population profile, it is noted that there was a lack of production and commercial activity. A market was removed in 1995 while no attempt was made to boost commercial activity on the residential ground floor and additional buildings; they are currently 75% vacant and therefore this aspect was considered to be a failure factor. With regard to the residential buildings, although the intervention plan included the entire estate, it ultimately focused exclusively on the high rise building (replacing two blocks due to structural stability problems), but less than 30% of the ground level buildings. This failed to create the homogeneous form that was originally intended, so this factor could be improved despite the difficulty that it entails.

In turn, although it is one of the most publicised and well-known low-income housing estates in Spanish architecture, the fact that it was never granted national heritage status meant that the intervention criteria could be left open. These criteria focused more on consolidating and repairing the building within economic constraints, without placing much emphasis on assessing its urban image, and less on following a process in line with that of the original work. Given its architectural interest, this practical side to the operation is considered a failure factor.

The overall assessment of items 04.11 Relationship between activity and residence, 4.12 Configuration of residential building and 4.13 Values intrinsic to the estate, highlight how an area that originally could have been improved become, after the intervention, a failure factor.

4.1.5 Typological Criteria

The types of housing, planned according to the needs of the typical family of the period, now have a very low occupancy rate: 2.43 inhabitants per home. This is due to the rehabilitation criteria that specified no intervention in private spaces to allow the inhabitants to remain in their homes while the work took place.

However, the intervention did address accessibility in the common spaces of the high rise building; the absence of an accessible route in the Original Condition represented a risk factor for the living conditions. Therefore, horizontal accessibility from the outside was improved, as was the condition of the entrances. Furthermore, communities that requested a lift had one installed. The decision, for financial reasons, not to replace the staircase was taken as a single criterion, applicable to the whole operation. This has meant that in places the entrances do not have sufficient clearance height, and when taking the lift to higher floors the lift must be exited at an intermediary level. So accessibility problems were only fully resolved in just 13.50% of the high rise housing, with 50% partially resolved and the remaining 36.50% unresolved. The ineffectiveness of this measure, added to the visual impact of the extra volumes added to house the lifts, establishes this aspect as a failure factor of the operation.

Regarding the bioclimatic conditions in the buildings, these retained their good original characteristics due the orientation of the buildings and cross ventilation. The rate of sunlight falling on the façades does not vary; very limited shade is cast on the high rise building or on the single-family housing. However, having analysed the façade openings, it is important to appreciate that, despite working with tightly closed volumes, the forward-thinking, personalised action of inhabitants anarchically changes the proportions of these gaps. Despite the change, the south-facing façade is still the most open; 27.6% of this façade consists of gaps. However, some types of housing have a greater number of north-facing gaps. The intervention provided an opportunity, which was partially seized, to improve the bioclimatic efficiency of the building; the configuration of the gaps was consolidated and protective elements were added to each orientation, helping to standardise the composition of the façades.

Therefore, items 05.04 Adapting the housing types, 05.15 Accessibility in communal spaces and 05.16 Bioclimatic conditions, are considered an area that could be improved that transformed to a failure factor of the operation.

4.1.6 Construction Criteria

Actions were performed that focused on resolving subsidence issues and structural pathologies; two blocks in the housing development were replaced. This succeeded in stabilising the building so is considered satisfactory.

With regard to the thermal envelope, prior to the intervention the building was in a critical condition because the roofs and parts of the façade had high transmittance, condensation forming on surfaces and thermal bridges, and leakage. The outer skin of façades and ending walls were often highly unstable while condensation and problems related to damp were found on the floors. These pathologies were satisfactorily resolved by fixing the material to the structure and adding a dry-lining system on the outside surface. This succeeded in giving all the elements transmittance levels that complied with current legislation, eliminating thermal bridges and resolving the leakage issue, so that the cladding complies with the standards established in the current *Código Técnico de la Edificación* (Technical Building Code).

With regard to the communal facilities, these had leaks and broke down due to their old age and were updated to current legislation. At the same time, gas was run from the mains to each home, but centralised heating or creating an additional, alternative energy source was not considered. Taking the above and bioclimatic factors into account, by comparing energy efficiency in the building in its Original and Rehabilitated Conditions, it can be seen that heating requirements decreased by 46.30% (from 99.64 to 53.35 kWh/m^2 year), primary energy consumption by 45.80% (from 240.37 to 130.22 kWh/m^2 year) and emissions by 73.60% (from 81.60 to 21.51 kgCO_2/m^2 year).

Therefore, items 06.17 Adaptation of structure, 06.18 Hygrothermic efficiency of envelope, 06.19 Adaptation of facilities and 06.2 Energy efficiency of building, progress from being a threat to being considered an area of success for the operation.

4.2 *Assessment of the Intervention*

4.2.1 Intervention

The operation was supervised by the joint action of the municipal, autonomous community and national governments, collaborating on the basis of signed agreements, and by the Neighbours' Association. The process was led and managed by a Managing Company established from varied sources, with involvement from the Autonomous Community, the Town Hall and a majority of owners. Its role was to manage the actions needed to develop and execute the operation, advise the owners and drive the intervention forwards via its Governing Board. Having one sole managing agent/developer, headed by the government, with economic forecasts (effective despite the financial limitations of the inhabitants) means that the management model represents a success factor for the operation.

The initiative to undertake the urban rehabilitation came from the inhabitants, who held a protest, led by the Neighbours' Association, focused on housing conditions, the lack of facilities and crime levels in the estate. The Neighbours' Association complemented its activism by participating in negotiations with the government prior to the process, in which favourable economic conditions and the opportunity to be involved for the duration of the operation were established. At the same time, a kind of participative assembly was created that was governed by the Community of Owners, where technical, management and financial aspects of each proposal were examined and disseminated to be approved by the maximum number of affected parties. This type of participation was successful, as was the fact that the works were carried out with the minimum amount of rehousing and inhabitants could continue to live in their homes.

Taking into account items 07.01 Development policy, 07.02 Intervention planning, 07.03 Management system and 07.04 Involvement of inhabitants, the Intervention Model is evaluated to be a successful area of the operation.

5 Results

This research has reached the following conclusions:

1. The initial threats to the estate were:
   - Exclusion factors present in the context (area 01) due to the estate's unsafe atmosphere; this measure could have been improved.
   - The socio-demographic and socio-economic vulnerability (area 02); these were not addressed so is therefore a failure of the operation.
   - Living conditions in the high rise dwellings due to a lack of accessibility via the communal spaces (area 05); this was not entirely resolved and is therefore a failure factor.
   - Unstable and weak construction of the buildings (area 06); better conditions than those requested were provided in new construction work, representing a success area of the operation.
2. The following opportunity factors were assessed:
   - Aspects related to the particular characteristics of the estate project (areas 03 and 04); their importance was not valued, and therefore in those areas where they are relevant they are considered to be failure factors.
3. The greatest success of the operation was the Intervention Model (area 07) which was based on collaboration between government administrations, on combined development and management that was exclusive to the process, and on active involvement from the owners. The fact it could be expanded to other operations means that this intervention model represents the greatest area of success for the intervention.

6 Conclusions

The enforcement of the assessment method to the case study demonstrates its usefulness and allows extracting the following conclusions:

- It is considered a wide scale action, understanding the housing estate as an urban piece affected by multiple thematic areas.
- Regulations are taken into account both at the time in which takes place the intervention, as at present to have a single threshold comparison with other study cases.
- For analysing the results it is taken into account that, depending on the case, priority areas will change.
- Quantitative and qualitative items are objectified and evaluated, being based on theoretical, methodological and legislative frameworks.

- Each item is compared, between initial and modified state, and in relation to others in the same or different areas, in order to obtain parallel readings.

Acknowledgements This paper is part of the frame work of Project BIA2013-44001-R: Integrated Design Model for Rehabilitating Social Housing and Urban Regeneration, selected by the 2013 call for R+D+i Projects from the State Research, Development and Innovation Programme focused on challenges to society by the Ministry for Economics and Competition.

References

Agencia de Ecología Urbana de Barcelona (2012) Guía Metodológica para los Sistemas de Auditoría, Certificación o Acreditación de la calidad y sostenibilidad en el medio urbano (Methodological Guidelines for Audit, Certification or Accreditation Systems of quality and sustainability in urban areas). Ministry for Development, Madrid

Castrillo M, Matesanz A, Sánchez D, Sevilla A (2014) ¿Regeneración urbana? Deconstrucción y reconstrucción de un concepto incuestionado (Urban regeneration? Deconstruction and reconstruction of an unquestioned concept). Papeles de relaciones ecosociales y cambio global 126:129–139

European Commission (1990) Green paper on the Urban Environment. Office for Official Publications of the European Commission, Bremen

Fernández Galiano L, Fernández de Isasi J, Lopera Arazola A (1989) La quimera moderna (Modern *quimera*). Hermann Blume, Madrid

Lefebvre H (1978) El derecho a la ciudad (The right to the city). Barcelona, Península

Leipzig Charter on the European Sustainable City (2007) http://www.fomento.gob.es/MFOM/LANG_CASTELLANO/DIRECCIONES_GENERALES/ARQ_VIVIENDA/SUELO_Y_POLITICAS/ACTIVIDAD/UE/

Rubio del Val J (2015) Potencial del nuevo marco normativo para el impulso de la rehabilitación y la regeneración urbana en los ámbitos autonómico y local (The potential of the new legal framework to promote urban renewal and regeneration at the regional and local levels). Informes de la Construcción 67:53–70

Tecnalia et al (2011) Metodología de análisis, diagnóstico e intervención en la Renovación del Parque Edificado de la CAPV (Methodology of analysis, diagnosis and intervention in the Building Park Renewal of CAPV). Document n. 000374-IN-CT-11/03.04 (v01). Basque Government, Bilbao

Vega Catalán L (2010) La rehabilitación en el Código Técnico de la Construcción (Technical Building Code). In: International congress rehabilitation and sustainability: the future is possible, Barcelona, Oct 2010. http://www.rsf2010.org/es/ponencias

Environmental Assessment and Energy Certification for the Sustainable Restoration of a Traditional Residential Building

Carmen Rodríguez-Liñán, Mª Jesús Morales-Conde, Filomena Pérez-Gálvez, Paloma Rubio-de Hita and Silvia López-Alonso

1 Introduction

The building sector represents a significant share of global energy consumption. As a result, the growing interest in Europe in sustainable development and the efficient use of energy generated has seen the implementation of policies to encourage a reduction in the energy consumption of buildings (European Parliament 2002) (European Parliament EC 2010) (EU 2012). In Spain, these EU Directives have been transposed into state legislation as regulations such as the Basic Document of Energy Saving (HE), the Technical Building Code (TBC 2006) and the Regulation of Technical Installations in Buildings (RITE) (Spanish Ministry of Industry, Trade and Tourism 2007). The implementation of these regulations and assessment of buildings' energy consumption by means of the Energy Performance Certificate (Spanish Government 2013) have been made possible by the development of tools (Lider and Calener VyP) that quantify energy consumption and CO_2 emissions, and indicate the necessary corrective measures to be taken. Each of these tools is aimed at evaluating a building typology, rating the building from an energy perspective.

However, when assessing the environmental impact of a building project, we need to consider not only the energy consumption and CO_2 emissions during its lifecycle, but also its overall impact on the local environment. A building should be understood as an element that can contribute to a more favourable local environment, and should be evaluated according to such criteria at all stages, from design, choice of materials, construction, operation and demolition (Llatas et al. 2010). In this sense, many public and private organizations have developed instruments to establish guidelines for "producing" buildings with a low environmental impact,

C. Rodríguez-Liñán (✉) · M.J. Morales-Conde · F. Pérez-Gálvez · P. Rubio-de Hita · S. López-Alonso
Universidad de Sevilla, Sevilla 41011, Spain
e-mail: rlinan@us.es

P. Mercader-Moyano (ed.), *Sustainable Development and Renovation in Architecture, Urbanism and Engineering*, DOI 10.1007/978-3-319-51442-0_16

such as the eco-labelling of products, guidelines for sustainable construction or tools for the analysis and assessment of the environmental impact of a building, such as LEED (from the USA, where its use is widespread), BREEAM (the UK) or VERDE (Spain), among others.

These tools aim to encourage the construction of buildings based on sustainability and high-efficiency criteria by evaluating various parameters to that end. LEED assesses five key areas of human health and the environment: site sustainability, water and energy efficiency, choice of materials and environmental quality. BREEAM evaluates a building according to nine categories: management, health and well-being, energy, transport, materials, waste, water, land use and ecology, and pollution. VERDE is based on a lifecycle assessment methodology that evaluates the building throughout its existence, from extraction of raw materials to reuse, recycling or transport to landfill once the building is demolished at the end of its useful life. A study of the site would analyse energy and atmosphere (covering the production of materials and their transfer to the site), natural resources, the quality of interior space, quality of service and socio-economic impact. The evaluation of each variable is used to rate the degree of compliance and determine the environmental impact of the building according to scale.

This work evaluates a rehabilitation project (Model B) for a building in the historic centre of Seville from an energy and environmental perspective. It is a standard rehabilitation project developed according to existing legislation and on a limited budget. As an alternative, this paper proposes a different scheme to improve this building's energy efficiency and reduce its environmental impact by the careful selection of more sustainable materials (Model C).

For the evaluation and selection of more sustainable materials, we rated materials used in the two models of the study. The scoring criteria only affected some of the issues that define sustainability. In this case, we did not rate all aspects covered by LEED, BREEAM or VERDE to assess environmental impact due to lack of data and because they were beyond the scope of this work. We did, however, consider those items that are most representative of an assessment of sustainability and the materials applicable to this project, and which are covered to a greater or lesser extent by these three assessment tools mentioned.

2 The Building Under Study

This paper examines three study models of the same building: Model A, of the building in its original state; Model B, the proposal executed by the architects based on existing legislation and with a limited budget; Model C, the alternative project proposed by the authors of this work.

2.1 Model A (The Original Building)

The studio building is located in the historic centre of Seville, and is a typical example of Sevillian domestic architecture of the 17th–19th centuries. This is a four-storey residential building, with a ground floor designed for commercial use. The plans show that the building has three bays between the party walls, with a small inner courtyard for lighting and ventilation in the second bay. The facade has the same pattern on the first, second and third floor, with 3 holes per floor all symmetrically deployed. The width of the facade is 8.60 m and the crown height is 14.23 m.

The construction system of the building is based on traditional brick walls, cast iron pillars and slabs of wood with beams measuring 15 × 40 cm and joists of 7 × 15 cm. The infill is wooden boards filled with alcatifa and the foundations consist of rubble and lime.

Some modifications were detected in the ground floor slab, resulting from the closure of the courtyard in later building work. In this case, a metal structure is placed in the second and third bays. The enclosure formed by the courtyard is not a load-bearing wall, so the whole support system reaches up to the same cantilever.

The roof of the building is a traditional Andalusian covering made of ceramic flooring tiles (Fig. 1).

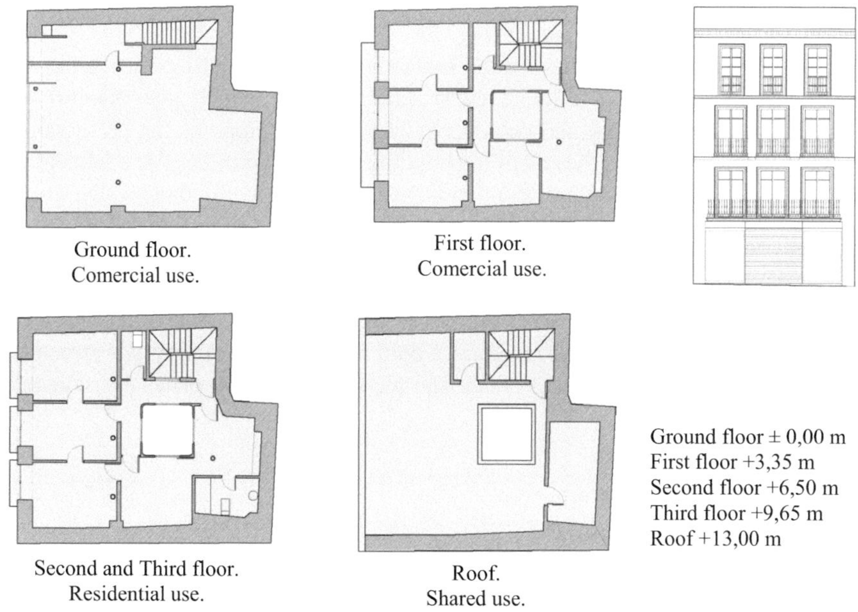

Fig. 1 Model A. Original state of the building before rehabilitation

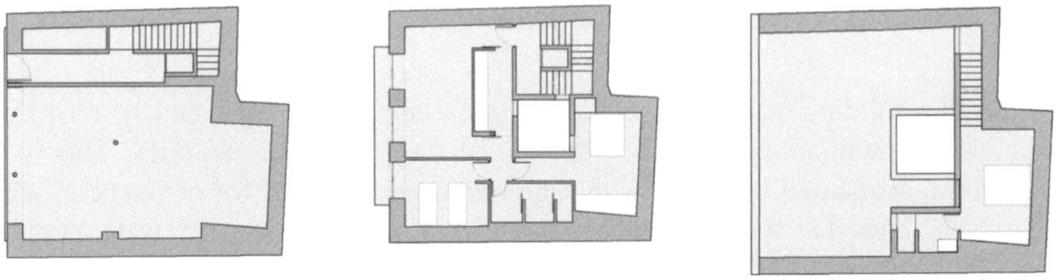

Ground floor. Commercial use. First, second and third floor. Residential use (2 bedrooms) Roof. Residential use. (1 bedroom of third floor house)

Fig. 2 Model B. Proposal executed by the architects

2.2 Model B (Proposal Carried Out by the Architects)

Public authorities are increasingly insistent on preserving the image of Seville's historic centre, thus obliging architects to retain the original facades of downtown buildings, as is the case with the building under study. They are permitted to strengthen and consolidate the facade but without modifying the composition of the holes and the general aesthetic. The architects' proposal led to the construction of a building similar to the original with four floors, and the ground floor designed for commercial use. In addition, part of the roof was modified to build a large apartment. The building now has 3 bays, with a central courtyard in the second centreline for lighting and ventilation (Fig. 2).

Current legislation provides for the aesthetic preservation of the façade but not for the continuity of the original construction system, which goes against the principles of sustainability and eco-efficiency. Model B proposes the preservation of the original facade and the rest of the building system is projected as a foundation slab that sustains the original foundation wall facade and dividing walls with a reticular floor slab-type cavity measuring 25 + 5 cm. This slab together with a minimum amount of concrete produces a ground floor that complies with municipal legislation. The facade walls and party walls remain untouched but would be covered with an air chamber and a thin brick wall, while the lintels of the holes are not modified. The rest of the structure would become a metallic pillar structure, discarding the original cast iron pillars and incorporating steel plates and concrete slabs.

2.3 Model C (Proposal by the Authors)

In contrast to Model B, the proposal in this paper (Model C) is an alternative which not only aims to comply with building regulations but also sets out to preserve the traditional construction systems of the building in order to minimize waste production in carrying out a sustainable and eco-efficient project.

Ground floor. Comercial use. | First, Second and Third Floor. Residential use (2 Bedrooms) | Roof. Residential use. Storage and facilities.

Fig. 3 Model C. Proposal by the authors of this work

The elements we consider worth preserving are:

- The vertical structure of the load-bearing walls and cast iron pillars.
- The horizontal structure of the slabs. We propose recovering the wooden structure of the slabs, except the heads embedded in the masonry walls; the beams and pairs are in good condition generally, and we have not detected any fungi or wood-eating insect attacks of importance. Neither did we observe any mechanical deformities that question the stability/functionality of the structure. We would replace the infill elements and wooden boards due to a high degree of deterioration.

The preservation of these structural elements accompanies our proposal designed according to sustainability criteria. Our plan provides the ground floor with natural light thanks to a glass coplanar with the first floor slab in the central courtyard. We aim to use the roof to create a collective private outdoor space accessible to all residents, with all dwellings provided with a storage area. The orientation of this phase of the construction will ensure that it does not cast shadows on the central patio. We also envisage a ventilated indoor area for clothes lines that does not occupy the community area or make it visible from outside (Fig. 3).

3 Environmental Assessment

The environmental evaluation of the proposals (Models B and C) was carried out by means of a study of the materials listed in the project phase, a common variable measured by all the most up-to-date assessment tools (Breeam, Leed, Verde …). We also took into account, although not specifically evaluated, the design criteria that promote the building's sustainability (provision of natural lighting on the ground floor, provision of outdoor spaces, enhancement of natural ventilation, orientation of volumes on the roof to favour natural lighting, incorporation of drying areas). The remaining variables are either not measured by all three tools or do not apply to interventions in other stages of the construction process, as in the case of management or waste (both of which are assessed by Breeam). Likewise, variables such as building site or contamination (evaluated by Breeam and Verde) are variables that come predefined by the type of building work, as is the case with a rehabilitation job.

We also assessed the potential environmental impact of the materials chosen by carrying out a preliminary study of their ecological footprint, considering the five following parameters (CTAV) (Agenda for Sustainable Construction 2002):

- Place of production (PP): a series of monitoring sites is established from the building site to the different production centres.
- Renewable raw material (RRM): the raw material from which the final material is obtained must be renewable and exploited responsibly; it should be repairable following each operation if necessary.
- Recycled Material (RM).
- Recyclable material (RBM): the material can be totally or partially recycled.
- Material Stability (MS): evaluation of the element's lifecycle and its safety properties when the work is commissioned, as such properties can affect the health of the workers on site.

These parameters are evaluated according to the following criteria (Table 1).

The materials studied were rated from 0 to 10 according to their sustainability. The materials were categorized for foundation and structure, masonry, insulation, finishes and facilities.

After studying the individual materials, we assessed Model B by examining the list of materials selected by the architects for their project (concrete foundation slab,

Table 1 Schedule of the environmental assessment of materials

Points	2.5	2	1.5	1	0.5	0
Place of production (PP)	Province	<250 km	<500 km	<750 km	<1000 km	>1000 km
Renewable raw material (RRM)	–	>75%	–	>50%	–	<50%
Recycled material (RM)	–	>75%	–	>50%	–	<50%
Recyclable material (RBM)	–	>75%	–	>50%	–	<50%
Material stability (MS)	–	–	Very good	Good	Average	Bad

masonry brick walls, metal structure pillars with steel plates and concrete slabs, XPS insulation for the roof and bitumen waterproofing sheets, stoneware tile finishing and plaster as indoor coating). In contrast, the materials in Model C were selected with a view to sustainable rehabilitation and low environmental impact (recycled EPS for lightweight concrete, concrete containing 50% recycled crushed aggregates, wooden joists reused from the building or from ecologically-managed forests, hemp ceramic block with hydraulic lime, insulation boards made of natural cork and/or hemp insulation boards). The materials chosen for use in Model B yielded a score of 29 for sustainability while the low ecological impact materials selected for Model C scored 57.5, rising to 77.5 when considering the improvement to the facilities provided by the system selected. By simply swapping some of the materials, a significant improvement in sustainability can be achieved according to the criteria applied (Table 2; Fig. 4).

4 Energy Analysis

We performed an energy efficiency analysis of models A, B and C using the Calener Vyp computer tool to rate the energy level of a building using $KgCO_2/m^2$ as an indicator. We quantified the CO_2 emissions from the building's heating and cooling systems, and from hot water production. The results are shown in Tables 3, 4 and 5.

4.1 Model A. Energy Analysis

See Table 3

4.2 Model B. Energy Analysis

See Table 4

4.3 Model C. Energy Analysis

As the tables show, the original building (Model A) has the highest level of $KgCO_2/m^2$ consumption, at 79.6 (energy rating G). When rehabilitated (Model B) the building improved its energy rating and decreased consumption to 21.0 (energy

Table 2 Materials study

	Material	PP		RRM		RM		RBM		ES		Total points	
		Fact	Points	F	P	F	P	F	P	F	P	B	C
FOUNDATIONS AND STRUCTURE	Concrete	Sevilla	2.5	<50	0	<50	0	>50	1	VG	1.5	5	5
	Structural steel[a]	Sevilla	2	<50	0	<50	0	>50	1	VG	1.5	5.5	
	Ceramic block[a]	Castellón	1	<50	0	<50	0	>50	1	VG	1.5	3.5	
	Cement mortar	Sevilla	2.5	<50	0	<50	0	<50	0	VG	1.5	4	
	Recycled EPS for lightweight concrete[b]	Cádiz	2	<50	0	>75	2	>50	1	VG	1.5		6.5
	Wooden beams[b]	Sevilla	2.5	>75	2	<50	0	>75	2	VG	1.5		8
	Natural wood preservatives[b]	Sevilla	2.5	>75	2	<50	0	<50	0	VG	1.5		6
	Lime coating[b]	Sevilla	2.5	<50	0	<50	0	<50	0	VG	1.5		4
	Hemp ceramic block[b]	Granada	2	>75	2	<50	0	>75	2	VG	1.5		7.5
INSULATION	EPS Panels[a]	Cádiz	2	<50	0	>50	1	>75	2	VG	1.5	6.5	
	Cork[b]	Granada	2	>75	2	<50	0	>75	2	VG	1.5		7.5
FINISHES	Stoneware tile[a]	Jaén	2	<50	0	<50	0	>50	1	VG	1.5	4.5	
	Recycled glass[b]	Barcelona	0.5	<50	0	>75	2	>75	2	VG	1.5		6
	Biosuro pavement[b]	Cáceres	2	>75	2	<50	0	>75	2	G	1		7
FACILITIES	Insulation. Synthetic rubber foam[b]	Barcelona	0.5	<50	0	>50	1	>50	1	VG	1.5		4
	Soundproof downspouts[b]	Barcelona	0.5	<50	0	<50	0	>50	1	VG	1.5		3
	Halogen-free powerwires[b]	Barcelona	0.5	<50	0	<50	0	>75	2	VG	1.5		4
	Low consumption cistern[b]	Valencia	1	<50	0	<50	0	>50	1	VG	1.5		3.5
	Low consumption tap[b]	Valencia	1	<50	0	<50	0	>50	1	VG	1.5		3.5
	Recycling gray water facility[b]	Cataluña	0.5	<50	0	<50	0	<50	0	VG	1.5		2
	Total model B											29	
	Total model C												77.5

[a]Model B materials
[b]Model C materials

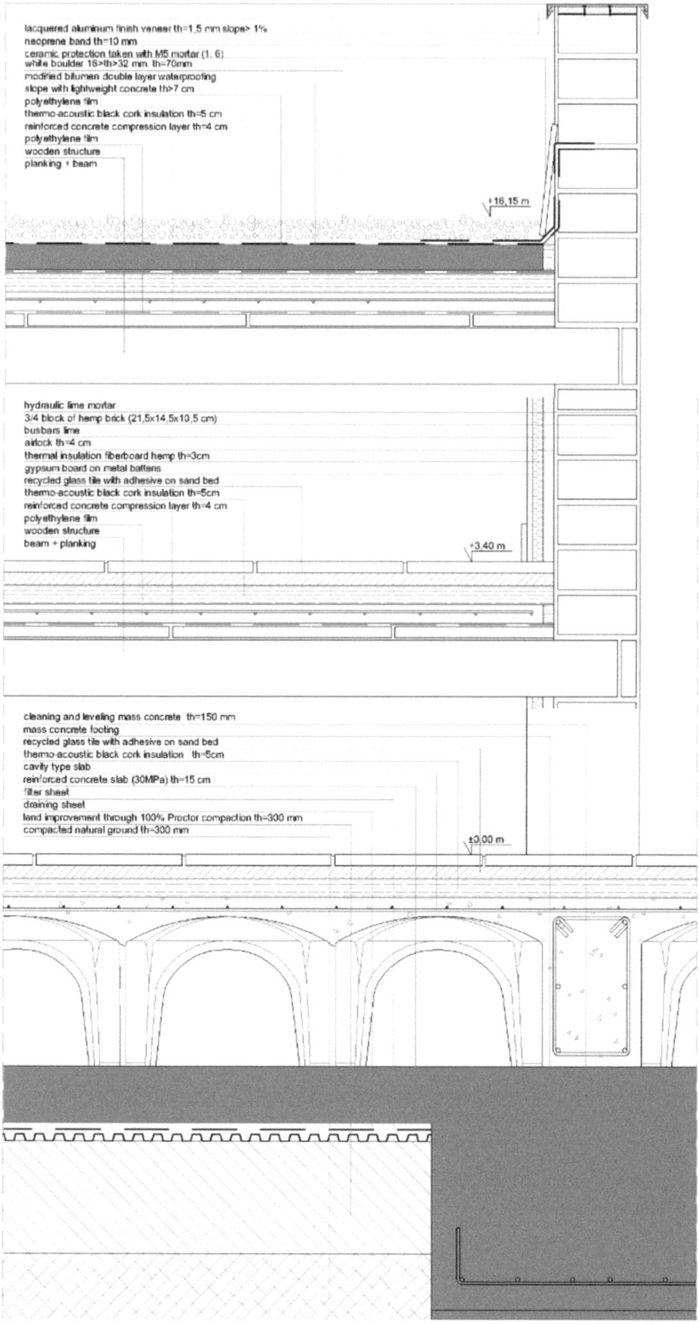

Fig. 4 Constructive section. Model C

Table 3 Model A energy qualification values obtained from Calener Vyp

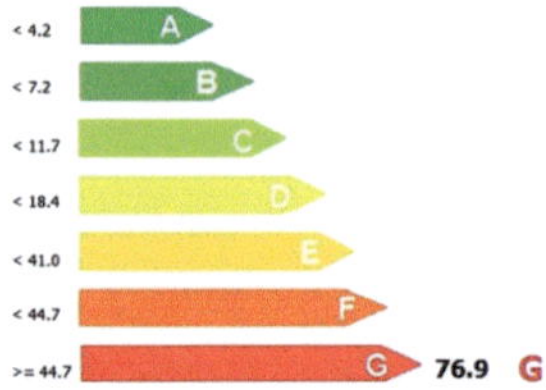

Table 4 Model B energy qualification values obtained from Calener Vyp

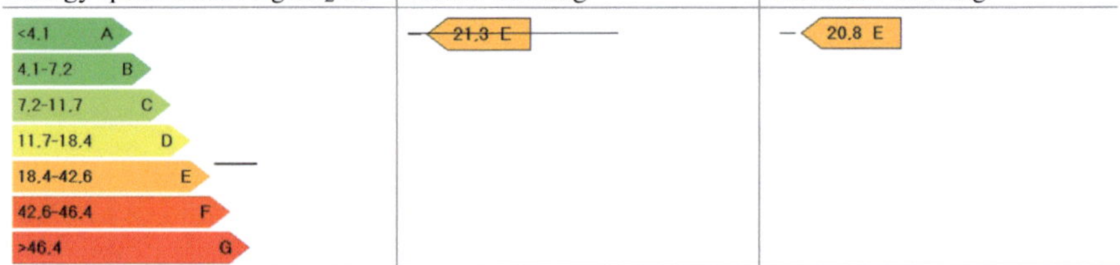

Energy qualification $KgCO_2/m^2$	Actual building			Reference building		
	Class	kWh/m^2	kWh/year	Class	kWh/m^2	kWh/year
Heating demand	E	44.9	10761.7	E	33.3	7998.3
Cooling demand	D	26.0	6246.7	D	22.8	5470.8
CO_2 emissions (heating)	E	12.2	2,926.0	E	10.7	2566.2
CO_2 emissions (cooling)	D	8.0	1918.7	E	8.7	2086.6
CO_2 emissions (water)	B	1.1	263.8	D	1.4	340.5
Primary energy consumption (PEC) (heating)	E	51.5	12354.4	E	48.4	11597.5
P. E C. (cooling)	D	32.2	7715.6	E	35.6	8534.4
P. E. C (water)	C	5.6	1338.9	D	5.9	1406.7

rating E) but still above the consumption rate for the building reference figure of 20.4. However, our proposed rehabilitation (Model C) reduces consumption to 14.4 $KgCO_2/m^2$ (energy rating D) (Table 5).

If we take into account demand for energy for heating, the difference in the improvement in energy saving between Model B, at 44.9 KWh/m^2, and Model C, at 29.6 KWh/m^2, is substantial, although energy savings on cooling are similar in both models.

Table 5 Model C energy qualification values obtained from Calener Vyp

Energy qualification $KgCO_2/m^2$	Actual building			Reference building		
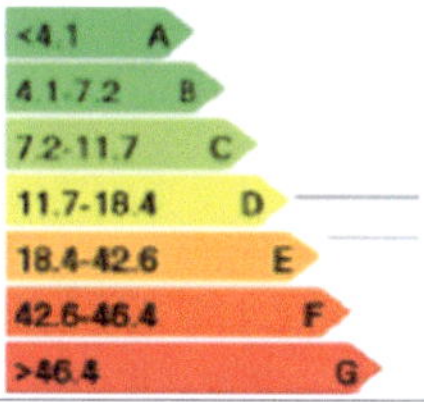	14.4 D			20.4 E		
	Class	kWh/m^2	kWh/year	Class	kWh/m^2	kWh/year
Heating demand	E	29.6	6709.9	E	32.0	7248.8
Cooling demand	D	28.0	6335.8	D	23.1	5244.3
CO_2 emissions (heating)	D	5.0	1132.7	E	10.2	2310.7
CO_2 emissions (cooling)	E	8.4	1902.9	E	8.8	1993.5
CO_2 emissions (water)	B	1.0	226.5	D	1.4	321.6
P. E. C. (heating)	D	22.4	5072.5	E	46.4	10510.8
P. E. C. (cooling)	D	33.8	7657.9	E	36.1	8181.0
P. E. C. (water)	C	5.1	1147.7	D	5.9	1328.7

5 Conclusions

Given the importance attached to the environmental impact of construction and its energy consumption, we need more studies to find new ways to protect the environment, cut energy consumption and also enhance the sustainability of future building projects.

This work has studied the rehabilitation of a historic building by comparing two different proposals (Models B and C). The former consisted of the rehabilitation of the building in strict compliance with current urban regulations whereas the latter applied sustainability criteria in the choice of materials, and aimed to retain the original building construction systems. Sustainability analysis was performed using a prior study of materials, and evaluating their ecological footprint within the following parameters: Place of Production, Renewable Raw Material, Recycled Material, Recyclable Material and Material Stability. The energy analysis was performed using Calener Vyp.

Studies show that the construction projects which apply sustainability parameters can significantly reduce a building's energy consumption, as our proposal demonstrates, raising its energy rating from E to D. Furthermore, sustainability levels increase significantly when using materials for rehabilitation that have a smaller ecological footprint. This study shows that our Model C was 1.98 times more sustainable than Model B.

Acknowledgements This research has been supported by the Ministry of Economy and Competitiveness of Spain (reference number BIA2013-43061-R). The author Mª Jesús Morales-Conde acknowledges the financial support of the V Research Plan of the University of Seville. Also, the authors acknowledge to the architects Marta Barranco and Adolfo Pérez for their special collaboration in this work.

References

Agenda for Sustainable Construction (2002) El Impacto Ambiental de los Edificios

Economic Committee of the European Parliament (2010) Directive 2010/31/ EU of the European Parliament and of the council of 19th May 2010 on the energy performance of buildings (recast)

European Parliament (2002) Directive 2002/91/EC of the European Parliament and of the council of December 16th 2002 on the energy performance of buildings. Office for Official Publications of the European Communities

European Union (2012) Directive 2012/27/EU of the European Parliament and of the Council of October 25th (2012) on energy efficiency amending Directives 2009/125/EC and 2010/30/EU, and repealing Directives 2004/8/EC and 2006/32/EC Text with EEA relevance. Official Journal of the European Union 315:1–56

Llatas et al (2010) Una aproximación a la eco-eficiencia en edificios. Herramientas básicas. In SB10mad Sustanaible Building Conference

Ministry of Industry, Trade and Tourism (2007) Regulation of thermal installations in buildings (RITE)

Spanish Government (2013) Royal decree 235/2013, of April 5th, approving the basic procedure for certifying the energy efficiency of buildings. Official Gazette, 13

TBC (2006) Spanish technical building code. Basic document of energy saving (HE). Royal Decree 314

Lider and Calener Vyp. Computer applications to verify limitation of energy demands and energy qualification in compliance with CTE-DB-HE1 standard

Territorial College of Architects of Valencia (CTAV) ICARO

Social Rent Housing Refurbishment Demonstrator of LIFE Project "New4Old" (LIFE10 ENV/ES/439)

Emilia Román, Gloria Gómez and Margarita de Luxán

1 Introduction

The aim of the intervention on the Demonstrator Buildings of the NewSolutions4OldHousing Project (LIFE10 ENV/ES/439), co-funded by the European Commission, is to analyse the actual results after defining the most appropriate methodology and the best practices for the rehabilitation of social housing with energy and environmental sustainability criteria. One of tasks performed has been the refurbishment of two buildings of social housing in Zaragoza, which act as a demonstrator in the project. In this way, it is possible to evaluate the measures implemented and the behaviour of buildings before and after the intervention. The institutions participating in the project are: the AITEMIN Technology Center, the Polytechnic University of Madrid (UPM), the Technology Centre of Ceramics and Glass of Portugal (CTCV) and the Municipal Zaragoza Housing Society (SMZV).

E. Román (✉) · G. Gómez
Architect cc60 Estudio de Arquitectura, Research Group Sustainable Architecture and Urban Planning (GIAU+S), Universidad Politécnica de Madrid, Madrid, Spain
e-mail: emilia.roman@upm.es

G. Gómez
e-mail: glogomu@cc60.com

M. de Luxán
PhD Architect Emerita Professor at the Upper Technical School of Architecture, Universidad Politécnica de Madrid, Madrid, Spain

P. Mercader-Moyano (ed.), *Sustainable Development and Renovation in Architecture, Urbanism and Engineering*, DOI 10.1007/978-3-319-51442-0_17

2 Objectives

The intervention on the two housing buildings has as specific objective to identify, apply and evaluate the most adequate project solutions to improve its energetic behaviour with regard to the climatic and social context in which they are placed and be able to measure the actual improvements resulting from it. The general objective of LIFE project is to define the appropriate methodology and best practices available for the renovation of social housing with criteria of energy and environmental sustainability, and also the application of innovative technologies to fight against climate change. Passive strategies in buildings are essential in social housing, especially when the tenants have very limited income.

3 Methodology

Regarding the Demonstrator Buildings, the main methodology is the analysis and comparison of diverse parameters of the construction states and results of their utilisation before and after the intervention.

There is no homogeneous corpus allowing to relate Demonstrators of the LIFE Programs in which antecedents can be established, neither specific states of the question that can be argued, since in the building field each case is unique finding itself in climatic, urban, constructive, technological, social even economic (concerning the money means to implement them); and even thematic as we are inside the Program that looks to care for diverse researches with very diverse Demonstrators.

The two buildings are located in the historic city of Zaragoza, the first of which (Building A) with access from the San Pablo St. 83-85 and the second (Building B) from the street Basilio Boggiero. Between the two buildings there is an inner courtyard. The buildings have, respectively, 18 and 12 homes for social rent. This represents a user profile with low incomes. Therefore, the working method to define improvement measures for the building, has paid special attention to the current conditions of use of housing and the possibilities of incorporating passive strategies to improve indoor comfort without increasing consumer costs. Thus situations of fuel poverty (Sanchez Guevara 2016) are avoided. Being rented dwellings, users are changing over time and therefore have chosen self-regulating solutions that do not require successive neighbors have to learn some management (Figs. 1 and 2).

For defining the solutions to implement in refurbishment, we analyzed the climatic and urban aspects of the area. A survey user to know their perception and their needs regarding comfort conditions of housing has been done. Monitoring the building before and after the performance has allowed evaluate the effectiveness and impact of the measures taken. An important starting data for the selection of solutions is that the contract budget was limited to 358,000 euros.

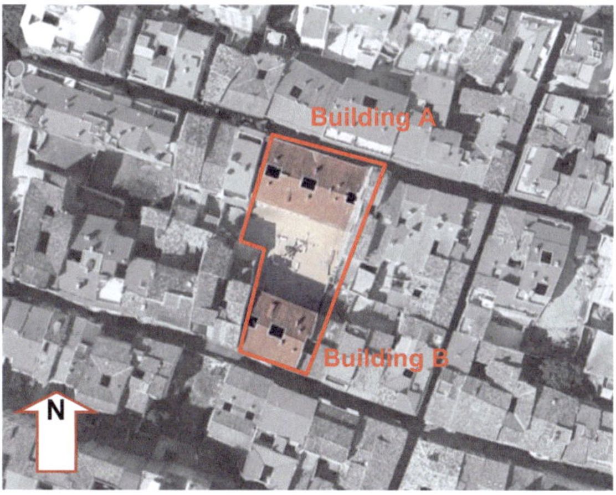

Fig. 1 Location of buildings

Fig. 2 North facade of the building in San Pablo St.

4 Initial Conditions Analysis

The buildings are owned by Municipal Zaragoza Housing Society (SMZV) and are located in one of the areas identified as vulnerable (Hernández Aja et al. 2011). They were built in the early 90s and are two blocks between medians with two façades oriented north and south. They have several inner courtyards of 3 × 4 m. Each building has three floors, in addition to ground floor. On each floor they are located up to 6 homes distributed in two staircases. The houses are oriented to the north or south and many of its rooms open onto interior courtyards. On the ground floor there is an access from the street, the stairs and the courtyard, and local occupied by neighborhood associations. The roof of the building is inclined gable, on a partitioned chamber currently has no ventilation.

4.1 Climate Analysis and Influence of Climate Change

Zaragoza has a continental mediterranean climate with warm summers. The average temperature of the maximum is around 21 °C and average minimum near 10 °C. In the winter months, the average minimum temperatures may be 5 °C and in summer the average maximum measurements can reach 30 °C. The relative humidity varies between 50 and 80%. The most frequent winds in winter correspond to the direction NW to SE and summer. The first winds are cold character and the second, warm.

Actions carried out with energy and environmental sustainability criteria, that want to get long-term results, should consider the effect of climate change. In the case of the refurbishment, the impact of these changes can be significant if the forecasts made for Spain (Olcina Cantos 2009) are met. According to this forecast, the summer conditions for inland cities, such as Zaragoza, will be modified by rising temperatures and reduced rainfall. Both issues, especially the first, involve changes in energy exchanges with the outside of the building and therefore its comfort conditions.

Bioclimatic comfort studies performed indicate that in the relatively near term, the number of days it will be needed to consider strategies summer, increases progressively. In the case of buildings, by refurbishment on existing buildings, it is necessary consider the progressive increase in average temperatures, especially in summer. This implies that although heating energy consumption will be reduced, the energy associated to cooling, if passive measures are not implemented, will increase.

In addition to the phenomenon of climate change, interventions in the urban context should consider heat island (Cuadrat Prats et al. 2005). Studies indicate again the importance of considering adaptation strategies to temperature conditions, since both the expected evolution of the climate and the phenomenon of heat island, they are longer necessary and will be of greater significance even in future.

4.2 Urban Contexts

After analysing the climate and identify interest in raising passive measures on buildings, it is necessary a solar study of the buildings and the constraints due its urban situation. Both buildings are oriented north-south. During the cold months the southern facade, which overlooks the central courtyard in the case of building A, is guaranteed to receive four hours of sunlight (10:00 am–14: 00 h) even in the month of January is the colder. This is due to the dimensions of the central courtyard and building height faced within the same plot in the north. Only the ground floor receives fewer hours of radiation. Aimed at San Pablo St., the north facade of this building would always be in shadow. The south facade of the building B, which is accessed from the street Boggiero, behaves like north facade in the cold months because urban arrangement prevents solar radiation reaches you, while in summer, is exposed to sunlight. The southern courtyards facades receive little direct radiation; only in the upper part.

During the colder months, the central courtyard is in shadow for much of the day. Only in the hottest hours of the months of October and February could receive radiation a small part of it, on the south facade (approx. 1/3 of its surface). The possibilities of natural ventilation are reduced due to the orientation and the high density of the urban fabric, but are allayed by the size, orientation and shape of the inner courtyard of apple and the chimney effect of the inner courtyards.

4.3 Users Characterization

After conducting surveys to users of the building, it follows that there are differences in the feeling of comfort among people living in dwellings oriented north and those living in the south oriented. In many of them, especially in the north-oriented, there is a problem of heating in the winter months. In the summer months overheating seen in all homes to south. In general, users consider getting warmer in winter and cooler in summer are the most important measures to achieve comfort. A significant finding is that 43% of users do not use the heating homes because they can not afford it, being an important indicator of the fuel poverty of some neighbours.

The above said makes that the mix of the energetic expenses does not match the usual average: thermal conditioning expenses are reduced, hence increasing the percentages of lighting and water heating.

5 Measures for Energy Refurbishment of Buildings

Following the completion of studies together with some previous articles (Luxán et al. 2009) and considering the budget for the work; in the project have been prioritized the following improvement measures in buildings after comparing and calibrating different possibilities.

Table 1 Measures to improve the thermal envelope of buildings A and B

Edificio	Element	Measure of improve	Before value	After value
Building A y north facades and courtyards in building B	Walls	External thermal insulation composite system 5 cm expanded polystyrene	0.50 W/m^2 K	0.27 W/m^2 K
Building A	Windows	Replacing glass maintaining existing woodworks	3.17 W/(m^2 K)	1.6 W/(m^2 K)
Buildings A y B	Roof	Incorporation 12 cm fiberglass cover. Improved ventilation	1.18 W/m^2 K	0.17 W/m^2 K
Buildings A y B	Floor	Incorporation of 5 cm fiberglass ceiling on outdoor spaces	1.16 W/m^2 K	0.43 W/m^2 K

5.1 Improved of Thermal Envelope

Is improved the thermal performance of building envelope to reduce energy demand. The buildings were built according to the Basic Conditions Thermal regulation CT-79, applicable at the time of the drafting of the initial project. The original building exceeds those requirements because the building had a heating electric wire radiant and this implied an improvement of the transmittances of these elements, required by the power company at the time of contracting the service. For reduction in energy demand has implemented the following (Table 1).

5.2 Hybrid Solar System for Hot Water and Electricity

The houses have an electric boiler to produce hot water. For budgetary reasons only it has built a solar system for hot water production in building A, which reduces energy consumption of the building. The installed solar panels are hybrids and are used for hot water production and for power generation for public areas and recharging points for electric car in basement.

5.3 Environmental Indoor Courtyard Conditioning

The courtyard between the two buildings serves as open for the use of residents and design of housing space. Before the refurbishment, it was characterized by the lack of shade. It has installed a pergola that acts as a sunscreen, which will minimize overheating soil and air in the warm months, and have relocated existing banks in areas of the most comfortable patio. The incorporation of vegetation boxes will be useful to raise the relative humidity.

5.4 Sunscreen in South Facade

Slats has been installed on the south facade of the building A for exterior shading (Figs. 3 and 4). Thus comfort conditions are improved inside the building in the summer months avoiding direct solar radiation on this side of the envelope subject to the biggest yearly variation in its conditions. It is a horizontal fixed sunscreen inclined slats, located on the bottom lines of the windows of each plant and continuously throughout the south facade of the building A. It has been designed to give shade to the windows and façade as well during the summer months and become minimal during the cold months without action from the users.

Fig. 3 South facade before renovation. October noon

Fig. 4 South facade after renovation. December morning

5.5 Improving Natural Lighting in Yards

This is one of the innovative solutions of this Demonstrator. As aforesaid, the percentage of electric consumption grows on the total as it is practically indispensable. To achieve energy savings in the building lighting, light ducts of the Solatube type, but redesigned to exterior-exterior use, have been incorporated in the courtyards of the building A for natural lighting and for energy savings in lighting the building.

This measure is important because the three courtyards of the building A have a dimension of 3.00 $\times$ 4.00 m and 70% of the rooms in housing and common areas get light through these patios. The ducts have been dimensioned to allow natural light to get into the rooms at least 2 m from the patio.

Measured in Spring 2016, the lighting in the lower plant of the courtyard was 600 lux without the duct effect (shutting the upper side captor) and 1.100 lux with the light duct in operation, what makes artificial lighting unnecessary most of the day for the rooms opening to these courtyards.

It has been installed motion detectors in common areas of the building illuminated by night from the hybrid panels' energy (Figs. 5 and 6).

Fig. 5 Courtyard before refurbishment

6 Monitoring Results

Monitoring of housing has been made by the team of the Polytechnic University of Madrid. The works of this energy rehabilitation of residential building were completed in November 2014. The monitoring was conducted homes since 2013, before the intervention in buildings, until the end of 2015, completion date of the project LIFE. This has allowed us to obtain and quantify the impact of interventions and to draw conclusions for the project objectives. The main monitoring data obtained were as follows.

According to data provided by the team, it has achieved a reduction of 71.42% in energy demand and 25.50% in the average real consumption. They were reduced by 76.32% CO_2 emissions.

Whereas the neighbours had previously made use of heating, it has been estimated household savings of 985 €/year. This implies a ratio of investment/savings:

Fig. 6 Courtyard after refurbishment

6.5 kWh/year reduction in energy demand per euro invested and an estimated return on investment of 12 years.

In the case of the demonstrator project, reducing overall energy demand is estimated at 266,000 kWh/year, reducing CO_2 emissions by 61,400 kg/year and economic savings 29,550 €/year for all 30 families benefit.

Natural lighting in the inner courtyards has raised the level of light at the lower floors from 600 lux to 1.100 lux, what makes the rooms opening to them not needing artificial light most of the day.

There has also been tracking production and waste during construction, resulting in a reduced consumption of resources required for the manufacture of construction materials by 180 tons, 18.48 and 68.37 tep tons of CO_2 emissions per household rehabilitated. For every inhabited housing they have reduced construction waste and demolition of 170 tons (Fig. 7).

Fig. 7 Building A after intervention facade insulation, incorporation of sunscreen, hybrid solar panels for hot water and electricity, solar lighting conduits for patios and prototype passive solar heating

7 Conclusions

The study of initial conditions residential building in Zaragoza and climate, urban, social and economic context has helped define a series of measures for energy refurbishment according to the characteristics and needs of its users. These measures seek to achieve greater thermal comfort for residents without increasing the economic cost of consumption, since they are people with low incomes, at risk of fuel poverty.

For this reason, the intervention in these two buildings has improved its passive behaviour, avoiding complex systems with high maintenance costs. Only by working in the building envelope, improving isolation and sunscreen hollow south, reducing energy demand it has reached more than 70% compared to the previous state.

As seen, passive strategies are very important when social housing rehabilitated, especially where social rental scheme where tenants live with very limited income. In that sense, the intervention on the building envelope is more durable and less costly to maintain and outlays for the user that the change of conditioning systems in homes and therefore more appropriate in situations like this.

The participation of residents in the building phase characterization has revealed the high percentage of users who can not use regular air conditioning systems for economic reasons. Hence the relevance of this work, which aims to propose actions adapted to urban, climatic and social conditions of these buildings.

Acknowledgements The authors wish to thank Roberto Diaz (AITEMIN) the project coordination and the Municipal Zaragoza Housing Society the collaboration in the project, especially to Juan Rubio, Fernando Albiac, Paloma Bozman and Margarita Garcia.

References

Cuadrat Prats JM, Vicente-Serrano SM, Saz Sánchez MA (2005) Los efectos de la urbanización en el clima de Zaragoza (España): La isla de calor y sus factores condicionantes. Boletín de la A. G.E., 311–327

Hernández Aja A, Vázquez Espí M, García Madruga C, Matesanz Parellada Á, Moreno García E, Alguacil Gómez J. y Camacho Gutiérrez J (2011) Análisis urbanístico de Barrios Vulnerables. [en línea]. Madrid. Disponible en: http://habitat.aq.upm.es/bbvv/bbvv.html

Luxán M, Vázquez M, Gómez G, Román E, Barbero M (2009) Actuaciones con criterios de sostenibilidad en la rehabilitación de viviendas en el centro de Madrid. EMVS pp 98–106, 50–65. ISBN 978-84-935719-8-6

Olcina Cantos J (2009) Cambio climático y riesgos climáticos en España. Investigaciones Geográficas 49:197–220

Sánchez Guevara C (2016) Propuesta metodológica de evaluación de la pobreza energética en España. Universidad Politécnica de Madrid, Indicadores para la rehabilitación de vivienda. Tesis doctoral

Using Mappable Indicators for Prioritizing the Refurbishment of Social Housing. A Case Study of Zaragoza (Spain)

Marta Monzón and Belinda López-Mesa

1 Introduction

In recent years, cities have undergone huge demographic and territorial growth. The United Nations states that the world's population that lives in cities and urban areas has increased from 750 million in 1950 to 3600 million in 2011. Housing all these people in a very short time entails the physical growth of urban areas, often with insufficient planning which, over the years, has resulted in physical and functional degradation in some cases. The selective depopulation and marginalization of degraded areas means that no-one invests in them, so they further degrade. Degraded areas in Spanish cities are concentrated generally in the old quarters and housing estates built during the Spanish Civil War. In Zaragoza, the old quarters of workers have begun to be studied and are called Urban Estates of Interest. They are groups of houses built according to some form of protection during the 1945–1965 period. In their current state, they are characterized by high-density housing, poor quality buildings, low environmental quality of public spaces and selective depopulation (Ruiz Palomeque and Rubio del Val 2006). Already some targeted investments have been made to refurbish and regenerate these estates with excellent results (Sustainable Buildings Challenge 2011).

Any interventions made within the urban regenerationframework of less favored areas should address this issue as a whole. Regeneration policies must be made to also address the problem as a whole; that is, integrated performance should be carried out. From Europe, mobilizing investments in the renewal of the built building stock is a pressing issue (European Parliament 2012), and the *Law on urban renewal, regeneration and renovation* refers to obtaining and updating maps and censuses of degraded areas to provide information to make such policies (Jefatura del Estado 2013). The government should promote urban regeneration;

M. Monzón (✉) · B. López-Mesa
Universidad de Zaragoza, Saragossa, Spain
e-mail: monzonch@unizar.es

P. Mercader-Moyano (ed.), *Sustainable Development and Renovation in Architecture, Urbanism and Engineering*, DOI 10.1007/978-3-319-51442-0_18

therefore, studies are required to diagnose built parks to direct public investment to the areas that need it the most (Lopez-Mesa et al. 2013).

Indicators are manageable pieces of information that help describe complex processes as they contain lots of synthesized data and serve as a guide for action. Indicators can measure and calibrate any progress made toward a goal, and can provide an early warning to prevent economic, social or environmental setbacks (United Nations 2007). Lots of indicators have been developed in several areas that aim to provide information, and this allows a given situation to be quickly and accurately analyzed, and to take measures and make decisions accordingly.

As public financial resources are limited, they should be allocated to areas most in need and, within them, to buildings that most urgently need rehabilitation. The purpose of this communication is, by means of a case study, to know the available indicators related to housing exclusion criteria that can help make policy decisions on the rehabilitation of buildings.

2 Methodology

The methodology involved studying the state of the art of existing indicators for the specific case of the Spanish city of Zaragoza. For this purpose, different systems of indicators used in this city were studied, and those related to the diagnosis of built parks were selected. Then the indicators found in two residential complexes were applied.

The indicators that prove useful to diagnose built parks, and can be also used to prioritize the refurbishment of buildings, and to focus investment policies and urban regeneration, can differ in nature. Socio-economic indicators, environmental indicators or sustainability indicators, among others, are the fields in which useful prioritization refurbishment indicators can be found.

3 State-of-the-Art Indicators for Diagnosing Built Parks and Prioritizing Refurbishment in the City of Zaragoza

In some Spanish cities, highly relevant indicators have been developed, but have not been applied to the city of Zaragoza. Among these, we find the urban indicators of Salvador Rueda (Rueda 1999) and the diagnosis indicators of intervention needs in renovating built parks of the Basque Country (Tecnalia 2011).

Agenda 21 is a pioneering program that defines indicator systems in general, including environmental indicators. This program was adopted during the United Nations Conference on the Environment and Development (UNCED) held in the city of Rio de Janeiro between 3 and 14 June, 1992 (United Nations 1992). The countries that signed Agenda 21 must promote environmental and socio-economic policies locally in order to achieve sustainable development. To monitor and evaluate these processes, a system of sustainability indicators was created to be used in accessible databases. Countries may adopt the indicators included in this system, which consists of several indicators that are periodically reviewed and have been modified over the years, and they can also use others that are specific for their own municipality.

Several indicators have been developed in Zaragoza since 2000 in the Agenda 21 context using the information available in both the Spanish National Statistics Institute and the Municipal Register, which provides data on the census scale. Among them, and from the point of view of prioritizing the rehabilitation of buildings, the Poor Housing Indicator is emphasized, which measures the number of people who live in homes that are in a poor state of conservation and habitability. Regarding noise pollution, another indicator provides information about the populations exposed to harmful levels of ambient noise (Ayuntamiento de Zaragoza 2014). In energy terms, two indicators report the final energy consumption and emissions of carbon dioxide (CO_2) and other greenhouse gases. These data are provided on the city scale and do not allow to differentiate the areas on this scale.

The Observatory of Urban Vulnerability is a long-term project of the Spanish Ministry of Public Works and Transport, which contains several studies on Urban Vulnerability, such as the Urban Analysis of Vulnerable Districts (Spanish Ministry of Public Works and Transport 2010) and the Urban Vulnerability Atlas in Spain (Spanish Ministry of Public Works and Transport 2012). The “vulnerable” concept used to conduct the Urban Analysis of Vulnerable Districts refers, on the one hand, to increased threats and the risks that affect individuals or social groups and, on the other hand, to the weakening of the mechanisms that address these risks and threats (Spanish Ministry of Public Works and Transport and the Juan de Herrera DUYOT Institute 2010). The aim of the report is to locate the districts represented in the census tracts, obtained from the Population Census, as the only source of extensive and intensive statewide data, which are also homogeneous and offer sufficient disaggregation (Spanish Ministry of Public Works and Transport 1996).

The developed vulnerabilityindicators consider four main areas: socio-demographic, socio-economic, residential and subjective vulnerability. Of these, the indicators that provide information to prioritize the refurbishment of buildings are the over-aging rate, the percentage of dwellings in buildings in a state of poor condition, the percentage of dwellings in buildings built before 1951, indicators of dwellings with a floor space that covers less than 31 m^2, and indicators of external noise. However, they all provide information on a census tract scale and not on a building scale, so relevant information can be diluted at the census tract level. In addition, these indicators are directly dependent on the way the census model is done. The model adopted by Spain to perform the 2011 census differs from that used to perform previous censuses. So results from different years are not comparable because the information is not that homogeneous, nor can it be disaggregated in the same way (Re-hab 2016; Bermejo Aguña 2014).

The city under study has undertaken the “Zaragoza block-to-block” project, which allows the location and accessibility to basic facilities and proximity services to be known. This information, developed by Ebrópolis and GEOT of the

University of Zaragoza, is shown on online satellite maps. Although this information could be useful to focus on urban regeneration strategies on the district scale, it is not relevant for the rehabilitation of buildings. Ebrópolis has also created a system of indicators on a city scale, with five areas marked in Zaragoza 2020: city of citizens, city partnerships, city of knowledge and communication, sustainable city and city territory.

The Regional Government of Zaragoza, like other cities, created the Special Data Infrastructure Portal of Zaragoza (IDEZar) in 2004. This is a basic set of technologies, policies and institutional arrangements that allow data and geographical attributes to be hosted in an environment where they are displayed, searched and evaluated (catalogs and map servers) (Ayuntamiento de Zaragoza 2004). It is a viewer that displays information on demographic and socio-economic aspects on a map of the city shown at the census tract level.

In order to obtain indicators that provide information about the physical characteristics of built parks, mappable Physical Vulnerability Indicators were developed at the block level while conducting a doctoral thesis (Monzón 2016). Thus in the same group of buildings, which presumably have the same socio-economic conditions, those that need to be renovated more urgently can be detected.

Among the existing indicators obtained from the above sources, those that can be used for the refurbishmentdecision making of buildings will be selected. From this point of view, it is considered that the relevant aspects that indicate the need for some residential blocks to receive emergency rehabilitation action as opposed to others can resemble residential exclusion criteria (Pablo et al. 2014). The social exclusion criteria that can be used to prioritize the refurbishment of residential buildings are:

- The elderly
- People with mobility and autonomy problems
- Overcrowding and overloads
- Constructive features. Households in a building that is ruinous or poor
- Basic facilities: homes with no heating
- Economic capacity or structural lack of income

4 Applying Indicators to Two Urban Estates of Interest

During the Spanish postwar period, groups of social housing for working-class people were built in Spanish cities. These housing estates were characterized, among others, by the unitary performance of these estates and by the traditional materials used in their construction. In 2001, the Zaragoza Land-Use Plan recognized many of these estates, and declared them to be Urban Estates of Interest according to the purposes of certain protective measures that this Plan included (Ruiz Palomeque and Rubio del Val 2006) (Ayuntamiento de Zaragoza 2001). Some of these groups today make up deprived areas of the city, with obsolete housing and populations at risk of social exclusion.

For this study, the Andrea Casamayor Group (the former Girón Group) and the Balsas de Ebro Viejo Estate were chosen. They are two estates with different years of construction, distinct construction types and varying number of dwellings. The Andrea Casamayor Group consists of two phases built between 1954 and 1961 and a total of 790 homes. The first phase consists in blocks of four floors and the second in blocks of five floors. The Balsas Set of Ebro Viejo was designed in 1964 and was built in 1975. It consists of 52 buildings spread over 47 blocks of five plants and five towers that are 12 floors high. They house 1260 homes.

Of all the indicators described above, those that allow decisions to be made about the rehabilitation of some blocks, as opposed to others, and in the same group and at the city level, were selected for the two aforementioned estates. Given the scale of detail, the Urban Vulnerability Atlas, the Special Data Infrastructure of Zaragoza and the Physical Vulnerability Indicators were used.

4.1 Urban Vulnerability Atlas

As stated above, the Urban Vulnerability Atlas provides indicators on the census scale. The Andrea Casamayor Group is divided into two census tracts, and each encompasses nearby buildings. The Balsas of Ebro Viejo Estate is divided into four census tracts in which no other buildings that belong to this group (Fig. 1) are included.

Figure 2 offer the graphical representations of the Vulnerability Atlas indicators of Urban Vulnerability with the 2011 census data. It shows how there are often different values in various census tracts of an Estate, which thus blurs information and makes decision making more difficult. The ratio of the average area per capita is similar in both estates, with a value of 0–75 m. The two census tracts of Andrea

Fig. 1 Representation of the limits of the census tracts that the estates house (*red line*) and that limit them (*orange line*)

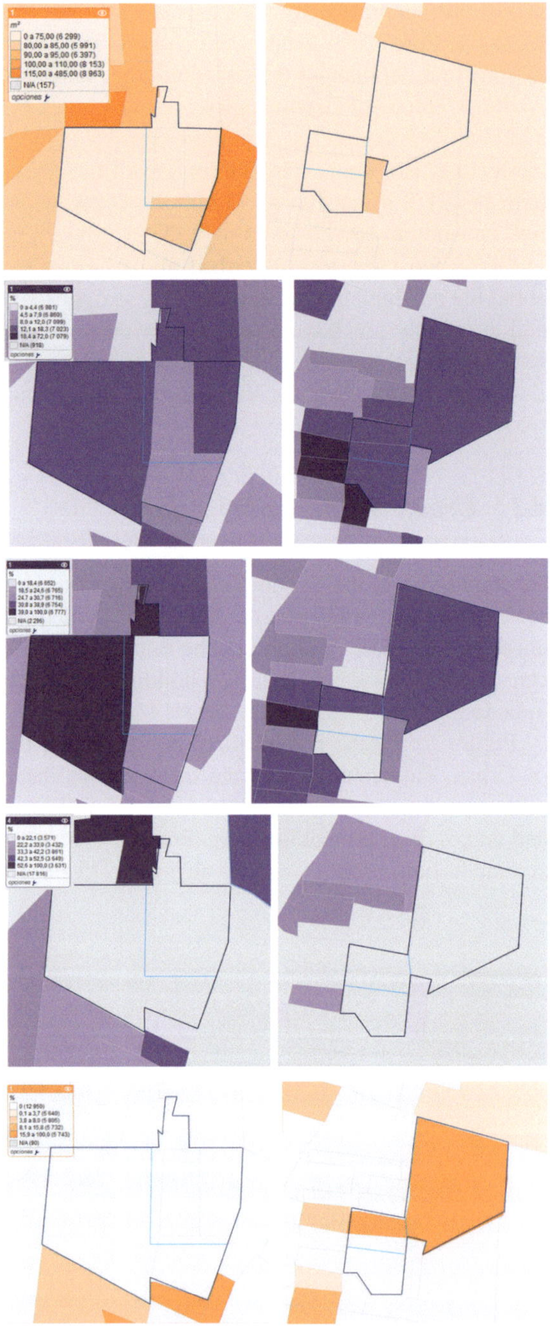

Fig. 2 From *top* to *bottom*, mean area per inhabitant, percentage of the population who has not completed their education, percentage of unemployed population, percentage of unemployed young population, percentage of homes in a ruinous, bad or poor state. *Left* Balsas de Ebro Viejo; *right* Andrea Casamayor. The *blue* limits the Estates

Casamayor have a population percentage who have not completed their education from 12.1 to 18.3%. In Balsas de Ebro Viejo, two of the four sections have the same percentage of population who has not completed their education, and the other two have a percentage of 4.5–7.9%.

The percentage of homes in a ruinous, bad or poor state takes the smallest value in Balsas de Ebro Old and in one of the Andrea Casamayor sections, while the other takes values from 15.9 to 100%. Note that from the Vulnerability Atlas, created with 2001 census data, data can be obtained on primary residences. This is a very valid argument for the rehabilitation of government-managed buildings. However, this information is not available for 2011.

4.2 *The Special Data Infrastructure of Zaragoza*

The Special Data Infrastructure of Zaragoza (IDEZar) provides indicators on the census tract or district scale of the demographic and socio-economic aspects of the city of Zaragoza. Among the residential exclusion criteria, it allows us to obtain information about the aging population and the average net income per household. Figure 3, top part, is the graphical representation of the population percentage aged 65 or more in the two studied estates on the map of the city, created with the 2014 data on a census scale. The four census tracts that form the Balsas de Ebro Viejo

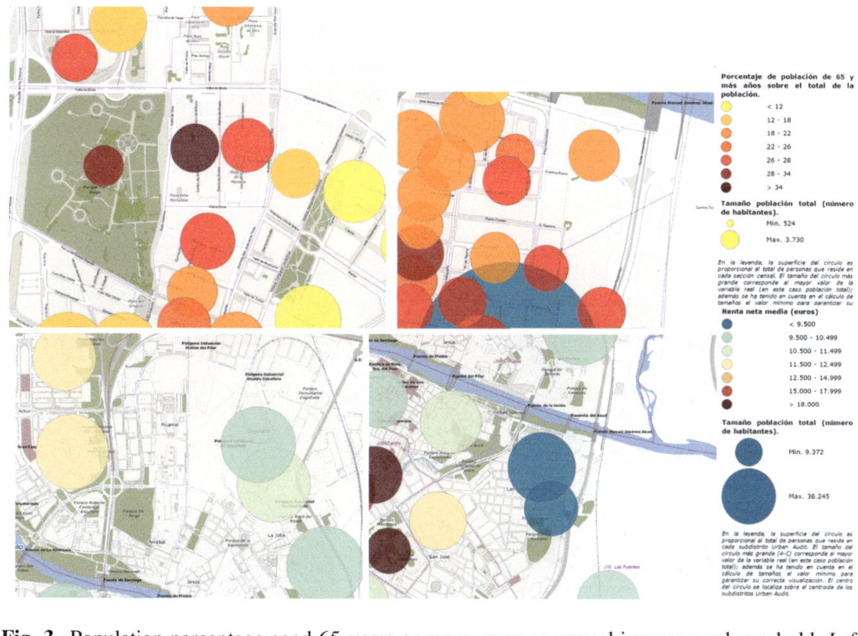

Fig. 3 Population percentage aged 65 years or more, average annual income per household. *Left* Balsas de Ebro Viejo; *right* Andrea Casamayor

indicators are red and dark red, with percentages that range from 26.38 to 35.32%. Both the Andrea Casamayor sections are shown in orange, with percentages of 19.81 and 20.34%. Figure 3, bottom part, is the graphical representation of the average net income, obtained with the 2013 data provided by the European Urban Audit Project at the district level [Spanish National Statistics Institute (INE) 2016], and shown on the map of the city. We can see how the average income in the Balsas de Ebro Viejo district reaches values of 10,300–12,000 net euros per year, and 9100 net euros per year in Andrea Casamayor.

4.3 Physical Vulnerability Indicators

The Physical Vulnerability Indicators described in (Monzón 2016) provide information at the block level on the physical characteristics of buildings in energy efficiency, acoustic insulation and accessibility terms. Figure 4, left, shows the indicator for lack of energy efficiency, and energy demand representing air conditioning understood as the sum of the heating and cooling energy demands. We can see how some Balsas de Ebro Viejo blocks have a higher energy demand than the rest. The Outside Noise Indicator represents the percentage of rooms of blocks that would meet the requirements of the Basic Protection Document Against Noise of the Technical Building Code (CTE), and the only change would be the hollow façade solution (Fig. 4, center). The rehabilitation of the blocks that do not meet the requirements and need other interventions should be prioritized, even when changing the window solution. Figure 4, right, shows the Physical Vulnerability Indicator through lack of accessibility, which has been configured by applying the installing elevators criteria in the Basic Security Document of the CTE for new buildings. Since the number of floors and homes per block is similar in all cases, we obtained the same values for this indicator, except for the towers because they already had elevators.

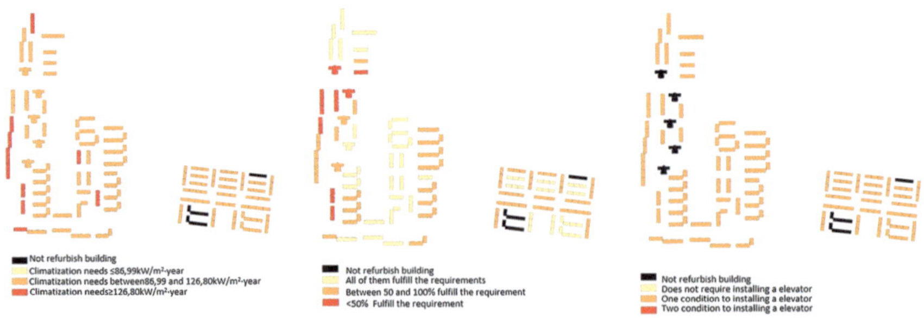

Fig. 4 *Left* physical Vulnerability Indicator for lack of energy efficiency, air conditioning demand. *Center* physical Vulnerability Indicator for outside noise, percentage of rooms that would meet replacing windows with 4-6-4. *Right* Physical vulnerability indicator for lack of accessibility

5 Discussion and Conclusions

Physical Vulnerability Indicators on the block scale show that the various blocks of the same housing group may perform differently. However, the information provided by indicators on the census or district scale offer fuzzy information when it comes to prioritizing the rehabilitation of some blocks as opposed to others because the information is diluted throughout the section.

After observing the values of the indicators, we reflect about prioritizing the intervention of both the studied estates, including the blocks that most urgently need to be rehabilitated.

Both groups obtained similar values for surface dwellings and the percentage of youth unemployment. So these data may not be relevant for decision making. Indicators of the population who had not finished their education, and who live in homes in a ruinous, bad or poor state, indicate higher percentages for the Andrea Casamayor group. As the values are higher only in one of the census tracts, this information may be ambiguous as they are nearby buildings.

The population is older in Balsas de Ebro Viejo, so installing elevators may be a more urgent matter. However, the average annual income is lower in Andrea Casamayor.

From the Physical Block Level Indicators, accessibility does not denote priority as all the blocks obtained the same value. This shows how the Energy Demand Indicator and the Exterior Noise Indicator highlight the underperformance of a few blocks in these areas, all of which are located in Balsas de Ebro Viejo.

References

Ayuntamiento de Zaragoza (2001) Plan General de Ordenación Urbana de Zaragoza

Ayuntamiento de Zaragoza (2014) Actualización de indicadores de sostenibilidad de Zaragoza 2014

Ayuntamiento de Zaragoza (2004) INFRAESTRUCTURA DE DATOS ESPACIALES DE ZARAGOZA (IDEZAR). Retrieved 10 Sept 2016, from http://www.zaragoza.es/ciudad/idezar/saber.htm

Bermejo Aguña A (2014) El impacto del Censo de 2011 en la investigación de carácter micro. El caso de Madrid. In *El censo de 2011 en el marco de la experiencia censal en España*. Jornada, Madrid: Universidad Carlos III de Madrid

European Parliament (2012) Directive 2012/27/EU of the European Parliament and of the Council of 25 october of 2012 on energy efficiency

Instituto Nacional de Estadística (2016) Indicadores urbanos (Urban Audit). Instituto Nacional de Estadística, Madrid

Jefatura del Estado (2013) Ley 8/2013 de Rehabilitación, Regeneración y Renovación Urbana

López-Mesa, B, Gairín Alastuey M, Monzón Chavarrías M, y Juan Rubio del Val (2013). Technical indicators to establish the order of priority for integral refurbishment of existing social housingPaper presented at 1st International & 3rd national Cong, University of Seville, Seville, 20–22 may 2013

Spanish Ministry of Public Works and Transport(1996) Análisis urbanístico de Barrios Vulnerables en España. Síntesis Metodológica y Créditos 1991

Spanish Ministry of Public Works and Transport (2010) Análisis urbanístico de Barrios Vulnerables en España. Introducción al Informe General 2001: metodología, estructura del Catálogo y Créditos. Madrid

Spanish Ministry of Public Works and Transport (2012) Atlas de la Vulnerabilidad Urbana en España. Metodología, contenidos y créditos

Spanish Ministry of Public Works and Transport, and the Juan de Herrera DUYOT Institute (2010) Análisis urbanístico de Barrios Vulnerables en España. Sobre la Vulnerabilidad Urbana

Monzón M (2016) Metodología para la obtención de indicadores de vulnerabilidad física de viviendas desde la perspectiva de la priorización de su rehabilitación integral. El caso de los Conjuntos de viviendas sociales de la posguerra de Zaragoza. Universidad de Zaragoza

Pablo, JM, Martínez SP, Eva María Tomás del Río, y David Pac Salas (2014) Exclusión residencial en la ciudad de Zaragoza. Hacia un sistema de indicadores. Cátedra Zaragoza Vivienda, Zaragoza

Re-hab (2016) Impacto de la metodología del nuevo Censo de Población y Vivienda 2011 en las investigaciones socio-urbanísticas http://www2.aq.upm.es/Departamentos/Urbanismo/blogs/re-hab/impacto-de-la-metodologia-del-nuevo-censo-de-poblacion-y-vivienda-2011-en-las-investigaciones-socio-urbanisticas/. Accessed 10 Feb 2016

Rueda S (1999) Modelos e Indicadores para ciudades más sostenibles. Taller sobre Indicadores de Huella

Ruiz Palomeque, LG, y Rubio del Val J (2006) Nuevas propuestas de rehabilitación urbana en Zaragoza. Estudio de conjuntos Urbanos de Interés. Zaragoza: Sociedad Municipal de Rehabilitación Urbana de Zaragoza, Zaragoza

Sustainable Buildings Challenge (2011) Grupo Girón 40 viviendas rehabilitación. Helsinki 2011

Tecnalia (2011) Diagnóstico de las necesidades de intervención en la renovación del parque edifado de la CAPV

United Nations (1992) Agenda 21. United Nations conference on environment and development

United Nations (2007) Indicators of sustainable development: guidelines and methodologies

Part IV
Minimizing the Consumption of Material Resources

Assessment of the Relationship Between Diameter and Tensile Strength of Piassaba (*Aphandra natalia*) Fibers

Cristian Balcázar-Arciniega and Francisco Hernández-Olivares

1 Introduction

Sustainable development aims to reduce the extraction and use of conventional materials through the use of regenerable natural resources from waste and by-products that can be recycled. The palms are considered one of the groups most economically important plants. In Ecuador is a great variety of species that provide some benefit from use of the fibers, leaves and fruits.

Piassaba is the name of the hard brown fibers, considered as a non- timber forest product, which can be obtained from three species of palm *Leopoldinia piassaba*, *Attalea Funifera* and *Aphandra natalia* (Kronborg et al. 2008).

Apandra natalia is an important resource for local communities from western Amazon. In Ecuador harvesting techniques are non-destructive, selective removal of other vegetation is made, even occasionally grown. The traditional uses range from consumption of their fruit, extracting fibers for sale and harvest leaves for thatching houses, its main product are the fibers from the leaf sheath and petiole with which brooms are made (Balslev et al. 2010; Pedersen 1992), see Fig. 1.

Piassaba of *Attalea funifera* and *Leopoldinia piassaba* have been characterized and used experimentally as reinforcements for polymer matrices, finding applicability in materials requiring good impact resistance and increase hardness (Monteiro et al. 2006; Shimazaki and Colombo 2010). On the other hand, so far there have been no reported properties physical, mechanical, chemical or the use as rein-

C. Balcázar-Arciniega (✉)
Department of Architecture and Arts, School of Architecture,
Universidad Tecnica Particular de Loja, 1101608 Loja, Ecuador
e-mail: cabalcazar@utpl.edu.ec

C. Balcázar-Arciniega · F. Hernández-Olivares
Department of Building Construction and Architectural Technology,
School of Architecture, Polytechnic University of Madrid, Madrid 28040, Spain

P. Mercader-Moyano (ed.), *Sustainable Development and Renovation in Architecture, Urbanism and Engineering*, DOI 10.1007/978-3-319-51442-0_19

Fig. 1 Main product developed fibers *Aphandra natalia*

forcement in composites of *Aphandra Natalia* fiber. In Table 1 shows mechanical and structural characteristics from different species of piassaba.

For many lignocellulosic fibers it has been reported a function of inverse proportionality between the tensile strength and the diameter (Carolina et al. 2015; da Costa et al. 2010; Monteiro et al. 2011a, b). The fibers of Sisal, Ramie, Curua, Jute, Bamboo, Coir, *Atelea funifera*, Burite, and pineapple had a hyperbolic correlation between the ultimate tensile strength (σ) and the equivalent diameter (d) expressed in the following shape function:

$$\sigma = \frac{A}{d} + B \tag{1}$$

where A and B are characteristic constants for each type of fiber. The reason for a general relationship specified by Eq. 1 is given in terms of flaws and imperfections, typical of a lignocellulosic structure. This inverse correlation has not been investigated for *Aphandra natalia* fibers (Fig. 2).

2 Experimental Procedure

The basic material investigated in this work was *Aphandra natalia* untreated, Fig. 2, harvested in the canton Sucúa province of Morona Santiago, supplied by the manufacturers of handmade brooms of the city of Loja (Ecuador). The diameter was calculated from a sample of 118 fibers randomly taken was assumed that its shape is cylindrical and 5 different points measured along its length, the tool used was a

Table 1 Mechanical properties and structural compositions of piassaba

Source piassaba	Tensile strength (MPa)	Elastic modulus (GPa)	Total strain at rupture (%)	Density (g/cm^3)	Cellulose (wt%)	Lignin (wt%)	References
Attalea funifera	134.8–142.9	1.07–4.59	21.9–7.8	–	31.6	48.4	d'Almeida et al. (2006)
Attalea funifera	86–123	4.499–5.449	4.00–3.00	1.1	–	–	Elzubair et al. (2007)
Attalea funifera	134.5	6.095	3.4	1.03	5.2	43.1	Martius (2015)
Attalea funifera	131 ± 36	03.8 ± 00.9	11.4 ± 3.6	–	–	–	Alves Fidelis et al. (2013)
Leopoldinia piassaba	22–23	1.116–1.447	5.00–7.00	1.13	–	–	Elzubair et al. (2007)

Fig. 2 *Aphandra natalia* fibers extracted from the pod and leaf petiole

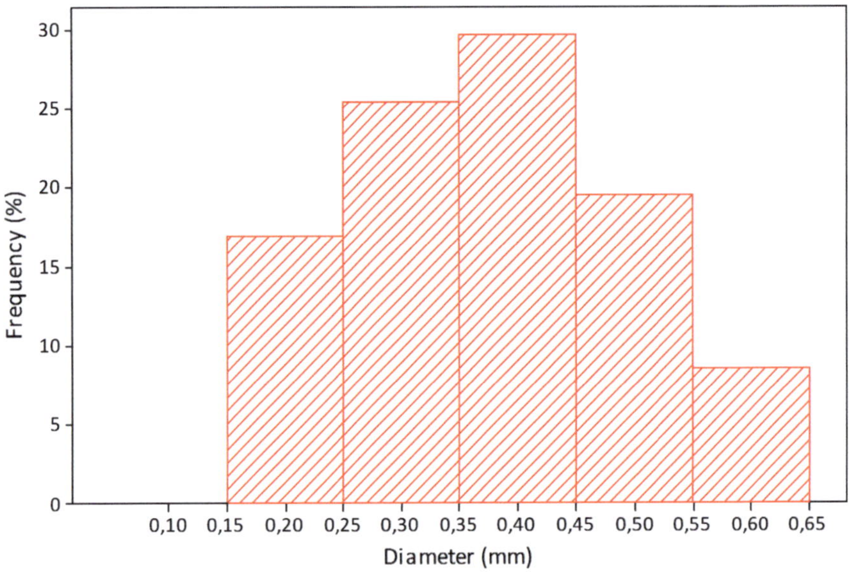

Fig. 3 Histogram of the statistical frequency of fibers within five conventional diameter intervals

0.001 mm micrometer precision. Figure 3 shows the histogram of the different diameters of piassaba, recital 5 intervals

Of the fiber diameters indicated in the histogram, we selected 10, two for each interval. To each fiber was performed of the analysis of tensile strength of 4–5 times at different points. The final tensile strength was calculated by the load ratio fails and diameter. The load was gradually applied and was measured with a dynamometer (50 N). The initial length between clamps was 20 ± 0.1 mm,

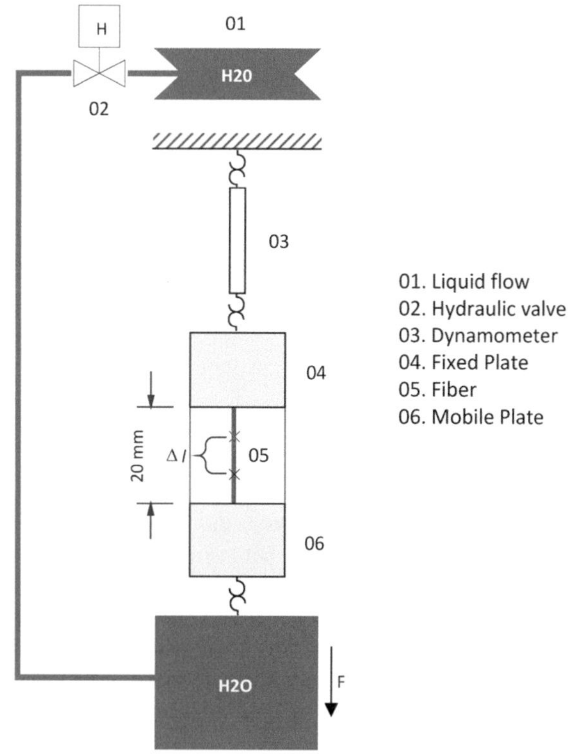

Fig. 4 Device for testing diagram tension

each fiber was marked at two points, which allowed obtaining the initial and final elongation length through a 0.01 mm gauge precision. Figure 4 shows the test device. The values obtained for the tensile strength, i.e. the last effort was statistically interpreted by a computer program Weibull analysis.

According with Carolina et al. (2015) and Nacional and Trabajo (2011), the Weibull statistical analysis is based on a cumulative distribution function F(x):

$$F(x) = 1 - \exp\left[-\left(\frac{\chi}{\theta}\right)\right]^{\beta} \tag{2}$$

where:

- χ position parameter which defines the starting point or origin of the distribution
- θ scale parameter, extension of distribution along, in our case represents the characteristic resistance
- β the shape parameter represents the slope straight de la describing the degree of variation of the failure rate.

The Weibull graph is functional scale graduated as follows: In the vertical axis you have In In$\left(\frac{1}{1-F(X)}\right)$ (neperiano logarithm double) and the abscissa In χ.

Any group of data following the Weibull distribution can be represented by a straight line, it can be inferred that the origin is well known and coincides with the experimental data. From Eq. 2, using logarithms twice, we have:

$$\text{In In}\left(\frac{1}{1 - F(x)}\right) = \beta \text{ In } \chi - \beta \text{ In } \theta \quad (3)$$

The breaking load was measured for each diameter range using data from 9 to 10 experiments. Figure 4, the computer program presented the linear graph of Eq. 3 and calculated Weibull parameters.

3 Results and Discussion

Aphandra natalia fibers have various diameters, including each fiber. Five characteristic estimated diameters, the largest relative frequency distribution corresponding to 29.6% sections between 0.35 and 0.45 mm, whereas the lower relative frequency distribution is 8.4% corresponding to the diameters between 0.55 and 0.65 mm. Similar research report higher relative frequency distribution of *Attalea funifera* originating piassaba 38 and 30% *Leopoldinia piassaba*, both where larger diameters are in the range of 0.4–0.6 mm (Elzubair et al. 2007).

In accordance with Carolina et al. (2015) and Pons (1995), and from the data obtained experimentally, Fig. 5, the tensile strength for each diameter range is determined by the following equation

$$\sigma = \frac{4L}{\pi d^2} \quad (4)$$

where:

σ tensile strength
L maximum load
d fiber section

The values of the tensile strength were analyzed by the method Weibull, 9 or 10 applied to experiments for each diameter range, Fig. 4. The Weibull analysis program provided diagrams, Fig. 6, that are characterized by unimodal. With a single straight line points are adjusted at each interval. This indicates a similar mechanical behavior of the fibers within the same diameter range.

Weibull analysis program also provided the characteristic resistance (θ) Weibull modulus (β) parameters for fine adjustment (R^2) and the average resistance and associated statistical deviations. The values are presented in Table 2.

In Fig. 7 the trend of the characteristic resistance (θ) is shown to vary inversely as the average equivalent diameter of the fiber, the values shown in Table 2 for the Weibull modulus (β) and fine adjustment (R^2) support the inverse correlation.

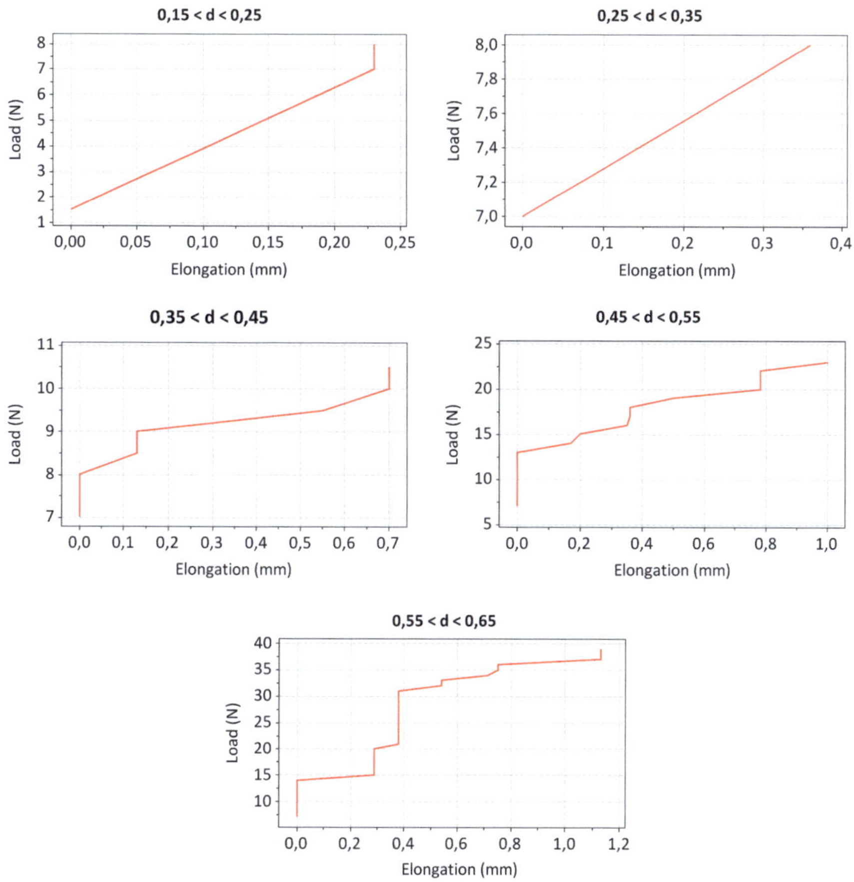

Fig. 5 Typical *Aphandra natalia* tensile load versus elongation for the different equivalent diameter intervals

Through a mathematical relationship of hyperbolic type, Eq. 1 adapted to fit the data in Fig. 7.

$$\theta\,(\mathrm{MPa}) = 36.01 + \frac{26.60}{\mathrm{d}} \quad (5)$$

Similarly, a hyperbolic equation fits the data of Fig. 6:

$$\bar{\sigma}\mathrm{m} = 38.01 + \frac{26.43}{\mathrm{d}} \quad (6)$$

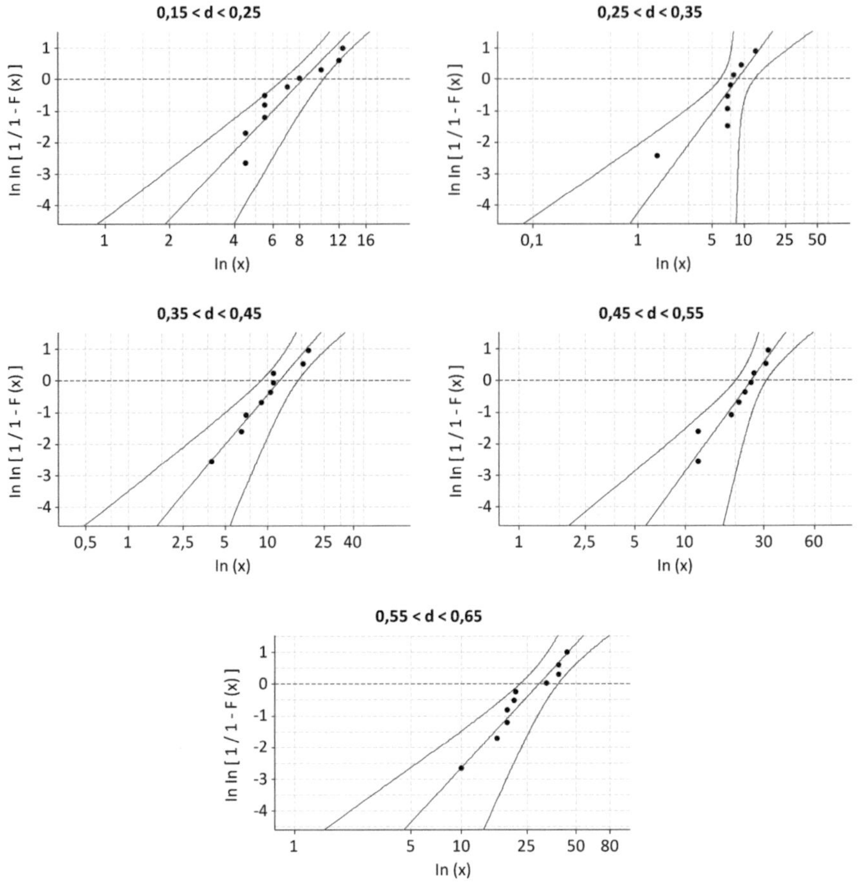

Fig. 6 Weibull diagrams for *Aphandra natalia* tensile strength in different intervals of diameter

Table 2 Weibull parameters for the tensile strength of *Aphandra natalia* fibers associated with different diameter interval

Diameter interval (mm)	Weibull modulus, β	Characteristic strength, θ (MPa)	Precision adjustment, R^2	Average tensile strength (MPa)	Statistical deviation (MPa)
0.15 < d < 0.25	1.428	184.834	0.983	183.152	47.7057
0.25 < d < 0.35	2.042	97.6595	0.888	105.43	58.4435
0.35 < d < 0.45	1.533	87.2399	0.983	88.3507	33.4141
0.45 < d < 0.55	1.549	105.424	0.963	104.923	30.7976
0.55 < d < 0.65	1.437	90.9011	0.979	91.7647	33.5328

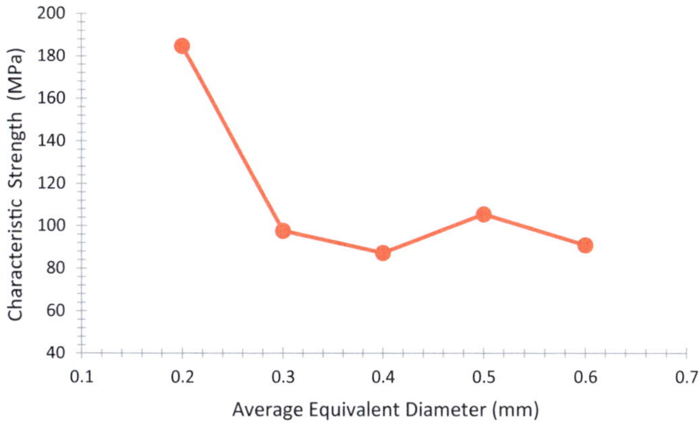

Fig. 7 Variation of the *Aphandra natalia* Weibull characteristic strength with the corresponding fiber equivalent diameter for each interval

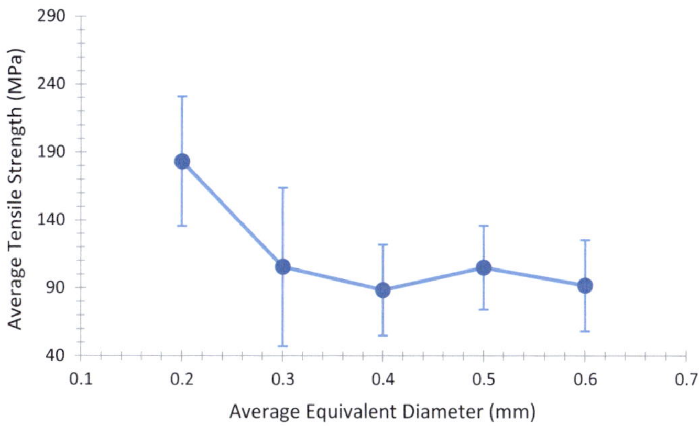

Fig. 8 Variation of the *Aphanda natalia* Weibull average strength with the corresponding fiber equivalent diameter for each interval

Equation 6 is very similar to Eq. 5, It indicates a coherence Weibull statistical analysis for tensile strength with dependence fiber diameter. The average tensile strength Weibull (σm) of Table 2 is plotted the diameter (d) in Fig. 8, a hyperbolic inverse correlation is also evident.

The microscopic views in the fracture of the fibers support the inverse correlation between strength and diameter demonstrated Eqs. 5 and 6. The thinnest fiber diameter 0.2 mm, Fig. 9a, is characterized by a uniform structure and simultaneous rupturing its fibrils. These fibrils are arranged in smaller numbers than in the thicker fibers and tend to be closer to reducing the gaps.

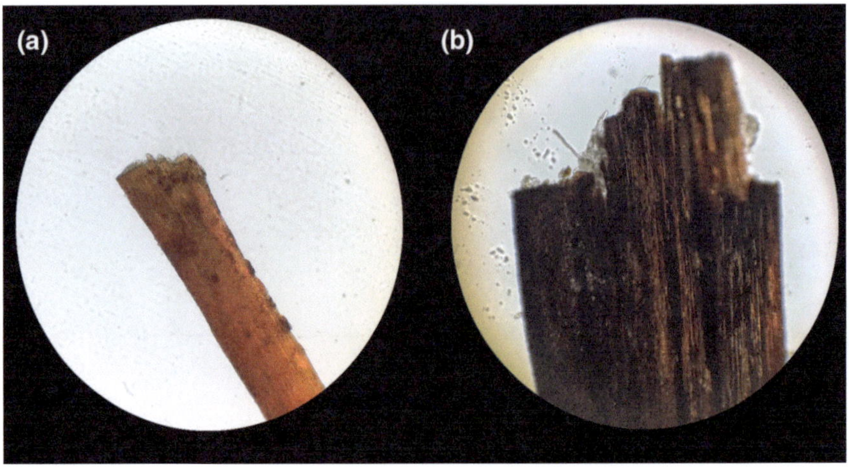

Fig. 9 Microscopic surface view, 100×, of *Aphandra natalia* fiber: **a** thinner fiber, d = 0.2 mm and **b** thicker fiber, d = 0.6 mm

On the other hand, the thicker fiber with diameter equal to 0.6 mm, Fig. 9b, shows more defects and reveals a heterogeneous fracture associated with a greater number of fibrils. The total rupture occurs when the first fibril breaks by propagating the defect all the structure. Statistically it is shown that the smaller diameter fibers will be stronger than the larger section. It is more likely that a thick fiber to a lower strength breaks down a thinner fiber.

As a final observation, in practice it is possible select the thinner fibers to develop composite materials with improved physical or mechanical properties.

4 Conclusions

1. An inverse interrelation was obtained between the resistance and the diameter of the fibers. The resistance of the fibers, they showed across a statistical analysis Weibull that the highest values are obtained for the fibers with smaller diameters.
2. Based on the results presented studies and other lignocellulosic fibers, it can be mentioned that a hyperbolic equation adjusts the inverse correlation between the strength and the diameter of the fibers.
3. Failure analysis and microscopic views statistically support that premature fracture of coarser fibers is due to the greater distribution of fibrils compared to the thinner.

Acknowledgements The authors thank the support to this investigation by: Ph.D. Sofía Sanz González de Lema.

References

Alves Fidelis ME, Pereira TVC, Gomes ODFM, De Andrade Silva F, Toledo Filho RD (2013) The effect of fiber morphology on the tensile strength of natural fibers. J Mater Res Technol 2 (2):149–157. http://doi.org/10.1016/j.jmrt.2013.02.003

Balslev H, Knudsen TR, Byg A, Kronborg M (2010) Traditional knowledge, use, and management of *Aphandra natalia* (Arecaceae) in Amazonian Peru. Econ Bot 64(1):55–67

Carolina M, Teles A, Oliveira G, Rodrigues G (2015) Evaluation of the diameter influence on the tensile strength of pineapple leaf fibers (PALF) by Weibull method. Mater Res 18(Suppl 2):185–192

d'Almeida JRM, Aquino RCMP, Monteiro SN (2006) Tensile mechanical properties, morphological aspects and chemical characterization of piassava (*Attalea funifera*) fibers. Compos Part A Appl Sci Manuf 37(9):1473–1479. http://doi.org/10.1016/j.compositesa.2005.03.035

da Costa LL, Loiola RL, Monteiro SN (2010) Diameter dependence of tensile strength by weibull analysis: part I bamboo fiber. Rev Mater 15(2):97–102. http://doi.org/10.1590/S1517-707620 10000200006

Elzubair A, Maria C, Bonelli C (2007) Morphological, structural, thermal and mechanical characterization of piassava fibers, vol 478, pp 37–41. http://doi.org/10.1300/J395v04n02 (April 2013)

Kronborg M, Grández CA, Ferreira E, Balslev H (2008) *Aphandra natalia* (Arecaceae)—A little known source of piassaba fibers from the western Amazon. Revista Peruana de Biología 15 (November):103–113

Martius A (2015) Artigo 38(2):161–165

Monteiro SN, Coeli R, Aquino MP, Lopes FPD, Roberto J, Almeida M (2006) Tenacidade ao Entalhe por Impacto Charpy de Compósitos de Poliéster Reforçados com Fibras de Piaçava, pp 204–210

Monteiro SN, Lopes FPD, Barbosa AP, Bevitori AB, Amaral Da Silva I L, Da Costa LL (2011a) Natural lignocellulosic fibers as engineering materials-an overview. Metall Mater Trans A Phys Metall Mater Sci 42(10):2963–2974. http://doi.org/10.1007/s11661-011-0789-6

Monteiro SN, Satyanarayana KG, Ferreira AS, Neto JB, Portela TG, Researcher V (2011b) Selection of high strength natural fibers. Rev Matér 15(4):488–505. http://doi.org/10.1590/S1517-70762010000400002

Nacional C, Trabajo DECDE (2011) NTP 331 : Fiabilidad : la distribución de Weibull, pp 1–13

Pedersen HB (1992) Uses And Management Of Aphandra Natalia (Palmae) in Ecuador. Bull Inst Fr Études Andines 21(2):741–753

Pons JM (1995) Método de ensayo para la determinación del módulo inicial de elasticidad de las fibras individuales. Annales Scientifiques Textiles Belge 2, 11. Retrieved from https://upcommons.upc.edu/revistes/bitstream/2099/6177/1/Article04.pdf

Shimazaki K, Colombo MA (2010). Gelatin/piassava composites treated by electron beam radiation, pp 303–308

Selection of Criteria for the Systematization of Technologies for a Sustainable Urban Water Cycle Management

Ana Prieto-Thomas

1 Introduction

Nowadays, the lack of sustainability in the territories and urban areas of the Mediterranean region is a fact as shown in a long list of national and international research institutions and projects[1], so that contributions that may allow a change in water resources administration, by means of a more sustainablemanagement of them, are becoming increasingly more important in our cities. That necessarily implies bringing to the technical actors the existing hydro-efficient strategies and technologies presently available, in a synthesized way, in order to their implementation on existing buildings and the free urban spaces that belong to the referred urban centers.

In order to propose changes in that management, the issue has been approached starting from a detailed analysis of the presently existing problems in building and urban spaces. The most relevant problems that have been detected are summarized as follows.

- Firstly, the urban models developed during the past decades along the Mediterranean region had a clear expansive character although they have been implemented, in many occasions, without an adequate control. Frequently, the problem is that these urbanizing processes have not taken into account their consequences on the systems that make up the hydrological cycle and have led to a generalized phenomenon of destabilization and unbalancing of those

[1]References on this matter are reflected in different research studies like Aqua-Riba (2013–2015) and Prieto-Thomas (2015).

A. Prieto-Thomas (✉)
Department of Building Constructions I, Higher Technical School of Architecture, University of Seville, Av. Reina Mercedes, 2, 41012 Seville, Spain
e-mail: apthomas@us.es

P. Mercader-Moyano (ed.), *Sustainable Development and Renovation in Architecture, Urbanism and Engineering*, DOI 10.1007/978-3-319-51442-0_20

abovementioned systems (Prieto-Thomas and Lara 2014). Some of the causes of those effects are:

- Watercourses channelling and undergrounding as well as the invasion riverbeds and riverbanks.
- Increment of impervious and sealed surfaces.
- Rise of demand and volumes of water that have to be supplied to urban centers and, subsequently, have to be treated.

Besides, it is also a proven fact that the huge increase of the world population has meant an important rise in the consumption of resources encouraged by a mentality, that appeared with industrialization and that has persisted during the 20th century, where the idea that humans could exploit the different resources present in nature without any limit predominated[2].

Starting from the abovementioned issues, solutions that defend the fostering of relations amongst territories, the city model, nets or infrastructures and water resources are proposed. An answer has to be given to the multiple consequences of urban expansion processes that have impacted on the hydrological system by means of revising the urban-territorial models. Furthermore, in the face of the expansive processes that mean the disproportionate and uncontrolled creation of the new city, it is proposed the renovation of the existing urban structure[3] with interventions implemented by approaches, strategies and innovative technologies that enhance all its possibilities and obtain the maximum benefit from resources, in this case, hydric ones.

In addition, it is necessary to change the view on the city. From now it must be considered as a living organism where advantage should be taken of all its existing resources—doing it along all stages—while the relationships and synergies amongst them are fostered. That will make it necessary to use *recirculation*, *reuse*, *recycling* and *regeneration*, amongst other strategies, as well as to develop material and energetic interchanges amongst the different areas. This will require a holistic vision of the city as a place of exchange of all type of fluxes[4].

All of the above must be accompanied by the implementation of new water management technologies which should be more environmentally friendly. At European Union level, as it is clearly shown in the Water Framework Directive,

---

[2]Miguel Delibes de Castro, in his admission speech when joining the Real Academia de la Lengua in 1975, already denounced the attitude of man abusing nature: “Nowadays, the abundance of technical means allows the transformation of the world to our liking, a possibility that has aroused among men a vehement dominating passion” (citated in Prieto-Thomas 2015).

[3]Such renovation must be considered as a necessary goal in the current context as an alternative to the production of the new city and, in fact, has become a main key in public policies within the present architectural and urban areas (Aqua-Riba 2013–2015).

[4]As an example of this trend, for several years, in northern European countries such as Sweden, this city model has been advocated with projects like *SymbioCity. Sustainability in Sweden* (SymbioCity 2015), that fosters “a holistic and sustainable urban development, tracking down potential synergies within urban functions and unlocking their efficiency and profitability” (SymbioCity 2008).

they already are very aware of the potentialities of sustainable technologies for water management compared to the traditional ones. Therefore, to implement these technologies in projects both to build new buildings and neighborhoods and to renovate them is a priority.

- Secondly, it has also been identified, through contact with the professional actors involved in Architecture and Urban Planning, the need of bringing closer to them, on a very simple and practical basis, updated and structured information on these issues in order to facilitate the implementation of these technologies on a real way in architectural and urban planning projects. Following an analysis of the current situation, with regards to its implementation in projects and works and in relation to the existing material on Urban Water Cycle management—including strategies and technologies—some information on the subject has been found, but most of it is very disperse and scarcely systematized, becoming of little use for architects and urban planners when it comes to dealing with specific projects in the field of cities and buildings. On the basis of this situation, it seemed interesting to create a document as a tool for the professionals and other stakeholders involved in building and urban planning that would allow the updating and deepening of knowledge about these aspects of a project in order to be able to approach the issues encountered more easily and safely.

  But the truth is that it is not enough to create that tool but it is also necessary to accelerate its implementation with mechanisms that facilitate the chances of its effective application on the ground. For this reason, the election of the criteria for its systematization is fundamental analyzing, in a very detailed manner, the organisation of the information offered to the agents involved in the subject. This is the main aim of this article, already approached in previous works, such as in the RDI Project Aqua-Riba (2013–2015) "Sustainable Urban Water Cycle Management Systems in the Integral Regeneration Plans for Districts in Andalusia (Aqua-Riba)" commissioned by the Consejería de Fomento y Vivienda de la Junta de Andalucía (Regional Ministry for Development and Dwelling; Regional Government of Andalusia) and the PhD Thesis of Prieto-Thomas (2015).

Therefore, an adequate management of the Urban Water Cycle, complemented with new effectively applied technological tools, is a key factor for the viability of these inhabited spaces and thus to be able to give an answer to the different problems that have been encountered.

2 Methods

This work has been developed following the *inductive method* where a research consisting of the analysis of methodological type data sources on the different existing organization systems has been carried out, with the systematization of hydro-efficient and sustainable technologies as a final aim. In this line, several European projects—such Eco-City (Carrasco 2009), WaND (Butler *et al* 2010), TRUST o SWITCH (2006–2011)—have been reviewed, highlighting the last one being a deep study on the issue that proposes very updated systematization criteria.

What follows is an enumeration of the issues analyzed within this work:

- Selection of criteria for the election of hydro-efficient and sustainable technologies currently available.
- Selection of relevant parameters for the characterization of technologies.
- Definition of the classification or systematization of technologies.
- Proposal for an outline for the organization and classification of the selected information.

3 Presentation of Results

3.1 Selection of Criteria for the Election of Hydro-efficient and Sustainable Technologies Currently Available

As a first step, a preliminary definition of the criteria for the selection of technologies has been considered indispensable on the basis of the specification of the strategic guidelines that are fundamental to be followed for the possible implementation of the new management models of the Urban Water Cycle.

3.1.1 Selection of General Strategies for the Joint Management of the Urban Water Cycle

Technologies should be targeted towards:

- Minimizing the potable water demand from the system and the energetic cost associated to it—as well as the economic costs derived from both of them—thus increasing the efficiency of the consumption and the use of alternative resources (rainwater, recycled or reused water, etc.).
- Reducing the water flow destined to urban sewage by means of decentralized systems for the treatment and reuse of water, as well as by means of the urban runoff reduction.
- Achieving the maintenance and recuperation of natural water fluxes, from both infiltration and surface runoff, fostering the recuperation of natural hydrological balance.
- Reducing the emission of CO_2, increasing its capture in urban centers, in order to fight against climate change, fostering the implementation of solutions, always under efficiency criteria, that boost the reduction of energetic consumption and/or favor the naturalization of cities.

3.1.2 Selection of Strategies Within the Different Fields[5] of Urban Water Cycle Management and Definition of Specific Measures

It has been considered essential to add within the definition of strategic guidelines a series of measures, which will complement those that are specifically more technological or *structural measures*. These other measures will favor the implementation of the technological ones and are related to the awareness of population in the need of achieving the previous objectives. They will conform what would be defined as *non-structural measures*.

- **Within the field of water supply**

Essentially, specific strategies are needed to help demand meet supply and thus, if possible, achieving other economic, social and environmental benefits. Below, the following are presented:

- Fostering, from the different areas of the society (legislative, political, social or educational fields), a new way of approaching the consumption of water, necessary for the optimization of the water cycle as a whole.
- Completing the incorporation of water-efficient devices and appliances in all buildings.
- Using resources alternative to traditional water sources that may reduce the pressure on natural ecosystems. In order to do this, the optimization of potable water consumption is proposed fostering the *adaption of qualities* to the different uses (*fit for purpose*), by means of water reusing and/or recycling taking advantage of water flows that were previously wasted (rainwater, gray-water, etc.).
- Increasing the control and maintenance of water service in order to reduce the demand due to waste or leaks.
- Bettering the efficiency of water use in outdoor spaces, by means of strategies within water efficient gardening applicable to open and green spaces in our cities. This will allow an increase of the naturalization of those spaces.

On the basis of these strategic working guidelines, groups of specific measures that have been defined within the field of water service and are presented. Within them, the different technologies will be framed in a future document:

- Social and educational measures.
- Economic and legislative measures.
- Water saving devices.
- Active management of leaks.
- Alternative water resources.
- Water efficient gardening.

---

[5]In this sense, it is essential to clarify that it starts from the proposal of dividing into three areas—water supply, rainwater and waste water—basic in current trends of research and implementation, against the division into two—water supply and sewerage system—that characterized conventional management models.

- **Within the field of rainwater**

The improvement measures proposed in this field take as a starting point an approach for the decentralization of rainwater fluxes and expect to minimize water collection from natural resources—something that will clearly reduce the pressure on the corresponding ecosystems—as well as the amount of water to be transported through water networks and to be subsequently treated in sewage treatment plants or its possible polluting effect due to uncontrolled discharge—essentially in the case of unitary networks.

In this regard, five basic strategic guidelines are proposed:

- Fostering, from the different areas of society (legislative, political, social or educational fields), a new way to understand rainwater as a source of wealth and not as another waste product that has to be immediately disposed of.
- Taking advantage of the enormous potential of rainwater as an alternative water source by means of the principle of *adaption of qualities* to the different uses (*fit for purpose*) by collecting and using it at its source, near collection areas, when possible. The collection of water in underground cisterns (*rainwater harvesting*) is also a basic strategy.
- Allowing solutions that will gradually incorporate rainwater to natural watercourses and aquifers in the surroundings of rainwater collection areas and thus avoid not only their deterioration but also the overload of wastewater networks. In this case, an essential solution is the implementation of drainage systems that are more ecological as well as better integrated in the natural systems of the area. This is the *Sustainable Urban Drainage Systems (SUDS)* philosophy (Perales, Andrés-Doménech and Fernández 2008), whose implantation is an absolute necessity in our settlements.
- Implementing solutions to reduce the water runoff and avoid its accumulation within urban spaces in the shape of waterlogging or flooding in the area. Also, SUDS have that purpose as well as *water harvesting* in underground tanks.
- Implementing solutions with the aim of reducing pressure on the sewage networks in order to not overload the sewage treatment plants and to prevent the possibility of polluting their surroundings with inconvenient discharges if their peak water flow is surpassed. The solutions that allow collecting or infiltrating rainwater in the areas close to collecting areas attain this function. Again, the suitability of underground cisterns and the SUDS in that regard is insisted on. Lakes and ponds can also perfectly carry out this stabilizing and laminating of water flows.

On the basis of those strategic work guidelines, specific measures, in which the following technologies are framed, have been defined within this field and are exposed:

- Measures in the field of urban planning and design.
- Measures with regards to the maintenance of urban spaces.
- Social and educational measures.

- Water harvesting and direct storage of rainwater.
- Sustainable Urban Drainage Systems (SUDS).

- **In the field of waste water**

Within this field, the improvement guidelines raised, also from the point of view of decentralization, are focused towards the reduction of environmental pollution given their devastating consequences on natural ecosystems (land, aquifers, natural basins, etc.) and in the field of human health. Those strategic guidelines are listed below:

- Fostering, from the different areas of society (the legislative, political, social or educational fields), the importance of the separation of waste water from rainwater, as well as the differentiation between the different components within waste water during the first stages of the cycle (solids and fluids).
- Achieving that all urban agglomerations have their water treatment system including solutions adapted to their size and nature. In the case of urban agglomerations of less than 2000 equivalent inhabitants, *non-conventional technologies* for water treatment represent an adequate solution in many situations.
- Increasing the control and maintenance of water networks in order to reduce the illegal discharges and/or the broken ducts along the pipelines towards the water treatment plants as well as discharge spots that may pose any direct or diffuse pollution.
- Using of specific types of water (regenerated urban waste water, etc.)—that have been hitherto considered as waste—as alternative resources including their corresponding treatments that may allow to give a different use to the water already used by means of reuse and recycle systems.

On the basis of these strategic work guidelines, specific measures that have been defined in this field are exposed:

- Measures in the field of urban planning and design.
- Measures with regards to the maintenance of urban spaces.
- Social and educational measures.
- Water separating systems.
- Water treatment compact systems.
- Pre-treatment of waste water.
- Primary treatments.
- Secondary treatments.
- Tertiary treatments.

- **In the field of water-energy relationship**

It must not be forgotten that the water-energy relationship is generated in both directions. Specific measures in the field of energy may reduce the water demand in the same way as the saving in water consumption generates energy saving. From there, three strategic guidelines of intervention have been defined:

- Fostering, from the different areas of society (the legislative, political, social or educational fields), the importance of developing synergies between both resources.
- Fostering measures to attain water saving and subsequently energy saving (controlling water pressure in consumption spots, limiting the distances of domestic hot water consumption, etc.).
- Fostering measures with the aim of saving energy in the field of water management (water networks differentiated to pressure, use of renewable energy sources, etc.).

The specific measures that have been selected in order to achieve a saving and optimization of both resources are the following:

- Measures in the field of architectural and urban planning and design.
- Measures with regards to the maintenance of urban spaces.
- Social and educational measures.
- Water saving measures in the field of energy consumption.
- Energy saving measures in the field of water consumption.

3.2 Selection of Relevant Parameters for the Characterization of Technologies

In order to facilitate the knowledge of and comparison between technologies, it has been considered essential to make a selection of the characteristics or parameters that are of most interest for the technical actors who will be final beneficiaries of the information. Since the beginning, those characteristics have also been clustered in a series of fields of information listed as following:

- Denomination of the technology.
- Description of the technology.
- Requirements, constraints and criteria of implementation of the technology.
- Predictable results of its implementation at an environmental and health level.
- Costs of the technology.
- Recommendations for its inclusion in projects.
- Other specific sources of information.

The first fields allow the technical agents to get in touch with the technology and to know it offering as well data that may be fundamental at the time of taking the decision of electing it or not. Once the decision to choose the specific technology is taken, the last ones are targeted towards, help to its implementation and/or allow elaborating on it, if necessary.

3.3 Definition of Technologies Systematization Criteria

The criteria that have been considered of most interest are the following:

- ***Area* to which the technology belongs to within the water cycle: water supply, management of rainwater, management of waste water or management of water and energy**

Since the beginning, it has been considered essential to present this systematization according to the areas of interest—with regards to the area or stage of the Water Urban Cycle where the technology is placed—following the tendencies shown in the most recent research works and projects carried out in this matter, such as the SWITCH Project (2006–2011) that divides the strategies and technologies within 3 fields of work in accordance with the different characteristics and quality of the fluxes with which they are working: water supply, management of rainwater and management of waste water. Besides, given its importance, the field of technologies that relate or associate water and energy has been added. Therefore, the classification of the technologies is structured according to the above mentioned division.

- **Type of *measures of intervention*: *non-structural* measures *versus structural* measures**

At the same time, it is appropriate to differentiate these clusters of measures according to their character, distinguishing if they are "prevention measures that do not require developing an infrastructure to be carried out, not being submitted to an specific location and that are closer to planning or even raise awareness of specific strategic work guidelines" or if they are "construction solutions that imply a material implementation". In this line, in order to classify the various measures it has been decided to use the following terms: *non-structural* measures and *structural* measures. This classification has been taken from scientific literature where they are being commonly used. The specific measures were listed in Sect. 3.1.2.

3.4 Proposal for a Classification Scheme of the Selected Information

Starting from the referred criteria, the classification of the selected information will be structured in a future document according to the following levels:

- Level 1: Areas or *fields* of the Water Urban Cycle.
- Level 2: *Measures of intervention* or strategic guidelines for specific actions.
- Level 3: Specific *technologies*.

The reasons for this structuration into levels are mainly two:

- Facilitating the location of the technologies according to the area that they belong to as well as to their main aim in a quickly and simple way.
- Distributing the selected information about the technologies according to their levels in order to avoid futile repetition of information.

From this structure, the proposed distribution of information on the various technologies, with regards to the coincidence or repetition of the approaches or methods, is the following:

- **Level 1: Areas or *fields* of the Urban Water Cycle.**

– *Specific legislation for implementation*: It is analyzed within a global level since most of the existing legislation affects complete areas of the Urban Water Cycle, with few exceptions that shall be commented in the relevant paragraph. In this case, what evidently has the greatest impact to each and every area has been selected. It will be extended with the information related to the water-energy relationship.

– *Listing of possible measures*: Starting from the necessary strategies in each area, a series of specific measures to be taken is defined. Besides, they may be of different character (*non-structural* measures or *structural* measures). These shall be described in Level 2.

- **Level 2: *Intervention measures* or strategic lines of action.**

In this case, all measures will be listed—*non-structural* measures as well as *structural* ones. But, while in the first case, they will only be explained in a general way, in the second, and given that it is within these *structural* measures where technologies are framed, specific paragraphs shall be detailed with common information to all the technologies included within the measure in question:

– *Definition and common aims of technologies*: All the technologies framed in an analyzed measure will be treated as a whole.

– *Requirements, constraints and implementation criteria of the set of technologies*: In this case, the paragraph is centered on the possibilities and limitations of the use of the set of technologies framed within a single measure.

– *Predictable results of their implementation at an environmental and health level*: Their benefits would be analyzed both for the environment (starting from a specific analysis, where possible, of the predictable water and energy savings, without forgetting the necessary material resources and the questions referred to waste production) and for the health of the users. The specific repercussions on the fields of architecture and urban planning (spatial, construction, etc.) merited their own particular approach given their importance.

– *General recommendations about design and calculations and other factors to be considered.*

- *Implementation technologies*: At this point are listed all that are framed within the measure in question that, finally, shall be detailed in the technological datasheets in Level 3.

- **Level 3: Specific Technologies.**

In this level, information will be gathered in the form of a *datasheet* following the referred model within the corresponding chapter. The explanation of each characteristic was also gathered in a previous chapter.

In short, and how it is represented below, 4 areas or *fields* of work have been selected to work on. In each of them a series of specific strategies is proposed in order to achieve a real change on management models. Subsequently, those strategic guidelines have been materialized in a series of 25 *measures*, 14 of those are *structural* measures in which 46 sustainable *technologies* will be permanently framed that it is the subject of another specific study.

WATER SUPPLY

- **NON-ESTRUCTURAL MEASURES**
  NEM 1. Social and educational measures
  NEM 2. Economic and legislative measures
- **ESTRUCTURAL MEASURES**
  EM 1. Water saving devices
  EM 2. Active management of leaks
  EM 3. Alternative water resources
  EM 4. Water efficient gardening

RAINWATER

- **NON-ESTRUCTURAL MEASURES**
  NEM 1. Measures in the field of urban planning and design
  NEM 2. Measures with regards to the maintenance of urban spaces
  NEM 3. Social and educational measures
- **ESTRUCTURAL MEASURES**
  EM 1. Water harvesting and direct storage of rainwater
  EM 2. *Sustainable Urban Drainage Systems* (SUDS)

WASTE-WATER

- **NON-ESTRUCTURAL MEASURES**
  NEM 1. Measures in the field of urban planning and design
  NEM 2. Measures with regards to the maintenance of urban spaces
  NEM 3. Social and educational measures
- **ESTRUCTURAL MEASURES**
  EM 1. Water separating systems
  EM 2. Water treatment compact systems
  EM 3. Pre-treatment of waste water
  EM 4. Primary treatments
  EM 5. Secondary treatments
  EM 6. Tertiary treatments

WATER-ENERGY RELATIONSHIP

- **NON-ESTRUCTURAL MEASURES**
  NEM 1. Measures in the field of architectural and urban planning and design
  NEM 2. Measures with regards to the maintenance of urban spaces
  NEM 3. Social and educational measures
- **ESTRUCTURAL MEASURES**
  EM 1. Water saving measures in the field of energy consumption
  EM 2. Energy saving measures in the field of water consumption

4 Conclusions

With regards to the criteria for the systematization of technologies, from a methodological point of view, the following is confirmed:

- The need to cluster the technologies by areas or *fields* of the Urban Water Cycle as well as in sets of *Intervention measures* or implementation strategic guidelines with common characteristics in order to obtain an easier location and understanding.
- The need to distribute, in each of the *Technological Datasheets*, the characteristics according to areas, beginning with those that allow a knowledge of technologies and help to make the decision of their possible use and ending with those that form more detailed recommendations, in case their implementation is decided or where it is desired to deepen their knowledge.

The study aims to provide the following contributions:

- A comprehensive study of the criteria to facilitate a listing, organization and systematization of the information related to technologies of the Urban Water Cycle currently available. It should allow, as a working tool, with a direct, practical and useful application for the technical agents involved in urban planning.
- With the document resulting from this systematization it is intended to go a step further in order to facilitate, in the coming years, the taking of decisions and the real implementation of these technologies in urban and architectural projects, from an up-to-date view and an overall picture of the available technologies and from the deep knowledge of each of them and their repercussions.

Acknowledgements To Doctors Jaime Navarro, Leandro del Moral and Laura Pozo from the University of Seville for trusting me, as well as to Juan José Sendra for staying with me along this long university journey together with the rest of my colleagues. Besides, to all the Aqua-Riba Project team for their great work and generosity. Especially to Ángela Lara for her many contributions.

References

Aqua-Riba (2013–2015) Sustainable urban water cycle management systems in the integral regeneration plans for districts in Andalusia (Aqua-Riba), RDI Project, Junta de Andalucía (coord.), EU-FEDER Programme (Andalusia-Spain) 2007–2013

Butler D et al (2010) WaND_Guidance on water cycle management for new developments. CIRIA. gow.epsrc.ac.uk/NGBOViewGrant.aspx?GrantRef=GR/S18373/01

Carrasco G (2009) Proyecto Ecocity. Manual para el diseño de Ecociudades en Europa, Revista INVI, nº 24, 197–200

Perales S, Andrés-Doménech I, Fernández A (2008) Los sistemas urbanos de drenaje sostenible en la hidrología urbana, en *IX Simposio de Hidrogeología*, Asociación Española de Hidrogeólogos, Elche (Alicante-Spain)

Prieto-Thomas A (2015) Systematization of technologies for a sustainable urban water cycle management. Spatial, constructional and economic repercussions in building and urban planning. PhD thesis directed by Navarro J, Del Moral L, Pozo-Morales L. University of Seville (Spain)

Prieto-Thomas A, Lara A (2014) Integration of sustainable urban drainage systems in the regeneration of urban spaces, Comunicación en Congreso "Greencities and Sostenibilidad.

Inteligencia aplicada a la Sostenibilidad Urbana", Área temática "Actuaciones Sostenibles del Espacio Urbano", Málaga (Spain) 2014
SWITCH (2006–2011) Managing water for the city of the future. Project under the 6th Framework Programme of the EU. www.switchurbanwater.eu. Accessed Dec 2013
SymbioCity (2008) *Take a deeper look*. The holistic approach to sustainable urban development, Swedish Trade Council Report, 24 pp, Stockholm. Accessed Aug 2015
SymbioCity (2015) Sustainability in Sweden, www.symbiocity.org/. Accessed Aug 2015

Study of Fine Mortar Powder from Different Waste Sources for Recycled Concrete Production

Eduardo Aguirre-Maldonado and Francisco Hernández-Olivares

1 Introduction

In recent years, has been evidenced different supporting sustainable development of the concrete industry. These advances based in initiatives apply in each phases of the life cycle of concrete, seeking to build a sustainable system that can be apply from different industries and reduce environmental impacts and improve the performance of concrete. One of the most important initiatives, is search for new or alternative materials to replace concrete components, without set aside ecology or concrete quality. However, even being an initiative with projection and continuous development, results in a gradual path. While sustainable options to replace fully concrete technology and materials, are generate, adopt and popularize, actually there are huge volumes of concrete waste material and even is estimated that present buildings at the end of its useful life, determines significant amounts of waste in the future.

In the context of construction and demolition waste (CDW) related to concrete, studies have focused to learn how to use all concrete waste components, and in this field has only normalize (with limits), the recovery of coarse aggregate for reentering the life cycle of concrete. An example of this, are the recommendations for the use of recycled concrete (use of recycled aggregates) showing the Annex 15 of the Spanish Instruction Structural Concrete (EHE-08 2008). Moreover, by recycling coarse aggregate, the remaining fine paste materials such as mortar adhering to the

E. Aguirre-Maldonado (✉)
Departamento de Arquitectura, Universidad Tecnica Particular de Loja, calle Marcelino Champagnat, 1101608 Loja, Ecuador
e-mail: ebaguirre@utpl.edu.ec

F. Hernández-Olivares
Department of Building Construction and Architectural Technology, Superior Technical School of Architecture, Technical University of Madrid, Av. Juan de herrera 4, 28040 Madrid, Spain
e-mail: f.hernandez@upm.es

P. Mercader-Moyano (ed.), *Sustainable Development and Renovation in Architecture, Urbanism and Engineering*, DOI 10.1007/978-3-319-51442-0_21

arid or RMF are a determining factor that influences in resistance of new concrete manufactured with recycled aggregates (Pedro et al. 2014), and due to characteristics of RMF, is difficult to search better options to exploits this material.

1.1 Recycling Fine Powder from Concrete Waste

From previous research of high temperatures influence on cement pastes (Zhang and Ye 2011; Castellote et al. 2004), also could be determined what happens with the burned cement pastes when subjected again to hydration. This has allowed notice the presence of similar setting characteristics as new cement setting, and somehow, recover the ability to display hydration products (Farage et al. 2003; HaGa et al. 2002; Xinwei et al. 2010).

Recent studies have confirmed that binder capacity of BRMF depends on the burning temperature and also has high pH value and water reactivity (Shui et al. 2009). Based on these results, application of BRMF as cement additive (Rui et al. 2013), alkaline activator (Shui et al. 2011) or even as partial replacement of cement (Xinwei et al. 2010) was evaluated, with good results. These advances have also allowed to rate the carbon emissions reduction in the manufacturing process of alternative cements, when CaO is obtained by dissolving portlandite and not through decarboxylation (Kwon et al. 2015). Despite this, BRMF characteristics has presented limitations and have unfavorable results when used as a binder directly, due to percentage of useful binder material in recycled paste, highly required water of standard consistency, low density, high surface area and high porosity (Shui et al. 2008, 2009). Moreover, the difficulty and the need to generate studies to improve the separation of fine aggregates of the cement components (Kwon et al. 2015) is recognized. Based on these advances, studies of use of waste concrete offer great opportunities, due to, assessment of this material allows the recovery of construction waste as well reducing waste concrete deposits and obtaining raw material for making new concrete. This is more important, when concrete is considered the most used and popular construction material nowadays, and in consequence of this, the amounts of material residues are plentiful and can be located throughout the modern world. In fact it is estimated that of all construction and demolition waste, from 20 to 75% are waste concrete (García 2012; Tam and Tam 2008). Also depending on the location, this percentage of concrete waste can increase considering that several countries, buildings uses concrete is most constructive elements and there are no efficient processes for handling waste concrete, for example, in Ecuador near the 76% of households use the concrete structures and masonry mortar systems (INEC 2010).

One of the different points that will help improve the knowledge to assess the RMF is the situation of ordinary waste, found in normal deposit places, different from those obtained in laboratory. It is recognized the handling difficulties, nature of concrete waste, and lack of technical certification for this product, due to material is deposited together with other wastes. Nevertheless, it is important to draw conclusions about which kind of concrete waste has better conditions for recycling and if there are significant differences between several sources and origins of the RMF related to its compressive strength, which it is one of the main characteristics of the concrete.

This research is focused on the comparison of experimental results of compressive strength determined for mortar specimens made based on the combination of BRMF and OPC, this will provide add one experience point for the selection of waste used in recycling cement raw material and impulse the manufacture of recycled concrete.

2 Experimental Research

2.1 *Materials*

The entire experiment was performed using Ordinary Portland Cement (OPC) GU Type I (31 Mpa), general purpose cement for construction created under the standard NTE INEN 2380: 2011 (equivalent to ASTM C 1157/C 1157M-11), the chemical properties are listed in Table 1. Three samples per each of the nine families of BRMF additions obtained from different sites are used (see Table 2), as well as fine natural aggregate (river sand) and distilled water. No additives were used.

2.1.1 Collection and Preparation of RMF

Concrete and mortar residues from architectural constructions: mortar coating, adhesive mortars and concrete samples, were collected (see Fig. 1). The description

Table 1 Chemical analysis of Cement GU type I and different sources of RMF dehydrated at 800 °C

% By mass	OPC	RMF dehydrated at 800 °C								
		M011	M012	M013	M021	M022	M023	M031	M032	M033
CaO	52.30	15.10	15.90	16.70	16.50	19.20	13.50	23.60	26.80	22.30
SiO_2	17.60	62.30	62.00	62.00	60.90	59.60	64.70	52.00	49.70	53.20
Al_2O_3	5.84	11.20	11.50	11.70	12.70	11.40	11.50	10.80	8.94	11.8
Fe_2O_3	4.15	3.10	3.32	4.09	4.47	3.97	4.07	3.89	3.94	4.13
K_2O	1.26	1.94	2.11	2.34	2.19	2.30	2.15	2.16	1.27	2.19
SO_3	1.75	0.65	0.65	0.68	0.79	0.70	0.53	0.85	1.49	0.80
MgO	5.69	3.86	2.28	0.22	>0.10	0.33	>0.10	3.95	>0.10	2.30
TiO_2	0.36	0.37	0.40	0.42	0.61	0.46	0.61	0.44	0.41	0.62
P_2O_5	1.78	0.90	0.92	1.04	1.03	1.12	0.89	1.26	1.85	1.24
LOI	2.70	7.30	6.60	9.50	11.7	9.00	8.60	27.70	24.50	22.00

Table 2 RMF types and approximate ages used in experimental study

Family RMF	Concrete waste origin	Approx. age (years)	Source	General composition
M011	Reinforced concrete	20	During demolition	Coarse and fine aggregates, OPC
M012	Mass Concrete	15	Waste construction fill deposit	Coarse and fine aggregates, OPC
M013	Ordinary concrete	7	In waste tank	Coarse and fine aggregates, OPC
M021	Cement mortar outer wall coating	25	In building area for deterioration of wall	Fine aggregates, OPC
M022	Cement mortar interior wall coating cement mortar	15	During demolition	Fine aggregates, OPC
M023	Cement mortar for brick adhering	22	Waste construction fill deposit	Fine aggregates, OPC
M031	Adhesive mortar for tiling	7	In waste tank	Fine aggregates, Portland cement, additives[a]
M032	Adhesive mortar for tiling	7	During demolition	Fine aggregates, Portland cement, additives[a]
M033	Adhesive mortar for tiling	10	Waste fill deposit	Fine aggregates, Portland cement, additives[a]

[a]According to general description of brands sold in the context of study, however manufacturers do not specify types of additive

Fig. 1 Waste construction samples: **a** concrete waste, **b** mortar waste, **c** adhesive mortar waste

of each group, origin, source description and approximate age is detailed in Table 2. The material has been obtained during the demolition process, from deterioration of buildings and of CDW deposit places, and by an inquiry into the building origin, was possible know the approximate age of each type of waste.

Samples has verified to avoid contaminated material with chemical or radioactive agents. Based on the analysis of the development line of concrete in the place of study, is possible mentioned that the cement used in all cases is the ordinary Portland cement. In reason that kind of binder is the only one that has been marketed in the city of study and is used for civil building, also in the first groups of families, concrete and mortar is mixing during the building, unlike third group where adhesive mortar is premixed in factory.

Each sample is dried RMF, and then proceed with two-stage crushing and one of pulverization. First crushing stage is performed manually releasing the coarse aggregate from concrete samples, subsequently the mechanical grinding was performed and finally use a disk mill for obtain a fine powder from the samples less than 70 μm. A subsequent step allowed get a finer material through desliming of samples, separating about 20% of sandy compound in samples. Finally, powder fineness can reach less 40 μm.

Dehydration processes were followed similarly to the methodology used in previous studies (Shui et al. 2011; Zonghui et al. 2012), applying a temperature of 800 °C during two hours in a range of 5 °C/min. This temperature was considered because after the 800 °C exist a tendency to decrease the compressive strength of BRMF in each curing phases (Shui et al. 2009), and this could be attributed to the transformation of partially dehydrated hydration phases over 800 °C (Okada et al. 1994). After the burning time (see Fig. 2), BRMF is rapidly cooled and placed in airtight containers stored in a dry environment. Chemical composition for each sample has obtained by XRF and listed in Table 1.

2.2 *Preparation of Specimens and Determination of Compressive Strength*

Determining the compressive strength of mortar specimens will be perform according to the Ecuadorian standard NTE INEN 488: 2009 (equivalent to ASTM C109/C109M-16), where a minimum value of 28 MPa is set at 28 days. Mortar

Fig. 2 RMF dehydrated at 800 °C samples: **a** concrete waste family, **b** mortar waste family, **c** adhesive mortar waste

specimens combining OPC and BRMF are manufacture-using BRMF at 30% of all binder mass (see Table 3). This percentage is established taking into account previous research (Oksri-Nelfia et al. 2016), because the collected samples are not separated completely from the original fine aggregate, making it difficult to harden when used in a percentage higher than 30% (Kwon et al. 2015). The water/cement +BRMF radius is 0.35. Mortar pastes were placed in molds 50 × 50 × 50 mm, compacted by vibration. After removing the mold, the specimens were cured at room temperature (21 °C) for 7 days. Finally, after 28 days compressive strength was tested, the results are show in Fig. 3.

Table 3 Mixing ratios of raw materials used in specimens and theoretical Ca/Si ratio

Sample family	%, BRFM	%, OPC	a/c	T-Ca/Si
M011	30	70	0.35	1.42
M012	30	70	0.35	1.43
M013	30	70	0.35	1.44
M021	30	70	0.35	1.46
M022	30	70	0.35	1.50
M023	30	70	0.35	1.37
M031	30	70	0.35	1.68
M032	30	70	0.35	1.76
M033	30	70	0.35	1.64

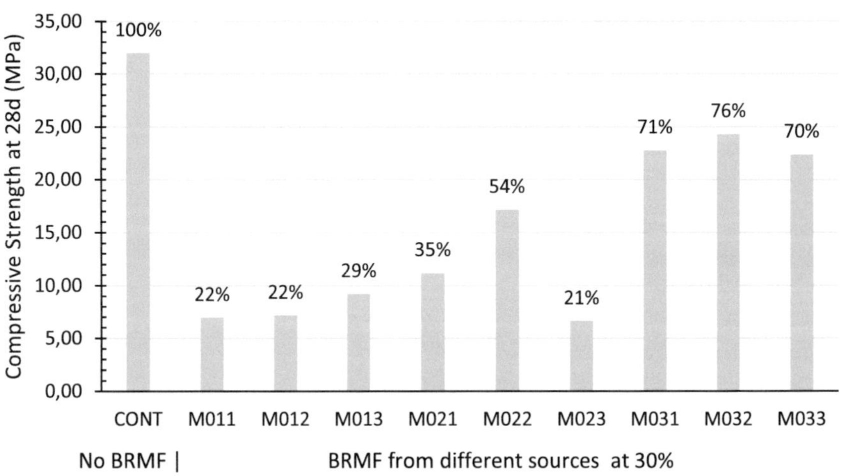

Fig. 3 Compressive strength of mortar specimens with addition of BRMF

2.3 Discussion of Experimental Results

The results tested at 28 days, show that in all cases the combination of cement and BRMF not improves or preserves the qualities of compressive strength of specimens. Compressive strength is less than specimens without BRMF and under the requirements in the standards. Although these data were predicted, because of BRMF was used directly in combination with new cement and wasn't separated most fine aggregate from BRMF, is possible appreciate certain particular behaviors in experimental results.

To discuss the results, the concept of theoretical Ca/Si ratio (T-Ca/Si) is used, apply in a previous study (Shui et al. 2011). In T-Ca/Si, CaO and SiO values are relates for cementitious material design, in present case concept is set to the variables and is expressed in Eq. 1

$$T = \frac{(md\ X\ d\mathrm{CaO} + mp\ X\ p\mathrm{CaO})/\mathrm{MCaO}}{(md\ X\ d\mathrm{SiO_2} + mp\ X\ p\mathrm{SiO_2})/\mathrm{MSiO_2}} \quad (1)$$

where T is the value of T-Ca/Si; md is the content BRMF; mp is the content of OPC; dCaO is the mass percentage of CaO in BRMF; pCaO is the percentage of CaO in OPC by mass; $dSiO_2$ is the SiO_2 percentage in BRMF by mass; $pSiO_2$ is the SiO_2 percentage in OPC by mass; and MCaO and $MSiO_2$ are the molar mass of CaO and SiO_2, respectively. The values of dCaO, pCaO; $dSiO_2$ y $pSiO_2$, are obtained from de la Table 1, for MCa and $MSiO_2$ is used 56.0774 and 60.08 g/mol, respectively. In this study, the value of T is calculated by a use of 30% of BRMF in all cases.

The values of T-Ca/Si of all mixtures obtained is indicated in Table 3. Specimens of the third group of samples made based on BRFM from mortars adhesive residues, presents results of compressive strength reaching 72% of the specimens that do not use BRFM, in other cases, only achieved a range from 21 to 54% of normal compressive strength. The results shows an increasing trend between the T-Ca/Si and compressive strength (see Fig. 4).

In other hand, regarding the assessment of waste from different sources for manufacturing recycled concrete, the results of the third family of samples in study, show highest values of T-Ca/Si relative to other groups, this can be attributed to type of concrete waste, for this case residues from adhesive mortar used in tiling. Prefabricated products undoubtedly have better processes to verify the quality of the material in addition to dosing and mixing the components are supporting by laboratory tested values, also they have advantage of having a higher content of raw materials rich in CaO based in own material requirements. For all other samples obtained in this study (concrete and mortar waste), in most the results tested differ slightly, this can be attributed in part to the manual preparation of concrete and mortar, because the mixtures dosage of materials is not as accurate. This last condition is not necessarily negative for the recycling of concrete and mortar waste, since due to having ratio of Ca/Si very similar could be considered within the same

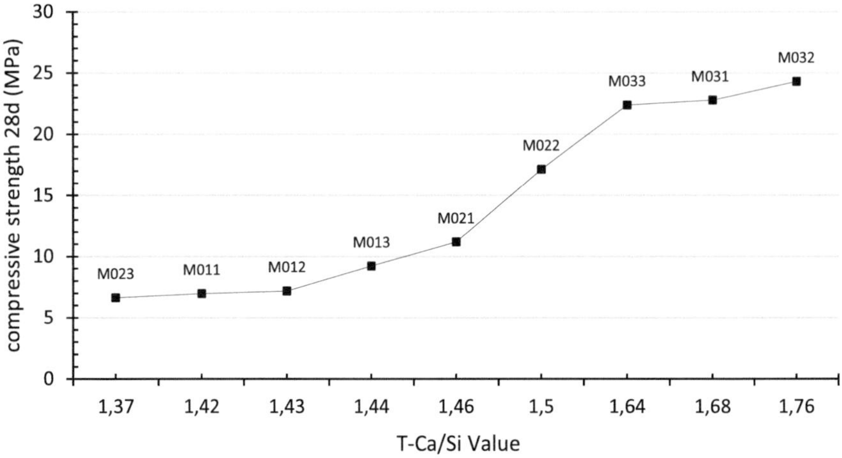

Fig. 4 Compressive strength in relation to value of T-Ca/Si

group and be handled as a single variety of waste concrete. Nevertheless, for the case of special mortars like adhesive mortars must be taken into account in selective demolition processes, because they are more easily able to respond positively to burned processes, although direct use is not recommended and like other types of waste concrete, should follow a mix design in combination with other components of cementitious properties.

3 Conclusions

This research has studied as the BRFM from different sources are combined with a normal Portland cement to observe its influence on the compressive strength of mortar specimens. The experimental study indicates the following conclusions:

1. Using directly BRFM in combination with ordinary Portland cement, in all case studies reduces the compressive strength of the specimens even when is apply in 30%, so it is not recommended for use directly in the mixture to make mortar or concrete recycling. Research continues to analyze the M031, M032 and M033 samples, whose results were significant related to compressive stresses, changing BRFM replacing ratio, from 30 to 20% in new specimens, the results of this new experimentation will be displayed in a new investigation.
2. Specimens made with BRFM from waste concrete and mortar, reach compressive strengths in range of 21–54%, but most remain in an average of 23%, is noticed that differences in compressive strengths have related directly to the CaO content in the waste.

3. In study was observed that the thermal processes applied in mortar and concrete wastes are related to the rate of T-Ca/Si, because the greater this ratio the compressive strength of the specimens is less affected.
4. Adhesive mortars wastes, for its higher content of CaO, have better qualities to be recycled in the concrete or mortar manufacturing recycled.
5. BRFM from concrete and mortar waste that were made manually, show similar conditions, therefore can be classified within a same group and being subject to recycling processes.
6. The heat treatment processes on RFM, should be adjusted considering the original Ca/si ratio. Samples with reduced values of this relationship could significantly impede recycling of this material for the manufacture of new mortar and concrete.
7. The process described in this communication shows experimentation with small samples in the laboratory. Once confirmed the results and value other alternatives ongoing experimentation, will proceed with a full cost analysis to make a comparison with other concrete recycling alternatives.

Acknowledgements The authors acknowledge the support and management to carry out this work, to the Ph.D. Sofia González Sanz, who served in the post of Department of Architecture and Art Director at the Universidad Tecnica Particular de Loja (UTPL). We are also grateful to the UTPL and its representatives for the support and provision of laboratories and equipment necessary for the conduct of the experimental phase of this work.

References

Castellote M, Alonso C, Andrade C, Turrillas X, Campo J (2004) Composition and microstructural changes of cement pastes upon heating, as studied by neutron diffraction. Cem Concr Res 34 (9):1633–1644

EHE-08 (2008) Instrucción para el proyecto y la ejecución del hormigón estructural. Anejo 15: hormigón reciclado. Ministerio de Fomento. España

Farage MR, Sercombe J, Gallé C (2003) Rehydration and microstructure of cement paste after heating at temperatures up to 300 °C. Cem Concr Res 33:1047–1056

García CT (2012) Hormigón reciclado de aplicación estructural, durabilidad en ambiente marino y comportamiento a fatiga. Dissertation, Universidad de Cantabria

HaGa K, Shibata M, Hironag M, Tanaka S, Nagasaki S (2002) Silicate anion atructural change in calcium silicate hydrate gel on dissolution of hydrated cement. J Nucl Sci Technol 39(5):540–547

INEC (2010) Instituto Nacional de Estadísticas y Censos. Censo de población y vivienda. Quito, Ecuador

Kwon E, Ahn J, Cho B, Park D (2015) A study on development of recycled cement made from waste cementitious powder. Constr Build Mater 83:174–180

NTE INEN 488 (2009) Cemento hidráulico. Determinación de la resistencia a la compresión de morteros en cubos de 50 mm de arista. Instituto Ecuatoriano de Normalización. Quito, Ecuador

NTE INEN 2380 (2011) Cementos hidráulicos. Requisitos de desempeño para cementos hidráulicos. Instituto Ecuatoriano de Normalización. Quito, Ecuador

Okada Y, Sasaki K, Zhong B, Ishida H, Mitsuda T (1994) Formation processes of b-C2 S by the decomposition of hydrothermally prepared C–S–H with Ca (OH) 2. J Am Ceram Soc 77 (5):1319–1323

Oksri-Nelfia L, Mahieux P, Amiri O, Turcry P, Lux J (2016) Reuse of recycled crushed concrete fines as mineral addition in cementitious materials. Mater Struct 49:3239–3251

Pedro D, de Brito J, Evangelista L (2014) Influence of the use of recycled concrete aggregates from different sources on structural concrete. Constr Build Mater 71:141–151

Rui Y, Zhonghe S, Jun D (2013) Using dehydrated cement paste as new type of cement additive. Mater J 110(4):395–402

Shui Z, Yu R, Dong J (2011) Activation of fly ash with dehydrated cement paste. ACI Mater J 108 (2):204–208

Shui Z, Xuan D, Wan H, Cao B (2008) Rehydration reactivity of recycled mortar from concrete waste experienced to thermal treatment. Constr Build Mater 22(8):1723–1729

Shui Z, Xuan D, Chen W, Yu R, Zhang R (2009) Cementitious characteristics of hydrated cement paste subjected to various dehydration temperatures. Constr Build Mater 23(1):531–537

Tam VW, Tam C (2008) Diversifying two-stage mixing approach (TSMA) for recycled aggregate concrete: TSMAs and TSMAsc. Constr Build Mater 22:2068–2077

Xinwei M, Zhaoxiang H, Xueying L (2010) Reactivity of dehydrated cement paste from waste concrete subject to heat treatment. In: Second international conference on sustainable construction materials and technologies (June 2010). Ancona-Italia: Coventry University and the University of Wisconsin Milwaukee

Zhang Q, Ye G (2011) Microstructure analysis of heated Portland cement paste. Proc Eng 14:830–836

Zonghui Z, Jinbang W, Decheng Z, Xuexu Z, Xiuzhi Z (2012) Study on preparing process for a recycled concrete. Proc Eng 27:357–364

Concrete Sustainable Light and of High Performance

Isabel Miñano, Francisco J. Benito, Carlos J. Parra and Pilar Hidalgo

1 Introduction

High consumption of natural resources, the production of large amount of industrial waste and the pollution of the environment require new solutions for sustainable development. The construction sector brings negative environmental impacts, such as excessive consumption of non-renewable natural resources for the extraction of raw materials (overfishing). Therefore the concrete industry, increasingly aware with the environmental aspects, tends to reduce the consumption of natural resources with the reuse of wastes, which minimize the environmental effects of the sector.

The increasing degradation of the environment and the concern of the industry to comply with the laws of waste make some of them of industrial origin, such as fly ash, increasingly used to replace part of the cement in concrete production. The achievement of all phases of this research has sought to deepen the knowledge of concrete of high strength (HP) and light (ECOHUL) for structural use, while more respectful with the environment.

I. Miñano (✉)
Department of Architecture and Building Technology,
Technical University of Cartagena, Group of Advanced Science
and Technology Building, Murcia, Spain
e-mail: ctac@upct.es

F.J. Benito · C.J. Parra
Department of Architecture and Building Technology,
Technical University of Cartagena, Murcia, Spain

P. Hidalgo
Department of Quality and Environment,
Cementos La Cruz, S.L, Murcia, Spain

P. Mercader-Moyano (ed.), *Sustainable Development and Renovation in Architecture, Urbanism and Engineering*, DOI 10.1007/978-3-319-51442-0_22

2 Characterization of the Materials

For this study is employed cement Portland CEM I 52.5 R, manufactured in the plants of Cementos La Cruz, S.L. La composition chemical of them ashes fly and the sludge of paper, employees as substitute of the cement were obtained by fluorescence of rays X.

The essential features that are looked up in the research is to use paper mill sludge, provided by the company Compost recyclable S.L, as active admixture to the cement.

The sludge is received with a 50% of humidity due to the process industrial by which is obtained as residue recycled from the paper. It was necessary to previously dried for 24 h in an oven at 105 °C. They then calcined in muffle at 750 °C for approximately 120 min, in order to remove organic matter and promote the Crystallographic transformation of mineralogical compounds that have. With the roasting was observed a loss in weight approximate 32.18%, calcium carbonate and other stable compound.

To investigate the effect of base paper sludge and calcined analyzed the resistance to compression (UNE-EN 196-1) standard mortar specimens (4 × 4 × 16 cm) with partial replacements of 1.33 and 2.66% within 90 days of age. The normalized composition of these mortars includes: natural siliceous aggregate (1350 g), water (225 g) and reference cement CEM I 52, 5R (450 g), as a reference, (REF 0%) mortar. Raised partial replacements of 1.33 and 2.66% of cement by sludge treatment (LP without treatment) and calcined sludge (calcined LP).

The results (Fig. 1) reveal that arrays of cement prepared with percentages of substitution of slurry of calcined paper equal to 1.33% in relation to the cement, increase compressive strength with respect to the reference mortar, being this greater than that experienced with mortars with substitutions of 2.66%. Sludge without heat treatment

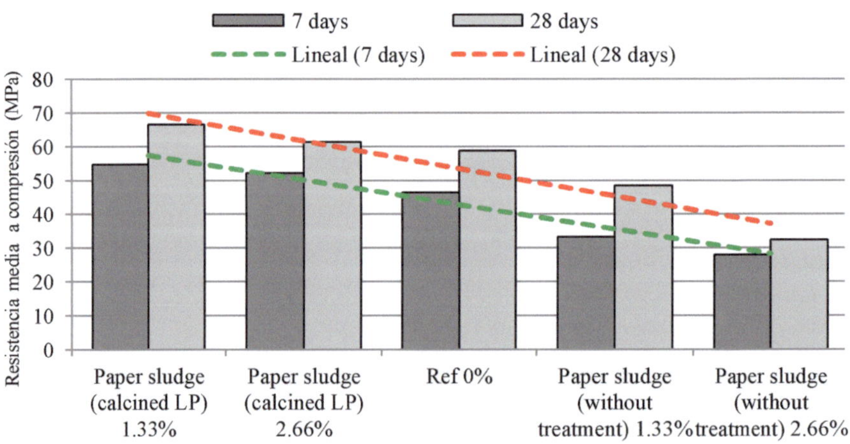

Fig. 1 Average compressive strength of mortars with charred paper sludge and untreated

tends to be lower than the control mortar resistance to compression. A decrease in the size of particular (Fig. 2) and a higher proportion of the SiO_2 vitreous (Table 1) increases the Pozzolanic activity of the sludge, due to greater contact between cement particles and paper sludge, what would justify partly the result.

To determine the distribution of sizes of particles very small is obtained the particle size using the Analyzer of particles Mastersizer 2000 of Malvern Instruments. Figure 2 shows a comparison between the granulometric curves of cement and additions used in this study. Can verse as the sludge of paper of split (represented in colour red), is the material more thick, with the peak of maximum concentration in 300 μm. With the heat treatment, the peak of maximum concentration descends to 4 μm, very inferior to the of the cement (11 μm).

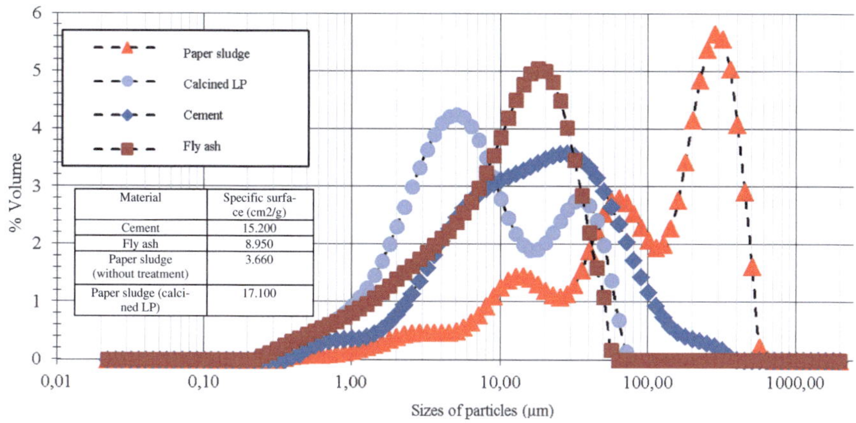

Fig. 2 Laser particle size analysis and specific surface of the cementitious particles

Table 1 Composition chemical of the components

Oxides	Cement	Fly ash	Paper sludge (calcined LP)	Paper sludge (without treatment)
Na_2O	0.372	1.24	2.84	1
MgO	2.52	1.98	2.82	0.698
Al_2O_3	4.09	23.47	4.77	1.56
SiO_2	16.89	48.38	3.68	1.19
P_2O_5	0.176	0.897	4.66	1.71
SO_3	4.061	0.958	2.09	1.36
Cl	0.142	–	0.17	0.17
K_2O	1.33	2.17	0.544	0.292
CaO	64.74	8.18	61.44	23.14
TiO_2	0.259	1.27	0.294	0.122
Fe_2O_3	3.507	7.502	0.624	0.151
SrO	0.104	0.2285	0.0434	0.050
BaO	0.053	0.278	–	0.014

In the investigation of Lorca (2014) indicates that when particles present a larger specific surface reactions occur with greater speed, since the energy required to activate the reaction is less. In general, minor used particle sizes are more reactive to have one higher specific surface, as in the case of paper sludge calcined. That is, the greater reaction surface in a particle causes hydration is faster and more effective.

Mineralogical analyses were made using diffraction of x-rays (DRX). For the operation of the equipment radiation CuK θ (1542 A) was used in the following conditions for all analyses: range of sweep of 10°–80° 2θ, for a passage of 0.03° 2θ and a time of incidence of 3 s per step. The identification of mineral phases present has been carried out by means of the JCPDS database (Joint Committee on Powder Diffraction Standards). The structure Atomic is a feature essential to concrete the properties cementitious of the material.

Cement replacement materials, as it is the case of fly ash and sludge of paper, requires a structural State amorphous—thermodynamically high internal energy and unstable and highly reactive chemically-(Hewlett 1998).

Fly ash is composed predominantly of mineral matter (65%) in the form of primarily vitreous particles with a small proportion of crystalline phase (35%). While the cement is composed predominantly of mineral matter basically Crystal (between 70 and 90%) (Table 2).

Table 2 Analysis structure of the part Crystal (% (w/w))

Phase	Cement	Fly ash	Paper sludge (calcined LP)	Paper sludge (without treatment)
Silicate	**79**	**9**	0	
Calcium silicate	79	9		
Oxides	**11**	**87**	**56**	
Quartz		32		
Mullite		49		
Corundum		6		
Oxide Ca, Mn, Fe and Al	9			
Ca and Al oxide	2			
Perovskite			29	
Calcium oxide			27	
Phosphates			**28**	
Hydroxylapatite			28	
Carbonates		4		**100**
Calcite				**100**
Dolomite		4		
Others	**10**		16	
Anhydrite	10			
Plasters	<1			
Portlandite			16	
Crystallinity (%)	**78**	**35**	**69**	**71**
Sum (%)	**100**	**100**	**100**	**100**

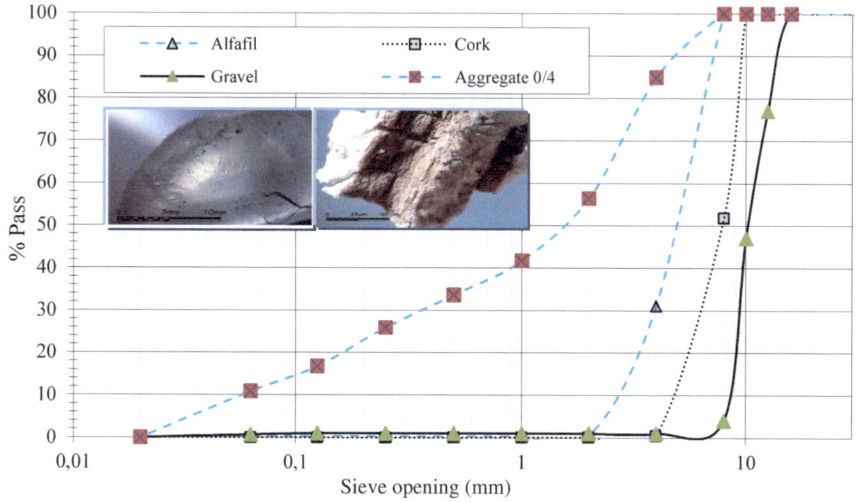

Fig. 3 Grading curve of different aggregates employees and micrograph at different scales of the arid Alfafil and Cork granules

The minerals more important found in them ashes fly and the cement is the: mullite, silica, aluminates calcium, anhydrite and oxide of calcium free.

Then collect analysis structure by diffraction of x-rays from the crystalline part, measured in % (w/w).

The best replacement materials are not 100% amorphous, but it is necessary that there is a small proportion of crystalline phase since they act as nucleation of the CSH gel sites and thus the compressive strength of these materials increases markedly. On this basis, we can say that the amorphous characteristics of the materials used in the research are in acceptable ranges and appropriate cementitious properties can be expected.

Is has effected the essay of particle size of the particles to them different aggregates employee in the research (Fig. 3), according to governs the standard UNE 933-1:2012. The Alfafil is a grain on the basis of propylene (PP) recovered, reinforced with ultrafine calcium carbonate 40%. And presents colour black and white.

3 Study of Dosages

In all the dosages is has employee a cement type CEM I 52, 5R to which is incorporates arid fine limestone crush of fraction particle 0–4 and aggregate thickness of fraction 4–10 mm. These aggregates are partly replaced by cork granules and similar granulometric fraction Alfafil. He is added to all the concrete a superplasticizer additive (MasterGlenium ACE 459 BASF) and nanosilica (MasterRoc MS 685 BASF).

Table 3 Dosages studied

	Dosages								
	Water (l)	Cement (kg)	Fly ash (kg)	$NanoSiO_2$ (l)	Aggregate (0–4) (kg)	Gravel (kg)	SP (l)	Alfafil (kg)	Cork (kg)
HP550	160	550	130	14	1015.16	507.58	11.00		
HP700	180	700	70	14	927.19	463.6	11.75		
ECOHUL700	180	700	70	14	593.39	129.82	11.75	189.49	19.77
ECOHUL550	160	550	130	14	646.94	142.27	11.00	206.16	21.5

In terms of the criteria for replacement of aggregates, is selection of the granulometric fraction of natural aggregate (gravel) for Cork granules and to replace natural aggregate 0/4 by Alfafil plastic waste, so that the total volume of aggregates was kept constant, since plastic waste were approaching more the aggregate size 0/4, although lack of fine fractions.

Mixtures were made with ratio water/cement material that remained constant 0.26 for the HP700 and the ECOHUL700. And 0.25 for the HP550 and the ECOHUL 550. We also studied the mechanical properties of the ECOHUL550 (L) and the HP550 (L) with additions of sludge of paper, around the 1.33%, since this was the best results obtained in the characterization of materials (Table 3).

4 Results and Discussion

Dosages of concrete were characterized by the following experimental plan.

4.1 *Study of Behaviour in Hardened State*

4.1.1 Resistance to Compression and Density

They are tested three specimens for each dosage, making trial as he governs the norm UNE-EN: 12390-2, at 7, 28 and 90 days for compressive strength and density (UNE-EN: 12390-7:2009).

It is defined as structural lightweight concrete that closed structure concrete, whose apparent, measured in dry conditions until constant weight, density is less than 2000 kg/m^3, but exceeding 1200 kg/m^3 and contains a certain proportion of arid light, both natural and artificial (Annexed 16, EHE-08). In this sense, in Fig. 6 are analysed to highlighting concrete delimits the area for structural lightweight concrete.

"In the case of structural lightweight concrete, the minimum strength is set to 15 or 20 N/mm^2 as maximum resistance depends on the type of arid light which concerned and the particular design of the mixture. While there are applications of lightweight high strength concretes, maximum structural lightweight concrete resistance is limited to 50 N/mm^{2}" (Annexed 16, EHE-08) (Fig. 4).

The evolution of resistance to compression is a very important parameter when it comes to study the mechanical properties. The least resistance of the ECOHUL is mainly due to the low bond between the surface of the Alfafil and Cork, with the paste of cement. In the micrographs (Fig. 5) the matrix presents micro cracks around Cork granules, which substantially weakened resistance to compression, in all the ECOHUL.

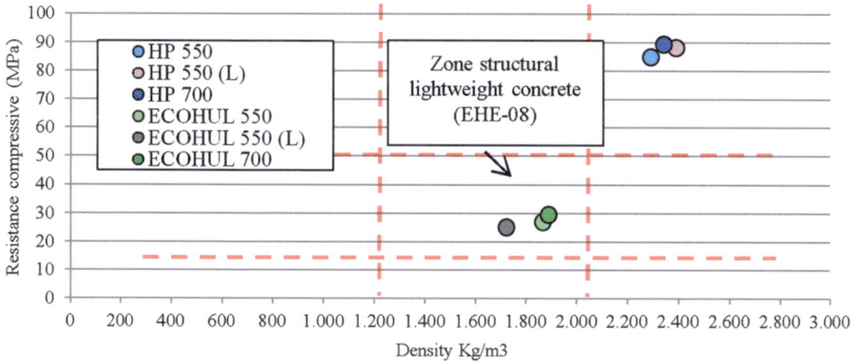

Fig. 4 Resistance average compressive to 28 days in relation to density

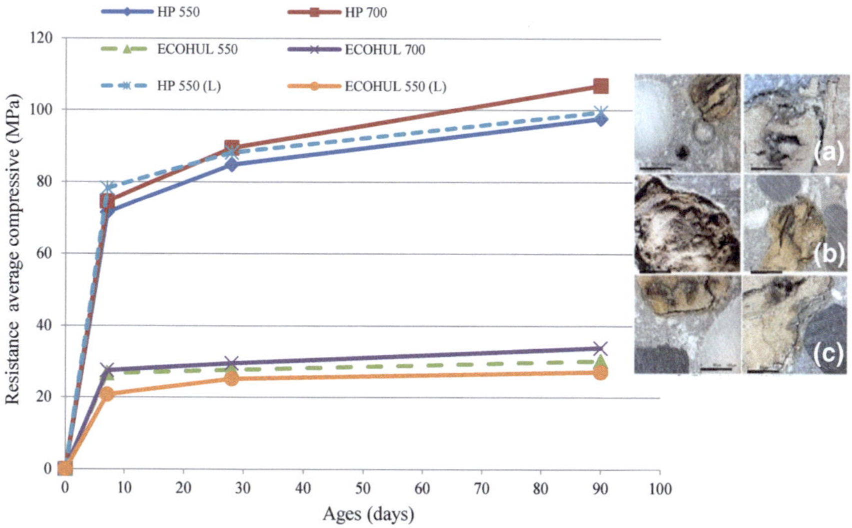

Fig. 5 Evolution of the resistance medium to compression and micrographs to the age 28 days **a** ECOHUL700, **b** ECOHUL550 and **c** ECOHUL550 (L)

4.2 *Study of the Durability*

4.2.1 Mercury Intrusion Porosimetry

The trial was conducted according to specification ASTM D4404-84-2004, validated by the technical literature for the characterization of the concrete, made at the age of 60 days. The Fig. 6 shows the pore size distribution.

All the concretes, the pore volume distribution follows a similar pattern, each curve has two peaks: one around 90 μm and another, pore sizes much more pronounced around sizes 0.05 μm and if spice of the ECOHUL, with Cork inside, stands a third peak in the 100 μm originated by the porous structure of Cork, with large pores. In addition, the greater porosity of the ECOHUL is clearly influenced by the high porosity of the Cork. The 550 HP clearly has a porosity of finer the 700 HP, to move the curve to the left (Fig. 6) and at the same time slightly has a lower total porosity (Table 4). Both results are consistent with results obtained mechanics. HP 550 lower porosity decreases the mechanical differences with the 700 HP.

As overview of all it indicated is highlights that with the use of them arid light, especially of the Cork, them concretes are more porous, with greater amount of Macropores but also with greater amount of pores of lower size, and not both in their properties resistant (Kumar and Bhattacharjee 2003; Mehta and Monteiro 2007) and thus has found in the investigation.

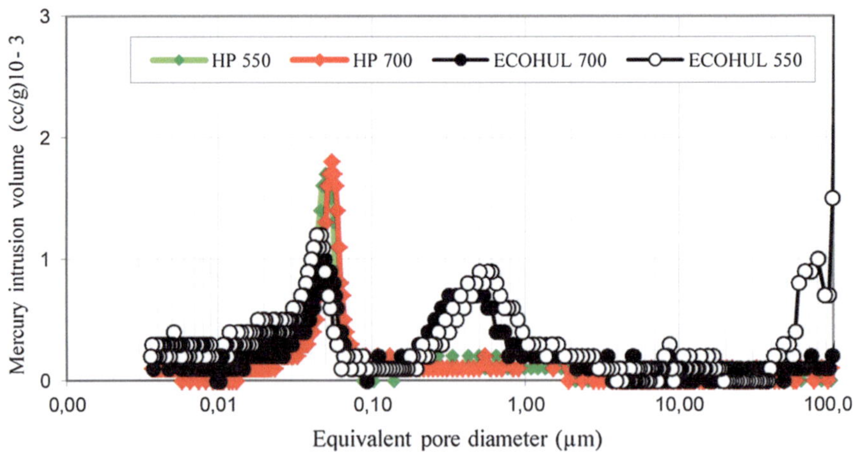

Fig. 6 Distribution by size of the porosity

Table 4 Porosity of concretes by mercury intrusion

	HP 550	HP 700	ECOHUL 550	ECOHUL 700
Total porosity (%)	9.831	10.012	17.763	11.846

Table 5 Absorption of the concretes tested in %

HP 550	HP 700	ECOHUL 550	ECOHUL 700	ECOHUL 550 (L)	HP 550 (L)
0.671	0.585	1.059	0.875	0.585	0.792

4.2.2 Absorption Capillary

The test is performed according to UNE 83982-08. He is observed an increase in absorption capillary concrete that contains higher amount of waste plastic and Cork than concrete containing natural aggregates only, the shape of the arid plastic and the interfacial union arid-pasta (Fig. 5) was most responsible for this behaviour (Table 5).

By this the ECOHUL-550 with a content of ALFAFIL of 206.16 kg/m^3, facing the 189.49 kg/m^3 that contain them ECOHUL 700, presented a greater absorption capillary (Albano et al. 2009). Clearly is stated that the concrete with a greater absorption is the ECOHUL 550 followed of the ECOHUL 700. This is due mainly to them arid light used of Cork (21.5 kg/m^3) that by its structure own porous and by them cracks that is created in its contact with the matrix cementing, create a road for the step of the water by capillarity. In the ECOHUL 700 occurs something similar available in your mix the same lightweight aggregates, but to a lesser extent (content of Cork 19.77 kg/m^3), but with more concrete and less content in Alfafil and Cork has been generated a denser cementing matrix and raincoats, which reduces its absorption by capillarity from the ECOHUL 550.

In the case of them concretes of high performance (HP 550 and HP 700), its absorption is much lower that the ECOHUL, by not have aggregates light as the Cork that facilitate the absorption. The HP 550 is 8.20% more absorbent than the 700 HP, for the improvement of the cementing matrix as discussed for the ECOHUL 700, with a 14% less absorbing than the ECOHUL 550.

In addition, displayed results, it can be concluded that the use of recycled lightweight aggregate makes the average 54% more absorbent concrete (58% for the 550 kg of cement and 50% for the 700).

In the case of the ECOHUL with sludge the trend is reversed, obtaining 35% less absorbing than the concrete of aggregate conventional. The most increased difference is obtained between the ECOHUL 550 and ECOHUL 550(L) with one further 80% which does not contain sludge.

4.2.3 Accelerated Carbonation Depth (UNE 13295) and Chlorides (AASHTO T259)

Results can be indicated that, in general, made concrete is have carbonated very little, with faces of specimens in which virtually no deepened the carbonation front. This happens in the 8 faces of the analysed HP 700. If you look at the media for each (Table 6) concrete results, it can be concluded indicating very good HP results, by their no carbonation. The HP 550, with 150 kg less than cement by m^3,

Table 6 Results of accelerated carbonation and chlorides

Dosages	Medium carbonation (mm)	Average penetration chlorides (mm)	
		28 dias	56 dias
HP 550	2.9	3.6	5.9
HP 700	0.0	5.7	8.6
ECOHUL 550	1.7	6.2	8.2
ECOHUL 700	4.0	7.0	8.6

medium carbonation front reaches 2.9 mm. For both ECOHUL, carbonation penetrates slightly more, possibly by their arid light of Cork, with a high open porosity, but they are very low values for this trial, facing other similar research.

The results obtained for the average penetration ion chlorides, at the age of 28 days and 56 days, and average carbonation (Table 6).

The excellent behaviour of the penetration of the ions chloride presenting concrete, partly can be attributed to refinement of the porous structure as a result on the one hand the use of additions pozzolanic and use of nanosilica which cause an effect of microfilled by making it more tortuous matrix.

This is coincident with the carbonation penetration tests, which may mean that reference for the analysed concrete dosage, including similar amounts of fly ash and nanosilica, is optimized to achieve concrete very waterproof and resistant to attacks foreign as carbonation or penetration of chlorides. This means a great step forward to increase the lifespan of the structures to give greater protection to their armour.

5 Conclusions

- In all the concretes, the obtained pore volume distribution follows a similar pattern. The HP 550 has a lower porosity total, but is more fine that the HP 700. The lower porosity of the HP 550 does decrease the differences mechanical with the HP 700.
- The ECOHUL-550 with a content of ALFAFIL of 206.16 kg/m^3, facing them 189.49 kg/m^3 that contain them ECOHUL-700, have a greater absorption capillary. In the case of concrete of high performance (HP 550 and 700 HP), its absorption is much smaller than the ECOHUL, for not having dry light. The HP 550 is a sauce 8.20% more absorbent that the HP 700, perhaps by the less dense matrix cementing. The use of aggregates light recycled makes to them concretes of half a 54% more absorbent (58% for those of 550 kg of cement and a 50% for them 700).
- In the case of ECOHUL 550 (L) the results are very good for improving the matrix produced mainly by its pozzolanic activity. The ECOHUL (L) is 35% less absorbent than the HP 550 and 80% less than ECOHUL 550.

- With the results the use of paper sludge is an addition with good performance to be used in base materials cement. The dosages analysed are optimal to achieve concrete very waterproof and resistant to attacks foreign as carbonation or penetration of chlorides. In the case of them ECOHUL, the great amount of materials recycled used in its dosage, them make very sustainable, contributing to reduce even more footprint of carbon of them structures of concrete.

6 Acknowledgments

Project CDTI-IDI-20130144. Contract research among the Group of research of science and technology advanced in the construction of the Polytechnic University of Cartagena and the company Cementos La Cruz S.L. our thanks to the company S.L recyclables Compost, BASF, and HORMICRUZ S.L. for your help and interest throughout the investigation.

References

Albano C, Hernández M, Matheus A, Gutiérrez A (2009) Influence of content and particle size of waste pet bottles on concrete behaviour at different w/c ratios. Waste Manage 29:2707–2716

Hewlett P (1998) Chemistry of cement and concrete, 4th edn. Butterworth Heinemann, New York

Kumar R, Bhattacharjee B (2003) Porosity, pore size distribution and in situ strength of concrete. Cement Concr Res 33:155–164

Lorca P (2014) Efecto de la adición de hidróxido cálcico sobre mezclas con alta sustitución de cemento por ceniza volante. Tesis doctoral Universidad Politécnica de Valencia. Dpto Ing Construcción y Proyectos de Ing Civil

Mehta PK, Monteiro PJ (2007) Concrete; microstructure, properties and materials. McGraw-Hill Education, New York

Part V
Sustainable Engineering

Project AURA: Sustainable Social Housing

Rafael Herrera, Paloma Pineda, Jorge Roa, Sebastián Cordero and Álvaro López-Escamilla

1 Introduction

Solar Decathlon is a competition for sustainable architecture that has a great international prestige. Technical Schools and Universities around the world take part in collaboration with institutions and private companies. The main objective is to design, build and operate a prototype of housing with the highest level of self-sufficiency and exclusively working with renewable energy.

Promoted by professors, researchers and students from the School of Architecture of the Universidad de Sevilla, the team is mostly composed by members of different areas from the Universidad de Sevilla and the Faculty of Engineering of the Universidad de Santiago de Cali. Thus, the "Híscali" team arises, from the union of the Spanish city of Sevilla (whose ancient name was Hispalis) and the Cali team (hísCali-Team, 2015).

The successful results obtained in the competition have served to verify the relevance and reliability of the project, ensuring the improvement and optimization of the results, in order to apply them to possible urban designs. In addition to being recognized with the 3rd Prize in the latest edition Solar Decathlon Latin America and the Caribbean 2015, it is worth highlighting the 1st Prize achieved in Communication and Marketing, as well as the 1st Prize in Comfort Conditions and the Honour Mention in Architecture.

R. Herrera (✉) · J. Roa · S. Cordero · Á. López-Escamilla
Department of Building Construction, School of Architecture,
Universidad de Sevilla, 41012 Seville, Spain
e-mail: herrera@us.es

P. Pineda
Department of Building Structures and Geotechnical Engineering, School of Architecture,
Universidad de Sevilla, 41012 Seville, Spain

P. Mercader-Moyano (ed.), *Sustainable Development and Renovation in Architecture, Urbanism and Engineering*, DOI 10.1007/978-3-319-51442-0_23

It is intended, from this experience and results, to activate social indicators (cohesion), urban indicators (occupation), economic indicators (work) and environmental indicators (efficiency), following the rationality of taking advantage of the available resources, and an ecological use in tropical climate environments.

2 Design Fundamentals

2.1 First Purposes

Instead of the urban development strategies that are designed prioritizing accessibility, road traffic, m^2 location and urbanization (including the usage assignments) the proposal offers an alternative: merging both the territory occupation and the way of life of the people from Cali. Thus, an efficient development of the dwelling place is proposed integrating the following basis: the social interest aimed at different communities and the place itself. The proposed strategy will preserve the intrinsic nature of community-place, preserving its character, resolving malfunctions and enhancing their activity in relation to the city and citizens from Cali (Fig. 1).

The focus is not an indiscriminate growth, and for that, and alternative process for the city development is proposed. As the process relies on the citizens from the very beginning, its viability in terms of cost and schedule is taken into account in a partial way (by means of scheduled interventions) instead of globally (responding to a foreseeable finalist picture) (Bonilla, 2015).

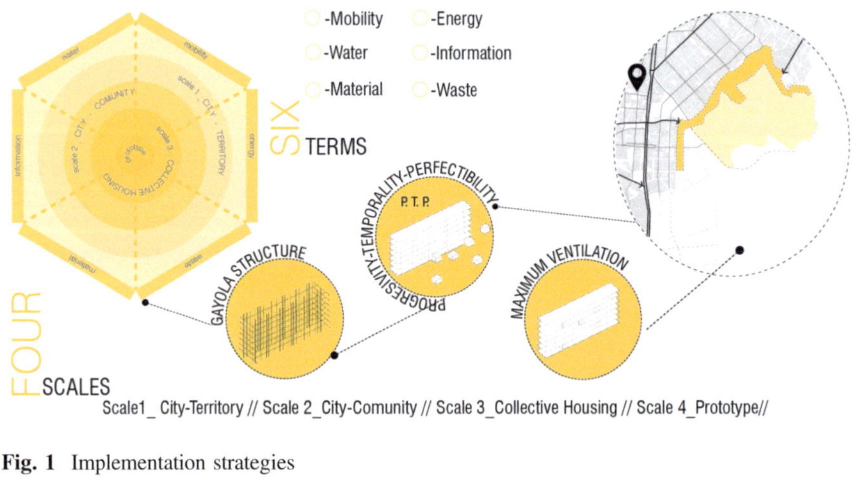

Fig. 1 Implementation strategies

2.2 The Dwelling Disbandment. Transition Spaces

In order to reduce the moisture content in a passive way, the cross-ventilation through the house is considered as a priority goal for the team. For that, the concept of traditional compact housing will be fragmented.

Transition spaces appear among the parts. Main interests of the team are those areas, as origin of new non-planned uses in the home or as cause of new situations that promote flexibility.

2.3 Self-build and Progressive Modular System

A very rooted concept of the Latin American building tradition, the self-build, has been assimilated into the construction process. It is essential that the dwellers know, feeling themselves identified and involved, the development and growth of the housing. That self-build method allows moving the starting time of the work. That time is overlapped to the remaining useful life, and, in this way, a real time adaptation to the domestic requirements is accomplished. Thus, from the structural and facilities core, "Gayola", the different spaces will "connect" to that nucleus. A housing, which is completely able of responding to the user requirements during the different stages of the family growth, is created (Habraken, 1979). That concept is inheriting the theories by Habraken, which were developed and very referenced at the end of the XXth century.

By means of adding prefabricated modular panels, and following pre-established rules for the assembly process, the family will grow over time, and in turn, the space will be occupied in an optimized way.

2.4 Progressiveness and Perfectibility

Within our social context, it is essential to understand and incorporate in the pre-fabrication processes one main issue: the difficulty that a family has to face when the housing is a rigid and complete unit, continuing over time. The family needs and opportunities are modified in a continuous and gradual way, being implemented within the design process. In a first stage, the users get through to the technical core of the house. From that core, a package comprising the needs of the initial housing arises. A second stage will bring two new living spaces, increasing the occupied area. A third stage will be dedicated to provide both quality and multi-functionality values to the house (Fig. 2).

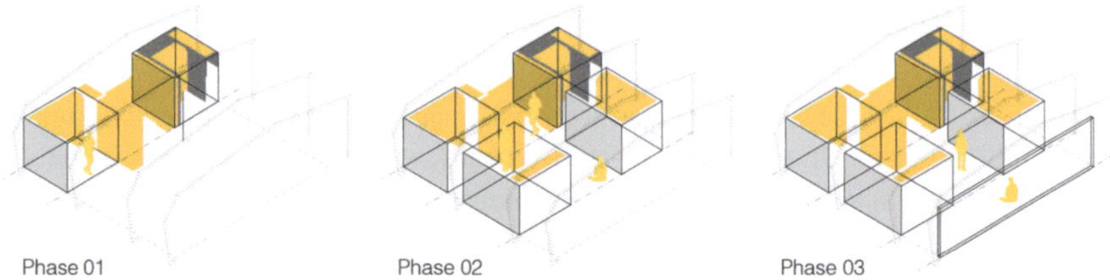

Fig. 2 Housing progressive growing

When the housing is completed, it will be able to improve itself by adding elements that contribute to increase the thermal comfort. Those elements also allow for a better response to the specific family difficulties. Thus, the housing will get better and, over time, will adapt to the family needs without be obligation to move to a different house.

3 Architectural Configuration

As aforementioned, Proyecto AURA is organized into two architectural pieces that respond to different purposes.

The former, the **MINIMUM HOUSING MODULE**, comprises the structural technical core ("gayola") and a progressive and spatial configuration of the dwelling based on the furniture.

The concept of carcass disbandment is taken to the extreme. Under that premise, the house is put together with a series of furniture pieces, which are equipped to provide multifunctional services to the rooms. The furniture comes from the idea of improvement and growth. Every piece of furniture has its specific use and, together with the different modules, generates new definitive equipment, which compartmentalise the space (Fig. 3).

For instance, the furniture-partition of the bedroom is the union of bed and storage space. Even it integrates the place for a possible improvement of housing conditions, by means of installing a small cooling machine. It is one of the primal furniture in the house, but it can evolve through the installation of air conditioning or through the placing of doors to compartmentalise the storage spaces. This furniture perfectibility together with the mobility of some of them (e.g. that of the living room) makes possible the multi-functionality of the spaces, via the furniture position and the home needs and growth.

The functional kitchen area is conceived as a minimum occupancy room that is linked to the initial "gayola". It is fully enclosed and hidden, being completely integrated into the living-dining room furniture.

The bathroom is also designed as a furniture piece, including all the necessary elements. Toilets, shower, storage, free up space for the washing machine, and cleaning supplies, are integrated as a whole. Due to the specific conditions, water resistant phenolic panels will be used.

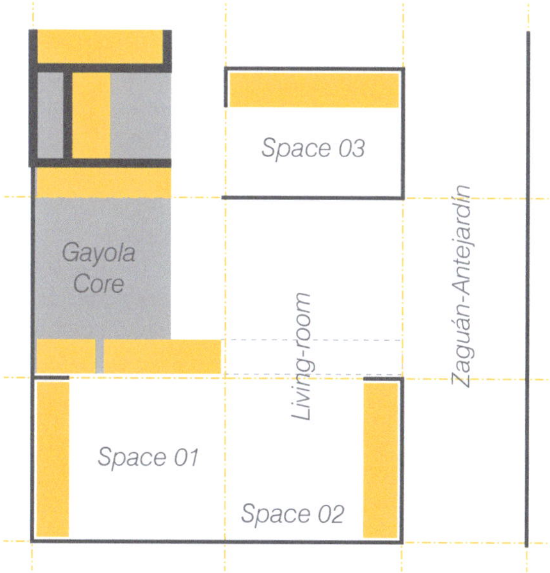

Fig. 3 Housing configuration diagram

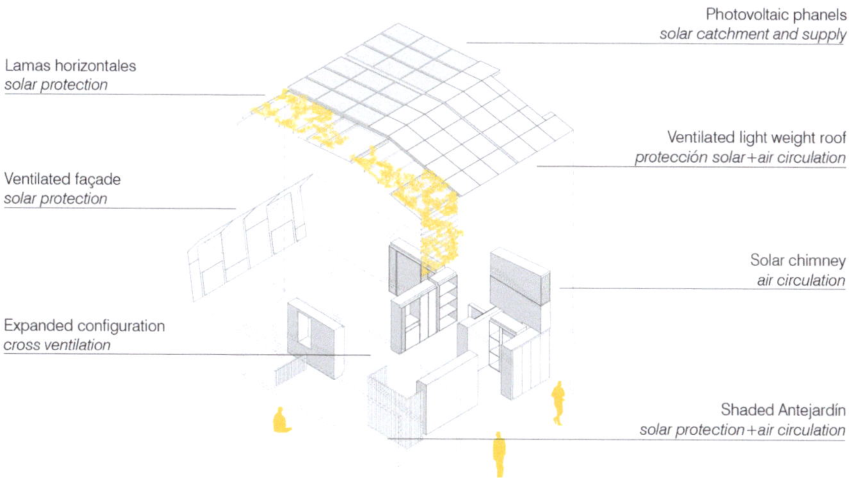

Fig. 4 Passive conditioning strategies

Finally, the passive bioclimatic technique is the last piece of furniture that, being part of the initial facilities, is introduced as an element of the collective dwelling. That technique is based on the solar chimney concept. Its form is related to the living room, kitchen and bathroom, and, working as a "vent", helps (in a passive way) to evacuate the aerial residual heat which is produced inside the house (Fig. 4).

The second architectural piece that organize the Proyecto is the **BIO-CLIMATIC MODULE/HALLWAY**.

The Proyecto AURA inserts a new element in the social housing, which is able to provide added values to the minimum housing module. Those values are not exclusively based on spatial occupation. It is a way of living, rescued from the vernacular architecture, which qualifies the spaces through an element that feeds and regulates novel (although basic) aspects, such as:

The "hallway (zaguan)" module works as a bioclimatic-damping element among the houses. Thus, cross ventilation, both individual and collective within the housing block, is guaranteed. This continuous air movement through those openings that connect the cold and warm areas contributes to reduce, in a passive way, the high humidity content, increasing the thermal comfort of the user (Fig. 5).

The second function of the hallway is the transition, or threshold space, between indoors and outdoors. The vertical opening provides a hazy light both for the collective entrance gallery and indoors.

In addition, this new element works as "damper" among the dwellings in relation to acoustic isolation. Due to its configuration, this space makes possible that each home has four independent facades. The facades are not dependent of the adjacent building, improving the acoustic behaviour of each one.

The task of this space is to serve both the home and its inhabitants, having the potential ability to be occupied, if it is required by the home itself, to maintain its own identity. It is an extension of the house that reminds the open process of continuous occupation, and being a covered and semi-open space, allows for neighbourly relations and community life. In this place the limit state between public versus private, between individual versus collective is lost.

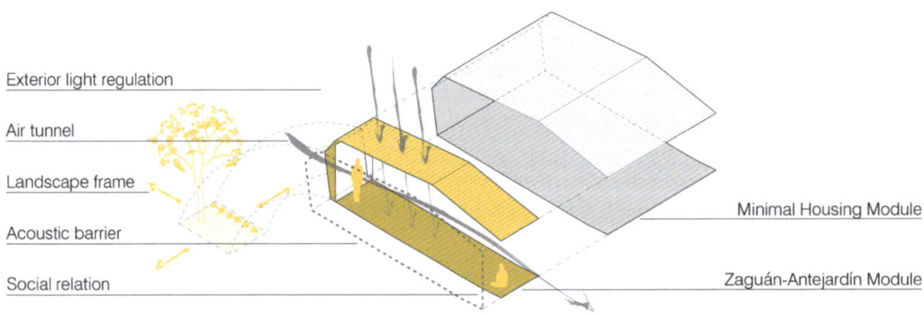

Fig. 5 Minimum hallway (zaguán) module

4 From the Theory to the Construction of the Room-Cell in Villa Solar

One of the most relevant complexities of this edition was translating and summarizing ideas from different areas and scales: urban and design scales, the housing and the collective level and an exhibition pavilion in Villa Solar. The difficulty was increased by the volume limitation requirements that were imposed by the rules.

One of the main objectives of the Híscali team was to show that the built sample is not a single and self-defined housing object, but an element of a more complex living ecosystem. To this aim, the north gallery of the original collective block is transformed into the entrance to the module. Thus, a connection among that element and both the vertical collective and the hypothetical neighbourhood is established.

Two ramps, pointed to East and West, were designed to overcome the +0.50 m height difference. Those ramps are covered with light elements, railings and lattices, showing that the prototype is not an isolated building, as it is part of a larger complex. Those dividing walls are adapted as solar envelope to emulate the adaptation of the collective project to the contest requirements (Fig. 6).

Regarding the aforementioned flexibility, the built prototype joined a number of adaptable elements, showing how the house could be used under different situations and family needs.

The mobile partition of the multifunctional room allows for two different options: to separate an additional bedroom or to maintain an open room that is connected to the living room in the centre of the home. That depends on the inhabitant requirements or on the time of the day. That partition could be used not only to generate a spatial relationship between living room, terrace and multifunctional room, but also to achieve high levels of cross ventilation in house.

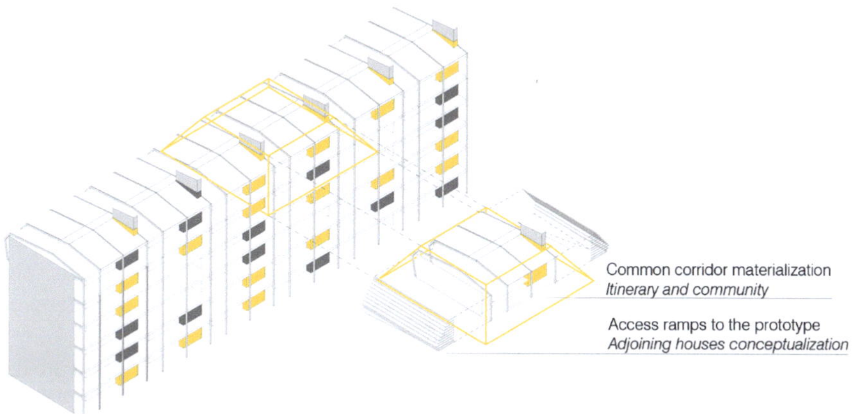

Fig. 6 Room-cell detail from the theoretical collective dwelling

Breathable panels or exterior glazing can also be opened for better natural lighting, if required. They also improve natural ventilation and dilate the view towards the exterior. It is important to mention that these protective panels mainly act as a screen to the predominant wind of Cali. That wind has a high content of dust and suspended particles. Therefore, the natural location of these breathable panels was to be closed to prevent the accumulation of dust and inside dirt in the house, without removing the unavoidable ventilation.

5 The Building Configuration

5.1 The Prototype Structure

From the beginning, the Proyecto AURA always worked around the idea of social housing, on the basis of the personal experiences of the team members who visited the city of Cali. The adaptation and translation of local practices and building techniques to prefabricated building systems are also central ideas.

As aforementioned, one of the concepts underlying the project is the progressive occupation of the main structural grid. That community part of the building consists of a structural grid of in situ concrete columns and slabs. The structural elements are built using simple and traditional techniques that a local company could perform without problems. The grid also distributes vertical communications and facilities in the building.

To build the prototype in Villa Solar, the structural grid is adapted following a three-dimensional system of rolled steel sections. That framework creates three completely built containers, responding to the requirement of relocation and assembly in Villa Solar. Thus, both assembly times and waste generation are reduced.

In the next stage, envelope and inside spaces, a space modular grid is performed. That grid is articulated to the main structural grid, where the communications are distributed horizontally and the facilities are vertically distributed.

5.2 Compartmentalization and Envelope

Within this coordination network, the bi-dimensional components are added using simple building techniques and a modular prefabricated system for the panels. The panels are multilayer and single sheet. In a second stage, they will be assembled by means of a dry board technique (Bernard, 1983).

Due to the promotion of self-build and to the inhabitant involvement in the housing growth, the panel fabrication could have a relative technical complexity.

But the vertical compartmentalization during the assembly stage must accomplish very specific requirements: lightweight as well as simple and repetitive assembly process.

From the aforementioned, one of the decisions is the use of very similar systems that, under the climate conditions of Cali, work both as inside partition and exterior envelope. This multifunctional character of the panels simplifies the assembly and solves the roaming during specific growing stages. Both functions will be carried out at different times.

Therefore, as the maximum simplicity is a main concern and objective of the project, a unique system of modular compartmentalization is designed. The aim is to solve all the singularities that were inside the main structural system, as well as the partition finish, by means of dry joint assembly.

The supporting structure of the panels is built with pinewood. That structure, used both in partitions and envelopes, is completed with the same composition, using OSB panels, 4 of 15 mm thick. The panels contain thermal and acoustic insulation. Thus, a double-sided sandwich is created.

From the specific characteristics of the Cali tropical climate, taking advantage of the low temperature gradient (day/night and winter/summer) a continuous breathable system is incorporated in the facade. That system consists of wooden slats, which allow cross ventilation in the house through the openings. To build those elements structural slab were used. Guadua boards, a local product, were easily fixed to the slabs.

For the facilities, the location of pipes was planned following, when possible, a horizontal distribution. Thus, the pipe networks could reach the supply points without joining to the vertical panels. Therefore, a very high percentage of the facilities works would be carried out in the hypothetical phase 0. That criterion is followed in the prototype in accordance to the primal design idea. The strategy of dissociating facilities from partitions and envelopes makes easier the self-build stage. The reason is that the panel assembly complexity decreases, as it is no necessary to prepare openings for the pipes.

6 The Work Process

Open prefabrication, proportion, module and flexibility are the central concepts that are used in the building process of the displayed room-cell in Villa Solar. Thus, two clearly defined stages are singled out:

Stage 1. Prototype fabrication, in individual and independent modules, with the objective of being relocated.
Stage 2. Module assembly—after relocation—in Villa Solar (Fig. 7).

Fig. 7 Module room-cell building, relocation and assembly in Villa Solar

7 Conclusions

Although the competition rules are well-defined in terms of location, volume and specific characteristics of the module, the AURA team proposes an intervention from the reflection on the urban configuration of the city itself.

The design fundamentals that are laid out for the further performance of the proposal are in balance between flexibility and diversity. Those factors are obtained analysing the city of Cali, its local building techniques, and its translation to pre-fabrication procedures.

The goal is to propose a systematic approach with the advantages of a designed action. That action is implemented by an innovative industrial process, which respects the adaptation to the boundary conditions of the setting.

In this way, the adaptation to climate and environment, the urban hallmark, the creation of public areas and neighbourhood units, among others, all should be part of the initial determinants of the proposal. Those are essential characteristics that distance from the initial idea of prefab prototype. A prototype that was absorbed by its own features, without considering the place.

That initial determinants of the place must be assessed and added, in order to strengthen the singular procedure of the design, aiming at obtaining a suitable and located proposal. The current automated building techniques will be implemented to obtain profitability in terms of quality and sustainability. The easy and speed of assembly–disassembly will be also considered, as it makes possible to incorporate the self-build process, and the gradual growth that makes easier a controlled and proportionate housing enlargement from the family needs.

Finally, after the performed analysis, one of the most revealing findings for the participants, professors and students, is the involvement in the process. That involvement includes not only the prototype assembly but also the funding search and promotion, the agreements, the learning in the fabrication labs, the timing control of a prefabricated work, and mainly, the management of such as singular project, which is located in an external environment.

8 Institutional and Staff Team of the Proyecto AURA

http://www.bienalesdearquitectura.es/index.php/en/muestra-de-pfc-e-investigacion/resultados-i-convocatoria-de-investigacion-investigacion/6018-investigacion-xiii-beau-investigacion-proyecto-aura.

References

Bernard P (1983) La construcción por componentes compatibles. Editores Técnicos Asociados, Barcelona

Bonilla M (2015) AURA, la propuesta colombo-española para el Solar Decathlon América Latina y el Caribe 2015, 11 Nov 2015. Plataforma Arquitectura. http://www.plataformaarquitectura.cl/cl/776978/aura-la-propuesta-colombo-espanola-para-el-solar-decathlon-america-latina-y-el-caribe-2015

Habraken NJ (1979) El diseño de soportes. Gustavo Gili, Barcelona

hísCali-Team (2015) "Proyecto Aura|Equipo hísCali" Revista Veredes. http://veredes.es/blog/proyecto-aura-equipo-hiscali/

Structural Refurbishment Projects. The Sustainability of Reinforcements Using Composite Materials

Victoriano González, Angela Barrios-Padura and Marta Molina-Huelva

1 Introduction

The feasibility and efficiency of an intervention of rehabilitation or repair is generally assessed based on issues related to the technical capacity of experts and economic of promoters (public or private). However, we must consider the complexity of operations and their impact on the environment, being determinant in most cases not to interrupt the use of buildings or infrastructure during the works.

Repairing a steel reinforced concrete structure requires really annoying works, and in many cases it is necessary to evacuate buildings or disrupt traffic in roads, in order to ensure welfare and safety of people.

They are complex and expensive works. The patching of concrete elements generates noise and dust in the atmosphere, and if the deterioration is severe, the structure must be propped during the process, making it difficult to access and operate. The works in structural elements must be done carefully, deciding in each case whether it is necessary to supplement or reinforce with steel bars or other structural systems. When the area is located in a zone difficult to access, or there are facilities, it is necessary to dismantle or demolish (wiring, water pipes, walls, windows, floors), which expensive works. Definitely, if analysing a work of strengthening of a steel reinforced concrete structure, from economic, environmental and social criteria, sustainability is questioned.

V. González (✉)
LABRUM Diagnosis and Specialized Consulting S.L, Seville, Spain
e-mail: victorianogonzalez@labrum.eu

A. Barrios-Padura
Department of Architectural Technology II, Sevilla University, Seville, Spain

M. Molina-Huelva
University Institute of Architecture and Construction Sciences,
Sevilla University, Seville, Spain

P. Mercader-Moyano (ed.), *Sustainable Development and Renovation in Architecture, Urbanism and Engineering*, DOI 10.1007/978-3-319-51442-0_24

There is an undeniable need to produce, in a sustainable way, in construction industry. Housing rehabilitation politics, preservation of buildings for its architectural interest or operation changes, and the need to extend the durability of infrastructures, address us to develop structural repair methods that "meet the needs of the present, without compromising the ability of future generations to meet their own needs".

However the implementation of CFRP reinforcement needs to follow a guideline for a proper execution, inspection and quality control (IVE 2008a, b).

The incidence of quality control of the implementation process in a structural reinforcement, is considered as a necessity since the 90s, in some studies and official documents such as Building Assessment Report of Valencia (Serrano 2014); works for a guide of interventions in concrete beams floor within the Subcommittee SC3, Building Quality Technical Commission of the Ministry of Public Works; documents as the "Procedure for action and intervention designed by the Institute of Housing of Valencia, IVVSA", "Recommendations for the systematic recognition and rapid diagnosis of concrete floors containing aluminous cement" of the Generalitat of Cataluña (Maña and Bellmunt 2005) "Pathology, repair and reinforcement of structures" (Rio 2008), and various instructions and guidelines (Basf 2011, 2012; Sika 2007; UNE-EN1504 2006; ACI 440.2R-08 2008).

The aim of the research presented in this paper is to assess the sustainability of the reinforcement of structures with composite materials, focusing on efficiency and limited environmental, economic and especially social impacts compared to other systems, and to value the need of quality control in implementation and maintenance during its lifetime stages.

The analysis is focused exclusively on the reinforcement used, considering that, for comparative purposes between different types of reinforcements, previous operations of preparing and repairing damaged structures is a common constant in any type of reinforcement.

This work it is a part of the research project "Nuremco" funded by the Ministry of Economy and Competitiveness, Feder Interconnecta Program, with the participation of the company VORSEVI and the University of Seville. The study continues among the research group Paidi TEP954 of the School of Architecture of Seville and the companies LABRUM and AZUL CONSTRUCCIÓN.

2 Sustainable Rehabilitation of Reinforced Concrete Structures

In conservation and maintenance of structures it is frequent to question whether rehabilitation is possible or we should proceed to demolish and reconstruct. In some cases it is necessary to implement a partial or a general reinforcement, according to the purpose of the project: the increased load due to change of use of the building, the need to improve operation conditions, changes in the structural system, adaptation to new regulations, the existence of defects of project, the existence of damages, etc. The three types of reinforcements most frequently used are: structural regrowth with steel reinforced concrete or structural mortars, overlap with metal profiles or with CFRP reinforcements.

In a sustainable rehabilitation, reinforcements of CFRP have important advantages, analysed from environmental, economic and social criteria, considering their lower weight and ease of transportation, better adaptation to the geometry of the structural elements, less interference with other elements and facilities, better resistance with reduced reinforcing sections, no corrosion, and its better performance in case of fire.

2.1 Environmental Aspects

To assess the sustainability of products or construction processes from environmental criteria, tools based on life-cycle assessment (LCA) are used, such as Environmental Product Declarations (EPD), which evaluate, at each stage of the life cycle (from extraction of raw materials to waste management), the impacts associated with global warming potential, depletion of the ozone layer, acidification, eutrophication, production of photochemical oxidants, depletion resources and waste production.

There are no DAP of CFRP reinforcements in construction sector, although some approaches to them can be found in wind turbine industry (Rivarola et al. 2011), or in automobile industry (Muñoz et al. 2008).

Due to the lack of data on LCA of laminated CFRP (Pereira 2015), the analysis of environmental cost is developed using a single parameter, the energy consumption in all stages of the manufacture of the product, to the treatment of waste and recycling at the end of its life cycle.

The first stage of analysis is the extraction of raw materials and the manufacture of the CFRP products. In Table 1 data primary energy consumed per kg of material in the production phase (Song et al. 2009) are collected.

As can be seen in the table, primary energy for the extraction of raw materials in the production of polyester and epoxy resins used in the manufacture of composite is low, but in the case of carbon fibres a high intensity of energy is needed (183–286 MJ/kg).

Regarding manufacturing processes, autoclave modelling is the one that consumes more energy and pultrusion the less (3.1 MJ/kg). However, in the first process of the analysis of the next stage, in the application, the high lightness of carbon fibre, produces a significant reduction in the energy required to transport in comparison with other heavier materials such as steel. As the difference of weight is about 60–80% compared to steel, at this stage of the life cycle, both reinforcements are matched in terms of energy consumption.

Table 1 Primary energy of the fabrication of composite materials (Song et al. 2009)

Materials	Energy intensity (MJ/kg)
Polymers	
Polyester	69–78
Epoxi	76–80
LDPE	65–92
PP	72–112
PVC	53–80
PS	71–118
PC	80–115
Fibres	
Glass	13–32
Carbon	183–286
Metals	
Aluminium	196–257
Steel	30–60
Stainless steel	110–210
Copper	95–115
Zinc	67–73
Manufacturing methods	
Autoclave moulding	21.9
Resin transfer moulding (RTM)	12.8
Vacuum assisted resin infusion (VARI)	10.2
Cold press	11.8
Pultrusion	3.1
Injection moulding (hydraulic)	19

The deduction **of energy consumption of transport** is carried out in accordance with the publication of CIEMAT (Lago et al. 2007), considering a midrange vehicle. Taking into account that gas has a calorific value of 42.9 MJ/kg, and that consumption is of 0.073 l/km = 0.0624 kg/km, restricting the transport of material to an 80% of the maximum capacity load, 489 kg of the reference vehicle, the energy required to transport 1 kg of a material would be 6.85×10^{-3} MJ/kg km.

$$(42.9\,\text{MJ/kg} \times 0.0624\,\text{kg/km})/(0.8 \times 489\,\text{kg}) = 6.85 \times 10^{-3}\,\text{MJ/kg km} \quad (1)$$

In a first analysis of energy consumption only referred to transport, if we compare the performance of a reinforcement for the same building with CFRP and steel, considering that both reinforcement materials weight are in a relation 1/8 (CFRP/STEEL), the unitary energy cost of the transport of carbon fibre is 6.85×10^{-3} MJ/kg km against 34.22×10^{-3} MJ/kg km of steel. It is necessary to note that the longer is transport distance between production and application site, the better is the compensation on energy difference and life cycles.

Table 2 Comparative study of CFRP and steel reinforcement

Stages	CFRP (MJ/kg)	Reinforcing steel (MJ/kg)
Obtaining materials	286	60
Manufacture	3.1	3.1
Transportation (500 km)	3.43	3.43
Application[a]	–	–
Maintenance[b], (Tjibaou and Piano 2010)	0	31.16
Extraction (unmounted)	2.03	2.03
Recycling	19	8.6[c]
Material recovered	−286	−60
Total final life	27.56	47.21

[a]Similarly energy consumption is estimated with CFRP reinforcements but require less auxiliary means
[b]Supposed a service period of 30 years, when repainted every 10 years
[c]Osorio (2011)

Regarding end of life phase of a product, waste production and recycling of carbon/epoxy composite materials, it must be noted that it is currently a booming industry, and very interesting for producers, considering the high energy and economic cost of the production of carbon fibre, being the most recent studies:

- Recycling by microwave. This study shows the great potential of the recycled carbon fibre with microwave irradiation (Jiang et al. 2015).
- Recycling by "solviosis". Method used to degrade the resin in two different varieties of carbon fibre with epoxy base using a mixture of acetone and water at a temperature of 320 °C, this process requires an energy consumption of 19 MJ/kg fibre, although the process has not been optimized it shows a great potential for the future (Keith et al. 2016).
- Mechanical extraction. Electric energy by milling as an option of carbon fibre composite recycling, reaching an energy consumption of 2.03 MJ/Kg (Howarth et al. 2014).

The calculation of the energy needed in the life cycle of CFRP reinforcement in comparison to a steel reinforcement is shown in Table 2. It is included recycling and recovery phases that optimize energy benefits at the end of life cycle.

Given the indicated ratio 1/8 of a CFRP/Steel, energy consumption of steel reinforcement is 377.68 MJ/kg, that is to say 13.7 times higher than CFRP. Therefore, from the standpoint of energy consumption it shows the great advantage of the reinforcing with composite materials than with steel.

2.2 Economic Aspects

Another of the misconceptions about the reinforcements of concrete structures with CFRP is that they are more expensive than other types of reinforcements. The cost

Table 3 Prices from CYPE database

Type of reinforcement	Description	Cost (€)
Screeds concrete	m pillar reinforcement by concrete screed	98.69
	m pillar reinforcement by shotcrete	101.42
	m beam reinforcement by concrete screed	43.93
	m beam reinforcement by shotcrete	19.37
Metallic elements	m concrete pillar reinforcement with steel	62.37
	Unity. Strengthening of capital and base of a concrete pillar with metal sheet	104.98
	m reinforcement of a concrete beam or joist with metal sheets	15.57
CFRP	m reinforcement of a concrete wall or pillar with carbon fibre laminate MasterBrace "BASF"	42.14
	m^2 of tying a concrete pillar with carbon fiber sheet MasterBrace "BASF"	90.29
	m reinforcement of beams and joists with laminated carbon fibre MasterBrace "BASF"	41.59
	m^2 Shear reinforcement beams, with carbon fiber sheet MasterBrace "BASF"	90.29

of raw material is higher, but as it requires less material, less auxiliary means and security measures, and less volume of waste from the demolition of walls, floors or facilities, etc., so overall costs are reduced.

In Table 3 are shown the prices of some reinforcements with concrete, metal and CFRP elements, collected according to the database of CYPE Corporation.

According to this table CFRP reinforcement elements are cheaper than those of concrete or steel. For example in the case of a reinforcement of a pillar, done with steel it would cost by metro 62.37 €, while 98.69 € with concrete, and 42.59 € with CFRP. To these costs there must be added maintenance costs at a minimum period of 50 years, then, much higher than in the case of metal reinforcements and concrete.

2.3 *Social Aspects*

An important aspect in the assessment of the sustainability of a structural reinforcement operation is the reduction of the impacts on people that can be achieved with CFRP reinforcements, both on users of buildings and infrastructures and operators.

In cases where it is not necessary previous repairing operations, in other words, buildings without any pathology where it is necessary to make an adjustment and put in value, reinforcements by CFRP are made with reduced intrusive operations, which do not require large volumes of demolition works or collection of construction material. These works do not need too much water, exclusively a preparation and cleaning of the surface with a sanding machine provided with a hoover, and they are really fast, so it is not necessary in most cases to completely move out buildings.

As they don't require many auxiliary means as concrete works, the use of the building or infrastructure is not interrupted, and it is reduced accidents risks on users. In most cases of bridge repair, the works are executed without the need of cutting or disrupt traffic.

The works cause fewer impacts on the neighborhood, reducing noise and air pollution, since they produce less dust and suspended particles. Also the production of waste is reduced as this material is commercialized with a reduced packaging volume and fully recyclable.

They are solutions that don't need much maintenance works, are resistant to corrosion and aggressive environments, so that costs of conservation to neighborhoods are reduced.

Furthermore, its lightness and ease of assembly reduce accidents risks of workers. They are safer works, faster, in which very skilled and conscientious operators work, regarding safety and execution quality.

2.3.1 The Work of the Companies Beyond the Technical or Economic Sustainability

The investigation has included some interviews with the company Azul Construccción, with the aim of, in a sample of representative works developed in the last 20 years, analyse environmental and economic aspects, and especially asses the importance of social aspects this kind of work.

Azul Construcción is an Andalusian construction company with over 25 years of experience recognized nationally and internationally in the execution of works of repair and reinforcement of concrete and steel structures.

Is a company that from its founding, commits to technological innovation and to provide a specialized service, with a highly trained professionals, highly skilled and experienced in reinforcement works and repair of structures.

Its interventions are developed with sustainability criteria, they study reinforcement systems that technically meet the needs of works, while not impacting on people, focusing, in relation to social aspects, in ensuring safety to workers, people, and avoiding problems in the use of buildings and infrastructure.

The experience of the company in road infrastructure works shows social actions, without which its viability would have been committed. Works in bridges normally have to be done without disrupting traffic, to prevent negative effects on the activity of communities. Representative cases that will be displayed at the conference are:

- Repair in a railway bridge of the Seville-Cadiz line P.K. 005 + 778-via I, on the SE-30 (Sevilla), whose repair and reinforcement was executed 100% without the need of traffic disruptions.
- Improvement of the bridge over the river Almanzora in the old C-323 road, in Purchena, Almería.
- Rehabilitation and reinforcement of the San Telmo Bridge, in Sevilla, performing structural repair works with mortars, crack injections with resins and reinforcement with CFRP.
- Repair and reinforcement of San Juan de Aznalfarache in Sevilla viaduct. In many cases in bridge works, you can find people living under them, and you must always respect their homes, building soccer fields, or even creating workshop schools to employ them.

A case of an intervention in a railway under a road in Ndjolé (Gabon) is an illustrative one, a difficult access place with poor infrastructure, in which the works were carried out without disrupting paths, which would have meant to isolate two cities and divert traffic by alternative routes with distances of more than 40 km.

3 The Importance of Quality Control in the Amortization of Impacts

The rating of the sustainability of CFRP reinforcements requires a study of an appropriate application that ensures its durability, so it is essential to make a proper quality control programme.

There are regulations such as UNE EN 1504-10 y ACI 440.2R-08 and publications such as DIT No. 572R/16 on the reinforcement system of concrete structures (MasterBrace System) BASF, different criteria and parameters are set to take into account in quality control plan, such as manufacturing, reception control and works control with the following indications:

1.1 The concrete should present surface traction strength greater than or equal to 1.5 MPa (Pull-off) UNE-EN 1542: 2000.
1.2 Some weather conditions are required during application, and a surface moisture less than 4%.
1.3 Substrate preparation is recommended with texture patterns (CSP 4–6) by the International Concrete Repair Institute (ICRI).

These documents define, in greater or lesser degree, parameters and tests for quality control executed during placement of reinforcement and throughout its useful life. However, these are exceptional and destructive tests, not applicable to a large area of reinforcement and therefore, they can not ensure enough quality and durability.

Detected this gap in manuals, regulations, recommendations and current research, in the research project NUREMCO and in subsequent studies of the Paidi TEP954 group of the University of Seville and the company LABRUM, it has designed a Quality Control Plan for the strengthening of structures with composite materials to ensure their quality, effectiveness, proper operation and durability, with the aim of extensive use of non-destructive testing correlated with destructive testing.

During this study a series of prismatic specimens were performed with two types of concrete, to which were applied a CFRP reinforcement with different roughness on surfaces and different curing environments (humidity and temperature).

Tests to asses reinforcement quality must be on one hand those focussed to find damages and imperfections, or and air bags in the CFRP through infrared thermography and ultrasonic testing (Fig. 1).

On the other hand, pull off; shear torsion and flexural strength tests to control adhesion (Fig. 2).

The realization of a suitable quality control of CFRP reinforcement guarantee durability, increasing life cycle extension and amortization of the impacts on production phases (Vilches 2014).

Quality control must be performed from the stages of preliminary studies, development of project, works, as well as maintenance operations projects.

Fig. 1 Non-destructive testing of infrared thermography and ultrasound performed

Fig. 2 Pull off and Shear torsion destructive testing

4 Conclusions

There are few studies, research projects and bibliographic references on quality control protocols and LCA of concrete structures reinforcements with CFRP. In this paper an analysis of the sustainability of this type of reinforcements considering from works, environmental impacts and on people, to issues related to quality control.

The comparative assessment made on energy consumption in life cycle between CFRP reinforcements and others, highlights the lower impact of them, being the overall cost of implementation and maintenance comparatively lower. The reduced volume of material and waste produced, together with the recyclability of products, place these systems among the more sustainable ones.

Based on the social aspects, it can be said that CFRP reinforcements are much more sustainable because of the lower impact of the works in inhabitants and workers. In building works it is not necessary to proceed to eviction, reducing disorders in people and cost increases for rehousing.

In infrastructure projects it is not necessary to cut or divert traffic, reducing disorder in communities. Being lightweight materials and not requiring auxiliary means, the risk of accidents is reduced. Workers tend to be more specialized, and with a greater awareness of the need to observe safety measures and quality controls.

These conclusions are valid as long as the durability of the reinforcement is guaranteed, being essential to ensure the proper application of reinforcement by performing a quality control programme that ensures an extended operation period with limited maintenance costs, as well as increasing impacts amortization by year.

Acknowledgements Acknowledgements to Nuremco Research Project funded by the Ministry of Economy and Competitiveness within the Feder Interconecta Program, to the V Research Plan of the University of Sevilla and to the company Azul Construcción.

References

ACI 440.2R-08 (2008) Guide for the design and construction of externally bonded FRP systems for strengthening concrete structures

Basf-Constructions (2011) Recomendaciones para el diseño de planes de control de calidad en ejecución de refuerzos con fibra de carbono. Basf Construction, Barcelona

Basf-Constructions (2012) Norma europea EN 1504. Guía simplificada ilustrada para todos los profesionales del sector de la reparación de hormigón. Basf Construction, Barcelona

Howarth J, Mareddy S, Mativenga PT (2014) Energy intensity and environmental analysis of mechanical recycling of carbon fibre composite. J Cleaner Prod 81:46–50

Instituto Valenciano de la Edificación IVE (2008a) Experiencia en Inspección de estructuras en edificios. Comunidad Valenciana 1991–2008. Conselleria de Medi Ambient, Aigua, Urbanisme i Habitatge, Valencia

Instituto Valenciano de la Edificación IVE (2008b) Guía para la Inspección y Evaluación Preliminar de estructuras de hormigón en edificios existentes. Conselleria de Medi Ambient, Aigua, Urbanisme i Habitatge, Valencia
Jiang L et al (2015) Recycling carbon fiber composites using microwave irradiation: reinforcement study of the recycled fiber in new composites. J Appl Polym Sci 132(41)
Keith MJ, Oliveux G, Leeke GA (2016) Optimisation of solvolysis for recycling carbon fibre reinforced composites. Paper presented at the 17th European conference on composite materials. University of Munich, Germany, 26–30 June 2016
Lago C et al (2007) Análisis del ciclo de vida de combustibles alternativos para el Transporte. CIEMAT, Madrid
Maña i Reixach F, Bellmunt et al (2005) Recomendaciones para el reconocimiento sistemático y la diagnosis rápida de forjados construidos con cemento aluminoso. ITC, Barcelona
Muñoz Marzá C, Vilar M, Garraín Cordero D (2008) Evolución del impacto medioambiental de los retrovisores de vehículos. Evaluación mediante análisis del ciclo de vida
Osorio Cardona JF (2011) El consumo sostenible de los materiales usados en la construcción de vivienda: la perspectiva sistémica. Dissertation, Universidad Nacional de Colombia
Pereira JJMG (2015) Bases de trabalho para a avaliação do ciclo de vida de vigas em betão armado reforçadas com laminados de CFRP de acordo com as técnicas EBR e NSM
Río Bueno A (2008) Patología, reparación y refuerzo de estructuras de hormigón armado de edificación. UPM, Madrid
Rivarola A, Torres L, Rivarola AM, Berrios G, Gardey MC, Gatica P (2011) Estudio y comparación de alternativas de fin de vida aplicadas a palas de aerogeneradores, empleando metodología de análisis de ciclo de vida
Serrano Lanzarote B (2014) IEE.CV Informe de Evaluación del Edificio, Comunitat Valenciana, Guía de Inspección. Conselleria de Medi Ambient, Aigua, Urbanisme i Habitatge, Valencia
Sika (2007) Procedimiento de aplicación. Sistemas de refuerzos de estructuras. Sika AG, Madrid
Song Young S, Youn Jae R, Gutowski TG (2009) Life cycle energy analysis of fiber-reinforced composites. Compos Part A Appl Sci Manuf 40(8):1257–1265
Tjibaou C, Piano R (2010) Sostenibilidad en la edificación. Centro Cultural Jean Marie Tjibaou
UNE EN 1504 (2006) Productos y sistemas para la protección y reparación de estructuras de hormigón
Vilches Such A, Lizana-Moral FJ, Barrios-Padura A, Díaz-heredia I, Serrano-Jiménez AJ (2014) Responsible management of construction resources. Amortization of the concrete's embodied environmental impact as a sustainable strategy. World Sustainable Building Conference 2014. Barcelona, Spain

Rethinking Ephemeral Architecture. Advanced Geometry for Citizen-Managed Spaces

Andrés Martín-Pastor, Amanda Martín-Mariscal and Alicia López-Martínez

1 Introduction

At the current moment of crisis, the traditional model of architectural production is strongly questioned due to its anti-ecological character and the negative imprint that this constructive dynamic has left us in the city. We refer to an architecture set by markets and whose processes are performed top-to-down at all levels. A trend that presupposes that the processes of innovation and technology resulting from them are applied mainly from the construction industry, then made profitable from the commercial side.

From this premise, the problems of the city are also understood—from top—as a set of transient imbalances that can only be addressed and resolved by accepting the mechanisms of the current system. Some problems that, according to Sennett (2008), are very complex to analyze and even more to solve, but whose consequences can be seen every day in our cities: degraded urban empty spaces, contamination due to unfinished works public spaces of impossible habitability, etc.

Precisely, to address this crossroad and from a totally different approach, citizen managing spaces offer a management system of organization that intended to answer to many of these problems—not all—we suffer in the current city. Social participation is important here, being the citizen who actively and collaboratively participates in the decision-making and projects to be undertaken. These groups

A. Martín-Pastor (✉)
Department of Graphic Engineering, Instituto Universitario de Arquitectura y Ciencias de la Construcción, University of Sevilla, Seville, Spain
e-mail: archiamp@us.es

A. Martín-Mariscal
Department of Architectural Design, Instituto Universitario de Arquitectura y Ciencias de la Construcción, University of Sevilla, Seville, Spain

A. López-Martínez
Department of Graphic Engineering, University of Sevilla, Seville, Spain

P. Mercader-Moyano (ed.), *Sustainable Development and Renovation in Architecture, Urbanism and Engineering*, DOI 10.1007/978-3-319-51442-0_25

have very diverse characteristics and different needs. Some contemplate the use of outdoor activities, since the main character of their activity takes place at precisely the outside—as the case of urban gardens—or because the activity itself necessarily involves the processing and use of that space according to the architectural action—as in the regeneration of urban empty spaces—architecture understood here as in its broadest sense of "transformation of the human habitat" (Martín-Mariscal 2016). In this sense it is interested to remember the words of Estalella (2015):

> The cities are mutating from one side to another of the globe. People that previously just transited through their streets or walked gardens, lodge in them to furnish squares and occupy lots. A form of citizen urbanism emerges, through which the urban space is redesigned on the street and from the street.

2 Self-building or Ephemeral Architecture?

The highlighted aspect, and which is linked to *sustainable construction processes*, is the formula already used by many of these groups called *ephemeral architecture*. This architectural style is suitable not only to solve the problem of lack of resources of promoters and managers, but also to reasonably cope—or perhaps in the unique way possible—legal problems arising from realizing an architectural/structural project on a floor without any license, either a disused land, public or other. The nature of ephemeral architecture—intentionally understood as simple furniture that is assembled and disassembled without damaging the environment where it is located—lacks of application of either mandatory compliance or constructive recommendation rules.

2.1 Digital Craft Versus Improvised Manufacturing

At the same time, we have observed that collective and agent drivers of these citizen initiatives frequently have few tools to promote the full phase of design, production and assembly of architectural facilities—ephemeral or non-ephemeral, simply self-built—optimized to develop their activities (Martín-Mariscal 2016). Scarce economic resources have led, in some cases, facilities almost as emergency mode, without planning, resources and knowledge to develop architecture according to different needs. Similarly, the character of innovation in architecture is appreciated and lived, for some groups, as something external and outside their means and possibilities, assuming—sometimes almost like a creed—improvising manufacturing and low technological character recycling as an environmental brand image even more outsider, rejecting everything that is not natural. Indeed, a CAD program, a CNC milling machine and a robotic arm could be understood as something artificial, alien to the human and to the craft. In contrast to this theory, *the digital craftsmanship and built digital materiality concept* (Chiarella and Martín-Pastor 2015), which later will be developed, is established.

The moment of crisis—or change—in which we live provides new opportunities recently unimaginable. One of these opportunities is access to technology and laboratories of digital production, allowing work on the basis of innovation, not only from the market sector, but also from the creation of knowledge produced in universities and researching groups associated to them. The outlook notes to how these new production processes, applied to architecture, can overcome some inherent challenges of the current time (Martín-Mariscal et al. 2016). Thus, numerical control machines, widely used in industrial design, are slowly incorporated into the architecture offering the possibility of manufacturing without intermediaries (Meredith and Kotronis 2012). These new-production architectural processes face the challenge of accompanying the complexity of the projects generated by informatics tools to prevent that complexity comes from execution besides complex geometries. Digital materiality starts from graphic thinking with digital processes, proposing new relationships between the architectural object and its representation (Chiarella et al. 2016).

3 A Project of Ephemeral Architecture

According to these social premises and framed within objective limits, the challenge we propose is to develop a model of design-production-assembly of a low-impact architecture in these citizen-managed spaces, using digital tools—Parametric Design and Digital Fabrication—, minimizing cost and impact of such architectural interventions. In short, perform a low-cost and high technological character architecture—*Low-Cost and High Tech*—involving low investment for their promoters but that maximizes efficiency and sustainability in all processes, incorporating optimized production and self-assembly forms. A productive logic that already exists in other fields, precisely in furniture design. The most obvious example is *Ikea* furniture, of which we have taken several references in the level of optimization and differentiation of teamwork: design, manufacturing, and assembly, the latter being accomplished by the end user.

Previous research on the use of surfaces (Casale et al. 2013) and its use in ephemeral pavilions, places us in an appropriate position to continue the research we have developed over the past two years, optimizing the design, reducing costs and impact of such facilities. Ephemeral architecture that proposed is composed of developable surfaces, those that can be deployed in the plane. Proposed prototypes are self-supporting skins made by a thin sheet of material—wood—approximately 6 mm thick. With parametric algorithms and CAD-CAM technology it is possible to deploy these complex surfaces in flat pieces. They are assembled on the ground, like a big puzzle, then curving from the plane into space to acquire its final shape and strength, thanks not the thickness but the geometry itself (Narvaez-Rodriguez et al. 2014; Martín-Pastor 2016).

The production of this type of architecture considers creating its own manual of *Ikea* type assembly, and part of the research is to soon become the architectural object into a real piece of furniture that can be self-built by any group of people. Part of the novelty of the proposal also influence on rethinking the use and enjoyment of certain spaces through so light interventions and so low impact, excluding the rules of architectural occupation, even true architectural structures.

3.1 Previous Experience

Previous experience of design, manufacture and installation of this type of ephemeral architecture consists of five experimental pavilions in the field of geometry research—architectural geometry—and digital fabrication (Fig. 1). These pavilions are: *The Caterpillar Gallery*, made in University of Sevilla ETSIE in February 2014; *The Cocoon Gallery*, at National University of Colombia (UNAL) Medellin in August 2014; *The SSFS Pavilion-Santa Fe*, FADU at Litoral University, Santa Fe, Argentina, from March 26th to April 27th, 2015; The Butterfly Gallery-*Helical Surfaces* in UFRJ at Federal University of Rio de Janeiro, Brazil, from August 4th to 14th 2015; and *SSFS Fablab-Pavilion Seville*, raised at Seville's Plaza Nueva in September 2015 (installation awarded with 2nd and 3rd Emporia National Award in the categories of "Innovation in Ephemeral Architecture" and "Green Stand").

Fig. 1 *Top-left* Caterpillar gallery. *Top-right* Cocoon gallery. *Bottom-left* Butterfly gallery-helical surfaces; *Bottom-right* SSFS-Pavilion Fablab Sevilla

This type of architecture, besides being an example of realization of Digital Fabrication processes described above, provides architectural and functional characteristics appropriate to be enjoyed in activities developed in the areas of citizen-managed spaces of our cities.

3.2 *Innovation and Sustainability. Architecture as Digital Fabrication Furniture*

As discussed above, we propose the development of a *furniture* type of architecture. Bringing these ideas from the world of industrial furniture to the world of architecture has clear advantages from the point of view of sustainability. From the point of view of design the classic concept of the construction process, where the architect is the author of the two-dimensional graphic information and a group of specialized craftsmen who execute changes are who execute—with more or less mastery—what was previously defined in the project. Gramazio and Kohler (2008), establish *digital materiality* in the existing complex interweave between computer programming, 3D construction—or digital modeling—, data and matter, at different stages of the architectural project. The new concept of digital craftsmanship emphasizes the figure of the architect as head of the interaction of all digital processes related to design and production (Fig. 2).

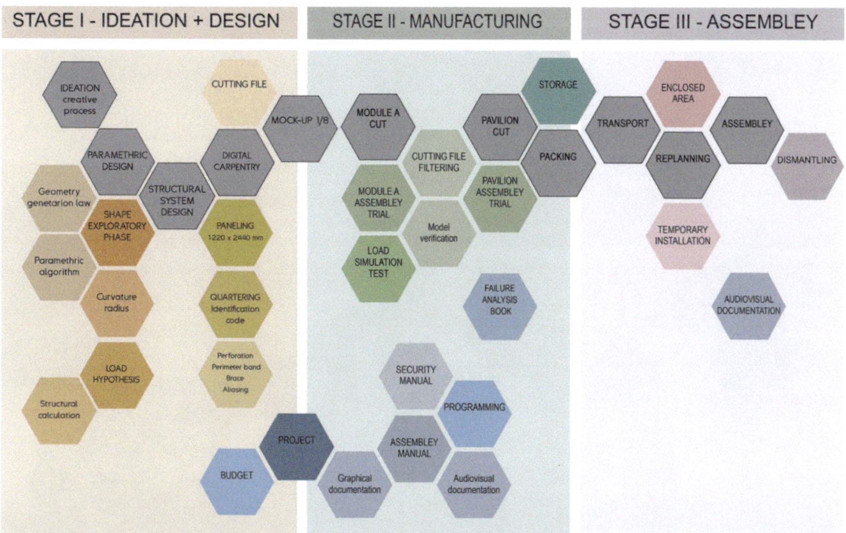

Fig. 2 This graph summarizes the three main phases of digital fabrication process: design, manufacturing and assembly, showing the various feedbacks that occur between them (image created by authors)

In the digital fabrication process, robotic cut machine—CNC or Robot Arm—handles the execution and machining work, minimizing the margin of error and getting cuts of very high precision. Human error is greatly reduced, focusing on monitoring and process optimization.

The optimized digital fabrication process will use the closer digital fabrication laboratory to the installation location, thus minimizing the energy costs and transport CO_2 emissions and giving preference to the local economy. Furthermore, a sustainable process must have priority in the use of materials of the area, with special attention to recycled and local products, thereby reducing the ecological footprint.

The architecture will be entirely reduced to a set of elements where does not exceed or lack any part or any screws. Thus, only necessary parts are produced, minimizing waste generated in the fabrication process (Fig. 3). Similarly, the packaging process ensures that only truly necessary material is transferred to the installation site, being a dry type construction that generates *zero* waste.

The assembly—or dismantling—is done by the end users, saving costs in the process both in the displacement of specialized personnel or machinery (Fig. 4). Maintenance spending in this architectural model is almost zero as it is a set of pieces defined by a code and replaceable by the user, who is in possession of the cutting file and the ability to reproduce it again in the laboratory of digital

Fig. 3 SSFS Pavilion-Fablab Sevilla assembly test in ETSIE building, and Fablab Sevilla

Fig. 4 Different stages of assembly of Fablab SSFS Pavilion, Plaza Nueva (Seville)

fabrication of his local network. Finally, the proposed architecture is completely removable, transportable and usable in other places, so no physical trace occurs in the place where it was located.

4 Results and Lines for Improvement

The proposed system tries to give architectural solutions from the logic of few material resources—a thin sheet of material—and minimum installation cost. The challenge of this project is to make progress on problems detected in previous experiences and to propose a truly useful model for use in public citizen-managed spaces.

Previous experiences faced problems such as the high number of workers required for assembly. This was due to the weight of each ring, oscillating around 80 kg. To move these pieces two teams of 7 people were needed, one (assemblers) to move the material and the second one (aid) to lift and bend it. The weight problem is due in part to the type of material used, wood and its variants, either on plywood or on medium density boards. Therefore, it would be interesting to investigate the possibility of using new materials to reduce this type of difficulties.

Managing to reduce the number of assemblers and aids involves research work in designing and optimizing geometries in material efficiency. We propose to investigate the use of recycled materials and recycled foams to reduce the density of the material while other properties are held constant as the material active tension or bending strength. In this regard, another difficulty has been the material unwinding over time, which has led to the appearance of deformations in the structure. We have thought to address this situation from two different perspectives: using materials that do not develop unwinding, and generating the meeting of surfaces to avoid bi-tangency points, due that this leads to weakness of the material over time.

Also, from the point of view of building systems, assembly method developed previously by nylon screws and flanges to suit unspecialized mounting personal must be improved. On the other hand, being objects dedicated to be installed in public space, these projects must be sensitive to the pavement. An important part of the research will consist on articulating support and tightening systems that allow this condition.

Therefore, it is proposed to advance the design of digital fabrication process from three different perspectives: (1) Design in material geometry, understanding it as flat sheets that are curved in space, seeking efficiency. (2) Design in purifying fabrication processes. (3) Design in the detailed description of each assembly stage.

This entails an exhaustive study of the project on several levels:

1. Materials, optimizing efficiency.
2. Advanced Geometry of developable surfaces.
3. Methods of structural calculation of self-supporting structural skins.
4. Investigation of shape and support systems.
5. Construction systems related to assembly and installation.
6. Process optimization.
7. Graphical expression, minimizing errors in assembly (Fig. 5).

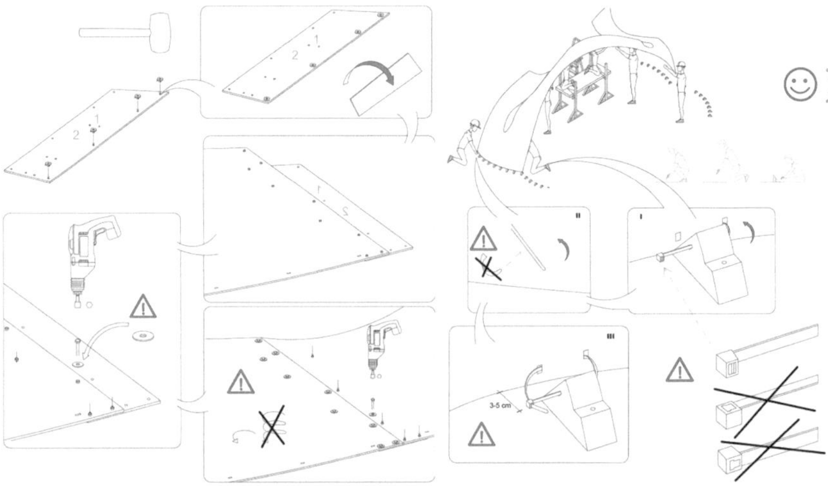

Fig. 5 An example of assembly manual proposed by Helena Santos Calvo, for SSFS-Pavilion. End of grade project illustration titled: optimization of the assembly phase. Ephemeral pavilions of digital abrication and lightweight materials. PFG 2015-2016, ETSIE, University of Sevilla

5 Conclusions and Discussion. A Reflection on the Role of the Architect

As discussed, it is proposed to define an easy to assemble, low cost and reusable model of ephemeral architecture, whose recipients are citizens and citizen-managed spaces. In this project the architect becomes the qualified mediator who uses his expertise to propose efficient architectural solutions adapted to the needs of a social collective. To do so, collective creativity or the possibilities for innovation or industrialized manufacturing are included, allowing access to digital fabrication laboratories. However, he refuses to be allied of an architecture based on work improvisation and low efficiency systems, while also does not demands results or unrealistic responses to people or unskilled collectives. Therefore, and sensitive to the underlying problems of these groups, a model of architectural production is proposed that leads to a product of unique architectural design. A *furniture* type, low-cost and high-tech architecture that exceeds improvisation and unskilled self-construction and proposes a real alternative to the architectural market.

> There are interesting opportunities for architects, mainly to reinvent the process of design and manufacturing, and to foresee the integration of technology at home so that it responds to the needs and values of its inhabitants and demographic, health and energy challenges of society. In the process of drastically improving the quality of the built environment, we can even create a more central and important role for the architect of today. (Bernstein and Deamer 2010, p. 113)

Acknowledgements *The Caterpillar Gallery*. Authors and project design: Roberto Narvaez-Rodríguez, Andrés Martín-Pastor, (In collaboration: Infante-Pereda Margarita Maria Aguilar-Alejandre). Manufacture equipment FablabSevilla. Assembly: Teachers and students of ETSIE Schools and ETSA, University of Sevilla. Collaboration: Department of Engineering University of Sevilla.

The Coccon Gallery. Project design Andrés Martín-Pastor and Roberto Narváez-Rodríguez. Manufacturing: Equipment FablabUnal Medellin. Execution and assembly: Architecture students UNAL UNAL Medellin and Manizales. Coordination: Es-strained Media Representation, Faculty of Architecture. Co-Unal Medellin Colombia. Collaboration: Extra. Grafica Engineering Equipment FablabSevilla.

SSFS Pavilion. Santa Fe Design and Project: Andrés Martín-Pastor, Roberto Narvaez-Rodriguez. Co-parametric design: Expósito Juan Bejarano. Academic Coordination and management: Mauro Chiarella. Master Program in Architecture, Fadu, Universidad Nacional del Litoral. Production and assembly: Students and teachers pro-Master of Architecture, Fadu. Collaboration: Extra. Graphic Engineering University of Sevilla, ETSIE Equipment FablabSevilla.

The Butterfly Gallery-Helical Surfaces. Project design Andrés Martín-Pastor, Juan Expósito Bejarano. Coordination: Maria Angela. Manufacturing: LAMO3D, Focus Design. Mounting: collaborative between teachers and students of the UFRJ. Collaboration: Extra. Graphic Engineering University of Sevilla, ETSIE Equipment FablabSevilla.

SSFS Pavilion—FablabSevilla. Project design Andrés Martín-Pastor, Juan Expósito Bejarano and Roberto Rodriguez Narvaez. Manufacturing: FabLabSevilla. Production Chief: Daniel del Valle. Mounting: collaborative between teachers and students of the ETSIE, FabLabSevilla and Edificalab. Collaboration: Extra. Graphic Engineering University of Sevilla, ETSIE Equipment FablabSevilla.

References

Bernstein P, Deamer P (2010) Building (in) the future: recasting labor in architecture. Yale School of Architecture, New Haven

Casale A, Valenti GM, Calvano M et al (2013) Surfaces: concept, design. Parametric modeling and prototyping. Nexus Netw J 15:271. doi:10.1007/s00004-013-0146-8

Chiarella M, Martín-Pastor A (2015) Pensamiento Gráfico y Diseño Colaborativo. Geometríasdesarrollables en composiciones plegadas Arquitectónicas. In: Proceedings of the XIX conference of the Ibero-American Society of Digital Graphics 2015 [Blucher design proceedings, v.2, n.3]. Blucher, São Paulo, pp 702–707. doi:10.5151/despro-sigradi2015-110107

Chiarella M, Martín-Pastor A et al (2016) Graphic and digital thought processes. Three cases of digital materiality built (COCOON/Colombia, BANCAPAR/Chile, SSFS/Argentina). In: Proceedings of the XVI international congress of architectural EGA 2016 [n.16]. Foundation University of Alcalá, Alcalá de Henares, pp 763–770. http://ega2016.com/wp-content/uploads/2016/06/Libro_EGA2016_Tomo_II.pdf

Estalella A (2015) Collective architecture: another urban sensibility. Retrieved from http://www.prototyping.es/destacado/colectivos-de-arquitectura-otra-sensibilidad-urbana

Gramazio F, Kohler M (2008) Digital materiality in architecture. Lars Müller, Baden

Martín-Mariscal A (2016) Collective creativity. Processes of contemporary architecture in contexts of complexity. Disertation, University of Sevilla

Martín-Mariscal A, Martín-Pastor A, López-Martínez A et al (2016) Advanced geometry for public spaces management. In: Proceedings of the XX conference of the Ibero-American Society of Digital Graphics 2016. Blucher, São Paulo (in press)

Martín-Pastor A (2016) A return to the fundamentals of geometry. The butterfly gallery, geometric strategies for digital fabrication. PROARQ notebooks. J Archit Urbanism 2016 [n.25]. UFRJ Post-graduation Program in Architecture, Rio de Janeiro, pp 2–30. http://cadernos.proarq.fau.ufrj.br/public/docs/cadernosproarq25.pdf

Meredith N, Kotronis J (2012) Self-detailing and self-documenting systems systems for wood fabrication: the BurjKhalifa. Advances in architectural geometry. TU Wien, Vienna, pp 185–198

Narváez-Rodríguez R, Martín-Pastor A, Aguilar-Alexandre M (2014) The caterpillar gallery: quadric surface theorems, parametric design and digital fabrication. In: Block P, Knippers J et al (ed) Advances in architectural geometry 2014. Springer, pp 309–322. doi:10.1007/978-3-319-11418-7_20

Sennett R (2008) The craftsman. Yale University Press, New Haven

NESS®, an Alternative System to Double Strand of Hot Water that Saves Water and Energy

M. Carmen Ladrón de Guevara Muñoz, Eduardo J. Dueñas Ladrón de Guevara, Marcos Ortega Rodríguez and Luis Martín Martínez

1 Introduction

The increase in water consumption by modern societies is one of the biggest challenges faced by regions worldwide (World Assessment Programme, WWAP 2012). The global water demand is projected to increase by 55%, by 2050 (United Nations World Water Assessment Programme 2015). A price increase of water consumption is expected by 2020 as one of the measures used to control or reduce water demand by different users. Even more, increased water-related problems due to Climate Change (UN-Water 2010) (droughts, floods, storm surges and landslides) will put additional strain on water resources management in the future.

In total water abstraction, public water supply accounts for approx. a fifth of water use across Europe and includes water supply to households, public buildings, small businesses and industries (European Environment Agency 2009). Analysing the information on the use of water from public water supply by economic sector in 2013, the main users of water in the EU were households, 2/3 of the total consumption, followed by offices and other buildings for diverse uses, c.f. Fig. 1 (European Commission, s.f.). Thus, major uses of water are carried out through and within buildings: 70–80% of water delivered from the public water supply system is used for domestic needs within buildings, or related to buildings' outdoor areas (Bio Intelligence Service 2009).

Domestic Hot Water (DHW), defined as human water needs is the dominating use of water in buildings and occurs in both residential (households) and non-residential (commercial, public and industrial) facilities, c.f. Fig. 1. The

M.C. Ladrón de Guevara Muñoz (✉) · E.J. Dueñas Ladrón de Guevara · M. Ortega Rodríguez
Métrica6 Ingeniería y Desarrollos S.L., Malaga, Spain
e-mail: carmenldg@metrica6.es

L. Martín Martínez
Hidrología Sostenible, Malaga, Spain

P. Mercader-Moyano (ed.), *Sustainable Development and Renovation in Architecture, Urbanism and Engineering*, DOI 10.1007/978-3-319-51442-0_26

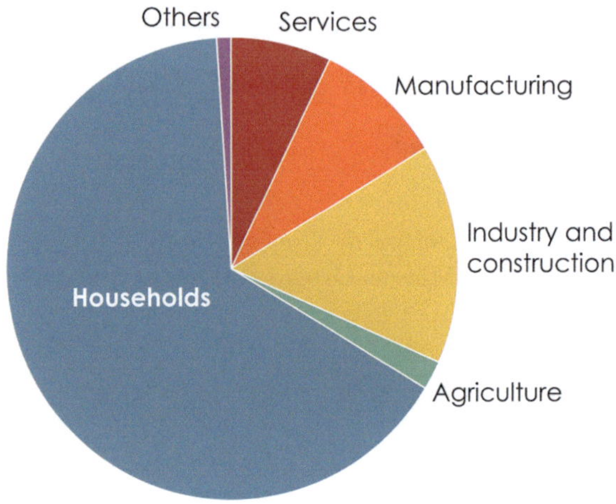

Fig. 1 Water use by economic sector

residential water use represents 72% of the total water use in buildings, and 28% for non-residential buildings.

The three largest sources of water use at home are showers and baths, toilet flushing and washing machine, where personal hygiene represents the 35% of the total consumption (Eurostat 2015), cf. Figs. 2 and 3.

In addition, Domestic Hot Water use is identified as a key issue from the building sector as it directly relates to the energy consumption of buildings. Inefficient use of hot water also leads to higher energy use, with extra financial and environmental costs; an estimated 10–20% of the overall energy consumption for heating purposes within EU is used for Domestic Hot Water (DHW) (Association of the European Heating Industry 2010).

Current water use in the EU is around 150 l/person/day (INE, Instituto Nacional de Estadística 2013). DHW consumption in buildings also depends on the number of occupants and it is strongly related to individual behaviour of its inhabitants.

Nowadays, a serious issue with hot water still exists: the wastage of water produced when trying to access hot water, leading to irremediably let a huge amount of cold water flow first. There are currently two systems for water saving in the market: on the one hand, the double strand of hot water, on the other hand, NESS® technology.

2 Methodology

To compare the efficiency of both systems, they are studied pursuing the same parameters: water savings, energy consumption, and costs.

Afterwards, results are compared and economic considerations are taken into the equation to open a discussion regarding energy and economic expenses as the main differentiation among them.

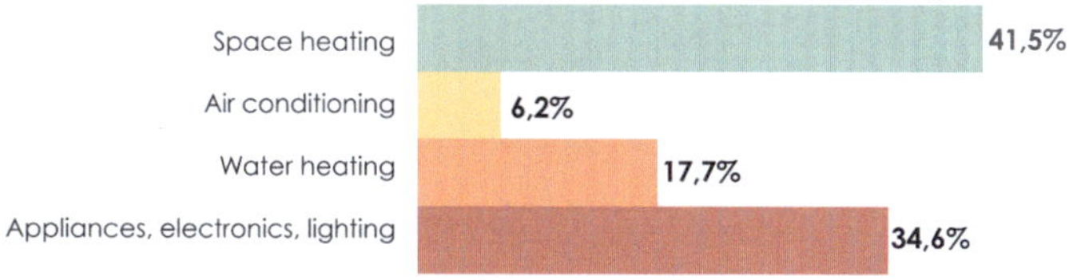

Fig. 2 Energy consumption by end uses

Fig. 3 Water consumption by end uses

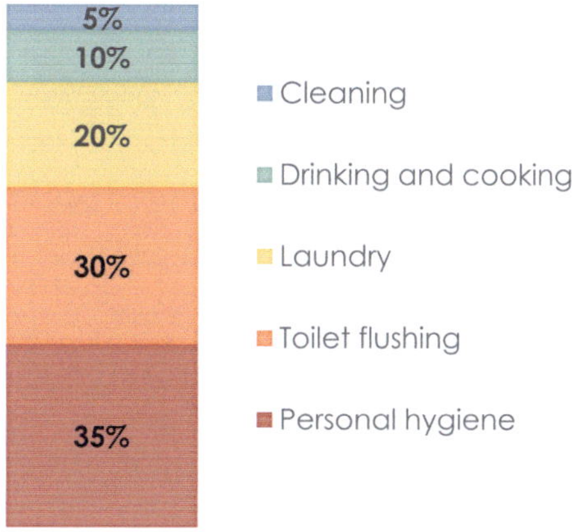

3 Conventional Facilities

Traditional facilities usually consist of one branch of cold water and one branch of hot water, departing from the cold water inlet supply and the water heater to all the consumption points distributed throughout the house or building.

According to actual water consumption statistics, the water disposal occurred every time someone intends to access hot water at home amounts approximately to over 800 l a month. This means around 2300 l wasted by the average family composed of 251 members (INE, Instituto Nacional de Estadística 2013).

3.1 Thermal Energy Disposal (Related to Hot Water Consumption)

After every use of domestic hot water, the temperature of the water stored in the pipes decreases to room temperature. This means that a part of the energy employed to increase the water temperature is lost in every use. In this paragraph, the energy loss is estimated; some assumptions such as the pipes dimensions or material must be done. The starting point is shown in the following Table 1.

Table 1 Parameters used during the calculations

Concept	Nomenclature	Value
Room temperature	T_{amb}	25 °C
Outlet temperature (heater)	T_{out}	60 °C
Water heat capacity	C_p	4180 J/kg°C 1163 Wh/kg°C
Water density (T_{amb})	ρ	1000 kg/m^3
Pipe inner diameter	d_i	20 mm
Hot water pipes length	L	15 m

The energy loss considered in this section refers to thermal energy since it is an amount of heat transferred to the atmosphere by radiation. It remains as follows:

$$E_{Th\ loss} = M_{water} \cdot C_p \cdot (T_{out} - T_{amb}) \cdot N_{uses} \tag{1}$$

where M_{water} is the water mass inside the tubes and N_{uses} refers to the number of times that hot water is used per day in the average household.

According to (Métrica6 Ingeniería y Desarrollos S.L. 2015), hot water is used 291 times a day, while there are an average of 251 inhabitants per household (INE, Instituto Nacional de Estadística 2013) which means the parameter N_{uses} in Eq. 1 equals 291 uses per person a day × 251 inhabitants per household = 73.04 uses per household a day. Then, the mass of water is obtained as displayed in Eq. 2.

$$M_{water} = \rho_{water} \cdot L \cdot \pi \cdot \left(\frac{d_i^2}{4}\right) \tag{2}$$

Thereby, applying Eqs. 1–2, the thermal energy loss results in 1.40104 Wh/day.

4 Double Strand of Hot Water (Facilities)

In this system there are two hot water pipes that distribute hot water around the house: one of them gives supply to each consumption point, and the other, which is actually an extension of the first, returns the water that has not been used to the heater. By means of this system, the hot water branch is always kept within a certain temperature range so that, when there is hot water demand, water is ready to be used without disposing any cold water, cf. Fig. 4.

In Spain, national construction regulations oblige the installation of the DSHW where the length of the hot water pipe exceeds 15 m (Ministerio de Fomento 2009).

However, maintaining all the water stored in the hot water pipe at a certain temperature regardless of demand periods translates into high energy costs. In order

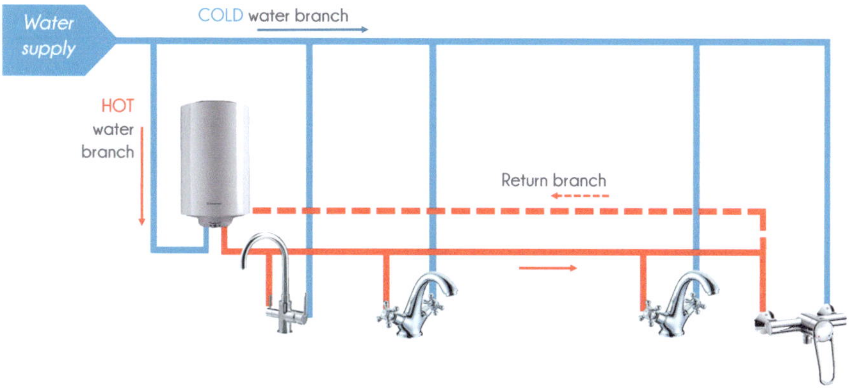

Fig. 4 Double strand of hot water

to quantify these costs, the thermal and the electric energy expenses due to the DSHW are surveyed in this section.

4.1 Energy Disposal

While water disposal is avoided through this system, there are other aspects to consider such as the energy that needs to be wasted to make it work. In order to quantify this amount of energy, the following aspects are considered:

- The DSHW activates when water temperature goes under 50 °C.
- Energy consumption depends on the number of times that the system activates along the day, making use of two different types of energy: electric (due to the activation of the pumping and control equipment) and thermal (as a consequence of the heat transferred by the heater to the water but not consumed by the final user).

4.1.1 Thermal Energy Due to Pipes Heating

In the case of the DSHW, the process of warming up pipes and water occurs every time water temperature decreases in 10 °C, reaching 50 °C. This means that no matter whether there is hot water demand or not, the heater will warm up the whole installation to be ready for the time a demand comes in. Parameters in Table 2 need to be considered together with Table 1 for the calculations.

To know the total amount of thermal energy transferred to the atmosphere by the DSHW, the number of cycles performed along the day needs to be known and can be calculated taking into account the time required to heat the total volume of water

Table 2 Parameters considered for the double strand of hot water

Concept		Nomenclature	Value
Double strand of hot water length		L	15 × 2 = 30 m
Boiler efficiency		η	90%
Pipe thickness		e	1 mm
Insulation thickness		e_{ins}	25 mm
Total diameter with insulation		D_{ins}	72 mm
Thermal conductivity	Copper	K_{cop}	398 W/mK
	Insulation	K_{ins}	0.04 W/mK

stored in the pipes, the time it takes to cool down and the actual amount of time during which hot water is used as Eq. 3 shows:

$$N_{cycles} = \frac{t_{eff.cal}}{t_h + t_c} \tag{3}$$

where $t_{eff.cal}$ refers to the amount of time that the DSHW (from now on, DSHW) is working without consumption or demand, namely, 24 h subtracting the amount of time during which water is actually being consumed:

$$t_{eff.cal} = 24 - t_{use} \tag{4}$$

Therefore, effective time of hot water use per day can be obtained as:

$$t_{use} = \frac{5\ \text{min}}{use} \cdot \frac{291\ uses}{pers} \cdot \frac{251\ pers}{house} \tag{5}$$

Applying Eq. 5, t_{use} equals 36.52 min, that is to say, 2191.23 s of hot water use per day in the average household, while the time DSHW is working without demand results in 84.20877 s distributed in several heating and cooling cycles along the day, t_h and t_c respectively, and are calculated as exposed in Eqs. 6 and 7:

$$t_c = M_{water} \cdot C_p \cdot R_c \cdot \ln\left(\frac{T_{out} - T_{amb}}{T_{act} - T_{amb}}\right) \tag{6}$$

$$R_c = \frac{\ln \frac{d_e}{d_i}}{2 \cdot \pi \cdot L \cdot k_t} + \frac{\ln \frac{d_{ins}}{d_e}}{2 \cdot \pi \cdot L \cdot k_{ins}} \tag{7}$$

Analogously, the heating time is obtained according to Eq. 8:

$$t_h = M_{water} \cdot C_p \cdot R_c \cdot \ln\left(\frac{P_{heater} \cdot \eta \cdot R_c + T_{amb} - T_{act}}{P_{heater} \cdot \eta \cdot R_c + T_{amb} - T_{out}}\right) \tag{8}$$

Once t_c and t_h have been resolved, the number of cycles, N_{cycles} can be calculated using Eq. 3, which results in 37.62 cycles per day.

The purpose of this section is to ascertain the thermal energy used in a day:

$$E_{th\ total} = E_{th\ eff} + E_{th\ loss} \tag{9}$$

Being $E_{th\ eff}$ the energy transferred by the heater to the water and pipes to increase the water temperature that is finally consumed by the user, while $E_{th\ loss}$ is the heat transferred to the water and pipes during the cycles where the water temperature decreases due to the lack of demand, and it is not either used afterwards.

However, the total amount of thermal energy employed by the DSHW can also be calculated as Eq. 10 displays:

$$E_{th\ total} = M_{water} \cdot C_p \cdot (T_{out} - T_{act}) \cdot N_{cycles} \tag{10}$$

This results in a total thermal energy expense of 412,363 Wh/day, of which 80.01 Wh are effectively used and 332,346 Wh are disposed to the ambient without attending any actual demand for hot water, cf. Eq. 11.

$$E_{th\ loss} = M_{water} \cdot C_p \cdot (T_{out} - T_{act}) \cdot N_{loss\ cycles} \tag{11}$$

where the number of cycles performed under no demand, $N_{cycles\ loss}$, results from $N_{loss\ cycles} = N_{cycles} - N_{uses} = 3762 - 251 \cdot 291 = 3032\ cycles$

4.1.2 Electric Energy Due to Pumps Functioning

In addition to thermal energy, the DSHW requires electricity to start the pumps in charge of moving the water stored in the pipes. Consequently, the pumping system consumes an amount of electricity in every cycle that needs to be considered among the energy expenses.

The pumping system power is estimated to be 90 W, and according to the previous section t_h = 157.03 s which means water requires 157 s to warm up from 50 to 60 °C. Hence, the electric energy consumed by the pump results in 1477 Wh per day according to Eq. 12.

$$E_{elec\ P} = P_{pump} \cdot N_{cycles} \cdot t_h \tag{12}$$

4.2 Effective Energy—Real Consumption

Although a great deal of energy is disposed through this system, not all of it is in vain: a part of it serves its purpose of heating water that is finally consumed.

Only 80.01 Wh/day of thermal energy are consumed effectively out of 412.36 Wh/day; meanwhile, the effective electric energy used relates to Eq. 12, applying 73 cycles (water consumption). Thus, 2867 Wh of electricity are effectively employed every day to impulse hot water through the pipes. These figures indicate that only 20% of the energy consumed by the DSHW is effectively employed.

5 The Alternative Technology: NESS®

5.1 *Elements of NESS®*

NESS® consists of a smart recirculating hot water system without any additional piping installation. It is composed of three separate bodies, each of them responsible for a different function: a pump module, a bypass or connection element and a switch. The three different modules communicate by a low frequency radio signal. Standard kit contains, cf. Fig. 5:

- PUMPING MODULE: Connected to cold water pipes before the water heater, the pumping module impulses hot water from the heater to the desired point of consumption.
- BYPASS MODULE: Bypass is installed at the location of usage where we want to enjoy instant hot water; the bypass is recommended to be located at the furthest point from the heater, so every faucet in between will benefit from NESS® cycle.
- SWITCH MODULE: The capacitive switch is the activator and indicator of the cycle. It emits a blue light when the water is cold and a red light when the water is hot. This module can be installed in any convenient surface.

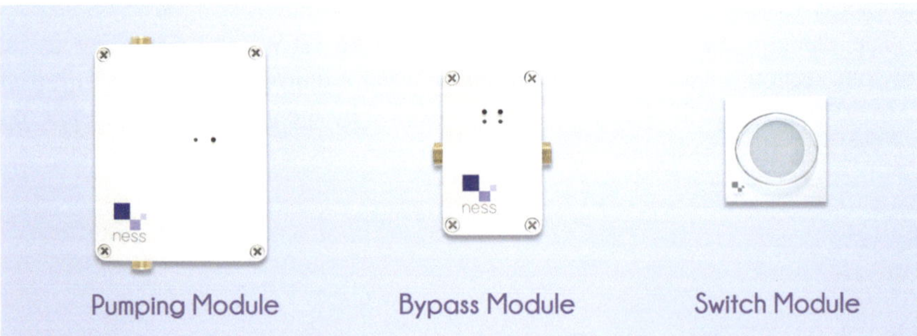

Fig. 5 NESS® modules

5.2 *Functionality of NESS®*

In order to understand how NESS® works, Fig. 6 shows the 3 main stages undergone by the device: first, the user needs to hold his/her hand in front of the activator for a couple of seconds, at this moment, with no need to turn on the tap, NESS® brings hot water from the heater to the faucet where the bypass has been installed in. Ultimately, may the water be hot in the desired point, NESS® shows a red light in the switch indicating that hot water is ready to be enjoyed in this tap and the rest in between.

5.3 *Energy Consumption and Savings Induced by NESS®*

The thermal energy loss using NESS® amounts the same than conventional installations without NESS®, that is to say, 14 kWh/day, cf. Eq 1 in Sect. 3. However, regarding the electricity required to power the system, the uptake of the 3 modules must be considered, c.f. Table 3.

Although Table 3 indicates the maximum power, modules do not use it to such extent constantly, just during specific stages of the recirculation period; most of the time they are in stand-by mode, waiting to receive the activation order from the

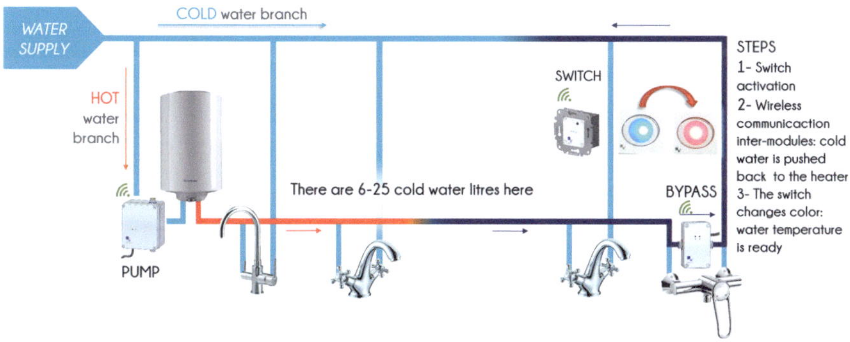

Fig. 6 NESS® functioning cycle

Table 3 Electric consumption of NESS® in one cycle/recirculation

Module	Max. power	Stand-by mode	During the recirculation
Pumping	90 W	$(24\ \text{h} - t_{cycle}) \cdot 0.68$ Wh[a]	$t_{cycle} \cdot 90$ W
Bypass	15 W	0.63 Wh[a]	Negligible
Switch	1 W	0.18 Wh[a]	Negligible

[a]According to the product CE marking

Table 4 Electric consumption of NESS® per day

Module	Mode	Electric consumption	Total
Pumping	Stand-by	16.07 Wh/day	$E_{pump} = 4894$ Wh/day
	Recirc.	32.86 Wh/day	
Bypass	Stand-by	1512 Wh/day	$E_{bypass} = 1512$ Wh/day
	Recirc.	0	
Switch	Stand-by	432 Wh/day	$E_{switch} = 432$ Wh/day
	Recirc.	0	
Total electric consumption of NESS®			$E_{elec\ N} = 6838$ Wh/day

user. The parameter t_{cycle} refers to the period of time during which the recirculation takes place. In this analysis, the recirculation is taken to last 3 min which represents medium size installations.

$$E_{elec\ N} = E_{pump} + E_{bypass} + E_{switch} \tag{13}$$

Here, each term is represented by the sum of the consumptions during the stand-by and recirculation mode shown in Table 3. The bypass and switch consumptions during the recirculation are negligible since the period of time during which they are actually working is short enough not to be considered as a significant electricity consumption. However, the pump is working during the whole period of 3 min' recirculation, consequently, it must be taken into account. Thus, the pump is propelling water during 3 min for N_{uses} = 73,041 times a day. Applying Eq. 13 and using the terms exposed in Tables 3 and 4 is obtained:

6 Overview of Both Systems: The Double Strand of Hot Water and NESS® Technology

This paragraph intends to synthesize results and gives the economic dimension of the study by bringing about current prices and costs for energy and materials to implement each system.

6.1 Results Chart

Regarding Table 5, NESS® only requires a 34% of the thermal energy expensed by the DSHW to achieve the same water saving. Accordingly, NESS® employs approximately half the electricity than the DSHW.

Table 5 Energy expenses of the DSHW and NESS® technology

	E_{th}—thermal energy expense (Wh/d)		E_{elec}—electricity expense (Wh/d)	
	Effective	Loss	Effective	Loss
DSHW	80.01	332.34	2867	11.90
	412.36		1477	
NESS®	140.10		6863[a]	
NESS®/DWHW	33.97%		46.47%	

[a]Average household with 251 inhabitants

Table 6 Electricity prices for 2016

Company	PVPC	EDP	Endesa	Gas natural	Iberdrola	Viesgo	Av.
€/kWh	0.11610	0.14521	0.13907	0.12980	0.14755	0.13466	0.135

Table 7 Energy costs according to different sources

Source	Biomass	Butane	Natural gas	Diesel	Electricity
€/kWh	0.0387	0.0962	0.0622	0.1136	0.163

IDAE (2007)

6.2 *Economic Dimension*

In order to get a further perspective, energy and electricity prices are taken into the equation. (Selectra España 2016) gives electricity prices for the six main suppliers in Spain, c.f. Table 6, in 2016.

Electricity average price in 2016 results in 0.135 €/kWh, which including taxes becomes 0.163 €/kWh. This price refers to domestic facilities, namely, below 10 kW. Most houses are within this power range.

Moreover, thermal energy expenses imply different costs and prices depending on the energy source employed, for instance, biomass, butane, natural gas, diesel or electricity, c.f. Table 7.

According to (CYPE Ingenieros 2015), the minimum price for the equipment and installation of the DSHW stands for 816 €, while the basic kit of NESS® costs 480 €, installation included; that is to say, NESS® technology can be installed and working for half de price than the DSHW.

NESS® energy costs imply only a 35% of the expense provided by the DSHW; Fig. 7 evidences such fact.

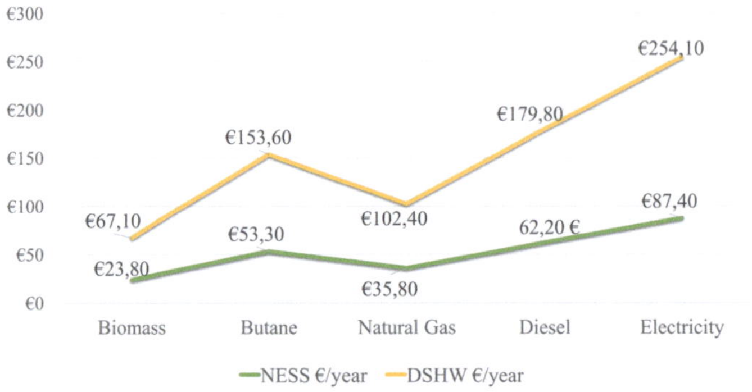

Fig. 7 Annual energy costs of both systems

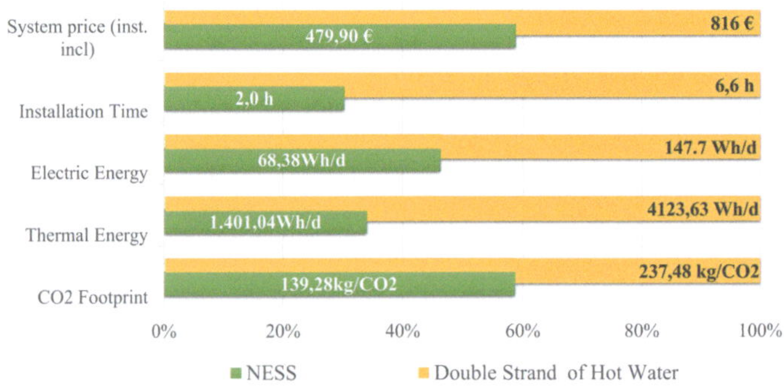

Fig. 8 NESS® compared to the DSHW

7 Discussion

Although with the DSHW there is no need to wait for hot water, NESS® achieves the same water saving at over a third part of the energetic costs produced by the DSHW. NESS® represents an energetic consumption of 35% with respect to the DSHW. In addition, material and installation costs are reduced to half the price required for 15 m of the DSHW. Figure 8 exposes such comparison.

In addition, NESS® can be installed with no need of works, that is to say, the DSHW needs to be installed in new or under refurbishment houses, while NESS® can be installed indistinctly in both.

8 Conclusions

There are two alternative systems to save the cold water that is wasted while waiting for hot water to arrive to the tap. The DSHW offers instant hot water without waiting, while NESS® system works under the user demand. This particular advantage of the DSHW against NESS® technology turns into the worst inconvenient since it brings about high energetic expenses to achieve instant hot water. However, NESS® solves this problem by using only a 35% of energy compared to the DSHW. NESS® complies with saving natural resources while reducing the energetic cost induced by the alternative solution.

The innovation cannot only be measured in terms of water savings but comfort to the user regarding everyday hot water usage routines. NESS® is able to achieve both, a significant reduction in water waste and monthly water bills. The system makes use of the recirculation technique in an innovative manner through its modular composition gaining in adaptability, low consumption, service notification without wasting a single drop, design and customization options, multipoint activation and the possibility to extend its functions. In addition, when used jointly with or as an alternative to the DSHW, it also saves energy and money.

A new and more economic, ecologic, efficient and viable alternative to the DSHW is identified and explained, in spite of not reducing the waiting time. Customers receive a new technology to increase their quality of life by saving water and money. Everyone in the region is a beneficiary of the system implementation, as less water wasted means more water for everyone, making it more accessible and decreasing its cost. A decrease on needed energy to treat and distribute water is also expected, making the whole water cycle more efficient in energy terms.

References

Association of the European Heating Industry (2010) ehi. Retrieved 26th Apr 2016

Bio Intelligence Service (2009) Study on water performance of buildings: final report. París. Retrieved 26th Apr 2016

CYPE Ingenieros (2015) Generador de Precios de CYPE

European Commission (n.d.) Eurostat. Retrieved 26 Apr 2016

European Environment Agency (2009) Water resources across Europe—confronting water scarcity and drought. Office for Official Publications of the European Communities, Copenhaguen. doi:10.2800/16803

Eurostat (2015) Household composition statistics. Retrieved 26 Oct 2015

IDAE, I. p. (2007) Guía Técnica 3: Diseño y cálculo del aislamiento de conducciones, aparatos y equipos. Madrid

INE, Instituto Nacional de Estadística (2013) Encuesta Continua en Hogares. Boletín informativo del Instituto Nacional de Estadística

Métrica6 Ingeniería y Desarrollos S.L. (2015). Estudio de ahorro y su impacto energético y económico. Málaga

Ministerio de Fomento (2009) Código Técnico de la Edificación. Documento Básico HS Salubridad

Selectra España (2016) Comparadorluz. Retrieved 6th Sept 2016, from http://comparadorluz.com/faq/precio-kwh-electricidad

UN-Water (2010) Climate change adaptation: the pivotal role of water. Retrieved Apr 2016

United Nations World Water Assessment Programme, W. (2015). The united nations water development report 2015: water for a sustainable world. UNESCO, París

World Assessment Programme, WWAP (2012) The united nations world water development report 4: managing water under uncertainty and risk. UNESCO, Paris. Retrieved 26 Apr 2016

Vegetation as a Design Element to Recover Green Areas in Settlements Developed on Contaminated Soils

Christhopher Contreras Lopez

1 Introduction

The mining sector, around the world, has played a central role as arranger and organizer of the territory, as it has guided the settlement, it has given rise to the construction of road networks and has been detonator other economics activities (Saavedra Silva and Sánchez Salazar 2007). However, the process of obtaining the precious metals has led to waste result as water sludge mine where transported solids and accumulate minerals who they are abandoned without any processing and are known as tailings (Hernández Ávila 2012). This mining wastes are the result of different processes of beneficiation of minerals. This wastes are placed in dams which are the site of final disposal, the construction and operation of dams occur simultaneously (SEMARNAT 2004).

In many countries there are abandoned tailings dams, as in South Africa, Australia, and Spain, the western of United States of America, off the coast of Chile and Peru and western of India (Tordoff et al. 2000). These dams cover hundreds of hectares, in these places the vegetation cannot developed, (Hernández Acosta et al. 2009) so that the soil remains bare and is moved by the wind, these dust devils contain heavy metals regarded as toxic elements that can be harm any living being (Hernández Ávila 2012). In Mexico, in the specific case of Pachuca, Hidalgo, of the XVI to XIX centuries, the mining wastes were thrown to the Avenidas river, that spread over a wide area several kilometers south of the city (Comisión del Fomento Minero 1959). In the XX century, the tailings were place in a planned dam in south of Pachuca, where the wastes joined the tailings scattered for four centuries. The tailings dam to the early 1980s occupied 391 ha with 65 million tons of wastes (Soto Oliver 1982).

C. Contreras Lopez (✉)
National Autonomous University of Mexico, 04510 Mexico City, Mexico
e-mail: christhophercl@hotmail.com

P. Mercader-Moyano (ed.), *Sustainable Development and Renovation in Architecture, Urbanism and Engineering*, DOI 10.1007/978-3-319-51442-0_27

In the 1980s, the city of Pachuca had a considerable urban growth southwards, towards Mexico City, the urban area grew so that first began to surround the mine tailings with construction mainly housing developments and late XX century and early XXI, were built on the tailings, two complexes of housing, covering almost half of the dam of mining wastes. Currently the tailings dam has an area of about 200 ha that are intended to be used to build dwelling, plus around this dam there are other housing complexes, shopping centers and urban facilities that were built due to the growth of the city to the south (Contreras López 2016).

Urbanization that surrounded and subsequently occupied the tailings dam has several problems, the most important is the contaminated waste difficult the development of vegetation, so that the soil remains bare and is removed by the wind, these dust devils contain heavy metals regarded as toxic elements, which can be detrimental to public health (Hernández Ávila 2012). On the other hand, changes are recorded in the physical and chemical soil parameters such as lack of organic matter and nitrogen, which are essential for plant development (Contreras López 2016). So in the spaces designated for green areas of urban development on tailings dam it does not grow vegetation, which makes these areas in inhospitable and isolated spaces without landscape quality.

2 Design of a Factorial Experiment

As previously cited, one of the main problems in mine tailings is that the development of vegetation is difficult, so that the soil is removed by the wind that generate dust clouds with dust particles, this particles contain toxic elements for human health. It is essential to avoid these dust storms and improve soil characteristics through revegetation process, so it is necessary to have vegetation able to grow in these soils and thereby confer habitability to urban areas on mining wastes. To achieve this, there are different types of treatments, including the technique of phytoremediation, which is a biological treatment that uses the ability of vegetation to survive in contaminated soils with heavy metals and simultaneously extract, accumulate, immobilize or transform the contaminants from soil. The success of this treatments is associated with the selection of appropriate vegetation that can recover the soil, and to the careful selection of improvements, such as organic matter or chelating agents, these improvements allow to development of soil properties and promote survival and plant growth (Ortiz Bernard et al. 2007).

Based on the above, we created a factorial experiment that allowed to observe over a period of nine months the development of seven plant species and its relationship with four soil concentrations. First, we perform an analysis of physical and chemical properties of soil to understand why not grow the vegetation, in addition to checking the contamination with toxic elements, which are a factor that interferes with the proper development of plants. Subsequently, by a bioassay, we measure the development of seven plan species: *Asclepias linaria*, *Carpobrotus edulis*, *Dietes vegeta*, *Hedera hélix*, *Sedum praealtum*, *Senna multiglandulosa* and

Trifolium repens; against four concentrations of soil: 0–100: it is a soil substrate composed of a mixture of organic matter, mineral matter and draining matter on thirds, which serves as a witness of the development of vegetation and topsoil; 60–40: it is a mixture of 60% of mining tailings plus 40% of the mixture 0–100; 80–20: it is a mixture of 80% of mining tailings plus 40% of the mixture 0–100; 100–0: 100% of mining tailings.

The experiment was performed with five repetitions and monthly measure of the development of vegetation on each substrate with three variables: height, coverage and vigor, with a metric ruler and monitoring book.

2.1 Physical and Chemical Properties of Soil

We need to understand why the vegetation does not developed on the wasted soils, so we perform an analysis about physical and chemical properties of contaminated soil, also we analyzed the tailings mixtures with topsoil and the topsoil that we use as a witness. Results are report in Table 1.

The pH is determined by potentiometer and the result is neutral to mildly alkaline (SEMARNAT 2003), the pH does not causes the development of vegetation; bulk density is determined by the method of measurement in a graduated cylinder (Flores Delgadillo and Alcalá Martínez 2010) and texture is determined by sedimentation test (SEMARNAT 2003), both are important to calculate the irrigation, this calculation says that in the dry season, which is from October to March, the need for water is up 4 l/m^2 and for the rest of year the irrigation is not required for this type of soil. The Soil Organic Matter is determined by the method of Walkey and Black (Flores Delgadillo and Alcalá Martínez 2010) and the result is from 0.27 to 0.87% in tailings mixtures, this means that it is associated with a poor soil structure and poor ability to withstand changes and contaminants, as well as inadequate reservoir of water and nutriments (Contreras López 2016).

For macroelements, the nitrogen (N) is determined by Kjeldahl method (FOSS Tecator AB 2003); the sample 0–100 has a high concentration but not to the other samples where the concentration is low, we determine that the concentration rises to improving soil. Phosphorus (P) is determined by Photo Colorimeter with molybdenum blue reaction, potassium (K) and sodium (Na) area determined by flame photometry, and calcium (Ca), magnesium (Mg) and Sulphur (S) are determined by X-ray Handheld XRF Spectrometer; the results are within normal ranges for soil and we observed that concentration increases with soil improvement (Allen 1989).

The microelements are determined by X-ray Handheld XRF Spectrometer, for chlorine (Cl) and iron (Fe), the results are within normal ranges for soil, for copper (Cu), magnesium (Mg) and zinc (Zn) the mixtures with tailings exceed the limits ranges for soils for what they are toxic (Allen 1989).

Besides heavy metals were detected in all samples with tailings. The analysis was by Atomic Absorption Spectroscopy and X-ray Handheld XRF Spectrometer. The results say there strontium (Sr), cadmium (Cd), lead (Pb), and silver (Ag), this

Table 1 Analysis of physical and chemical properties of soil

Analysis	Types of soil			
	0–100	60–40	80–20	100–0
Physical				
Texture	Sandy loam	Sandy clay loam	Sandy clay loam	Sandy clay loam
Bulk density (g/cm^3)	0.76	1.15	1.23	1.34
Chemical				
pH	6.8	6.95	7.05	7.19
SOM (%)	3.77	0.87	0.67	0.27
Macroelements				
N (%)	0.2	0.04	0.03	0.02
P (mg/100 g)	2.38	1.28	0.62	0.50
K (mg/100 g)	109.4	48.2	30.2	19.4
Na (mg/100 g)	41.4	26.2	22.8	13.6
Ca (mg/100 g)	2.72	3.18	3.24	3.22
Mg (mg/100 g)	2.01	1.98	1.89	1.87
S (mg/100 g)	0.29	0.58	0.74	0.78
Microelements				
Cl (%)	0.03	0.03	0.02	0.02
Cu (%)	0.00	0.01	0.01	0.01
Fe (%)	5.00	3.58	3.41	3.12
Mn (%)	0.09	0.61	0.73	0.79
Zn (%)	0.01	0.2	0.24	0.25
Heavy metals				
Sr (%)	0.00	0.03	0.03	0.03
Cd (%)	0.00	0.00	0.00	0.01
Pb (%)	0.00	0.07	0.09	0.11
Ag (%)	0.00	0.002	0.006	0.01

elements are included on the list of the Environmental Protection Agency (EPA) of United States of America, as a Potentially Toxic Elements (SEMARNAT 2012).

According to the interpretation of results, the problems on the tailings are: low organic matter, deficiency of nitrogen in samples with tailings, this element is essential to life, as it stimulates the growth of vegetation. The high levels in microelements which become toxic, besides the presence of heavy metals in all samples with tailings.

According to these results, the plant species should be able to allow uniform soil cover to mitigate dust that contains toxic elements, be invasive, low requirements, besides being able to grow in poor, thin and rocky soils. Moreover, it was observed that the toxicity levels are reduced with topsoil and the essential elements to sustain plant life increase with the mixture of tailings and topsoil.

2.2 Selection of Plant Species

Although, the tailings obstruct the development of vegetation, as already discussed, there are many plant species that were established in mining wastes. The plant species reported in tailings from Pachuca belong to families compositae: *Chamaemelum fuscatum*; graminae: *Agrostis capillaris*; cruciferae: *Brassica juncea*, leguminosae: *Medicago polymorpha*; solanaceae: *Solanum corymbosum* and chenopodiaceae: *Atriplex suberecta* (Contreras López 2016).

Also around the world it have studied many plant species, which can be established on tailings for example in the municipality of Aznalcollar, Seville, Spain, following the spill of the tailings dam, there studied species of *Brassica juncea* and *B. carinata* for use in the recovery of contaminated soils, besides *Amaranthus blotoides*, *Beta bulgaris* and *Convolvulus arvensis*, plants that are accumulating heavy metals (De Haro et al. 2003). In Encartaciones, Vizcaya, Spain, it tested plant species that are tolerant to heavy metals accumulation like *Thlaspi caerulescens*, *Jasione montana*, *Rumex acetosa* and *Festuca rubra* (Becerril et al. 2007).

Many plant species has been tested that are able to accumulate heavy metals like *Pistacia terebinthus* in Cyprus, *Bidens humilis* in Ecuador, besides *Atriplex lentiformis* and *A. canescens* in United States of America, with very promising results (Mendez and Maier 2008). In Mexico has been tested *Juniperus depeana*, *Celtis reticulata* and *Prosopis juliflora*, which they are species that resisted soil contamination (Puga et al. 2006).

Although there are variety of trees, shrubs and herbs that can grow on mine tailings and can assist in the rescue of contaminated soils, has only been investigated its contribution to environment rescue, the studies omitted much of these soils are already urbanized and it is vitally important to consider plant species of ornamental interest. Some plant species that have been studied can meet the requirement to be ornamental like *Brassica rapa*, *Gazania splendens*, *Lepidium virginicum*, *Sanvitalia procumbens*, *Montanoa tomentosa* and *Solanum eleagnifolium*, but only we can buy in mexican stores *G. splendens*. So it is important to propose an ornamental plant palette, which can be used in green areas and they may face the environmental problem in addition to the formal and aesthetic requirements and it can buy in stores.

Therefore we tested the species *Carpobrotus edulis*, *Dietes vegeta*, *Sedum praealtum* and *Asclepias linaria*, herbaceous plants with fast growth, easy reproduction and low maintenance, which are useful to form vegetable beds of different height, able to cover the ground, avoid dust devils and give structure to space. *Hedera helix* is fast growing and produces a high level of coverage, it can grow as a vine or ground cover, which allows options for the design of green areas as it is able to grow with adventitious roots on walls that have no nutriment elements. *Senna multiglandulosa* is a small tree capable of producing a light shade and Trifolium repens is mainly spread by seed and stolon and allows for very fine green carpet texture. The latter two belong to the family of leguminosae and indicates that they are species associated with nitrogen-fixing bacteria that can grow in inhospitable soils lack of this fundamental element.

All species were chosen to be tested in experimental design because they allow uniform ground coverage, are invasive and low requirements, in addition to being able to develop in poor, thin and rocky soils, and the other hand all meet aesthetic requirements and they can be purchased at the store.

2.3 Bioassay

Each plant species was tested in four different types of substrate with five repetitions, so that 20 containers were assigned to each plant. The total experiment had 140 containers with seven plant species and four substrates, with five repetitions as already mentioned.

The planting density for each species was calculated based on the number of individuals that fit into an area and a plantation for real framework in which an individual has against another, both horizontally and vertically and is calculated based on adult height of plants to ensure full coverage of the tray, avoiding competition, the number of plants per container is in Table 2 (López de Juambelz 2015).

We measured the development of vegetation in each substrate with the following variables: height, measuring the vertical size of the plant from the substrate surface; coverage, measuring the sides of the surface covering the plant to become an area; and vigor, counting the number of live plants in each substrate. Each measurement was performed on a monthly basis with a metric ruler and monitoring book as presented in Table 3.

With the monthly record we obtain results of each species in each type of substrate. Once we obtained all monthly measurements, the data are subject to a descriptive statistical management, once the data of different variables captured, we obtain an average value and standard deviation for each species and for each variable. Finally we perform a monthly average and comparative trend is performed by each species in each substrate for each of the variables, throughout the experiment.

Table 2 Plants per container

Plant	Number of plants per container
Asclepias linaria	5 plants per container
Carpobrotus edulis	9 plants per container
Dietes vegeta	9 plants per container
Hedera hélix	16 plants per container
Sedum praealtum	9 plants per container
Senna multiglandulosa	5 plants per container
Trifolium repens	20 plants per container

Table 3 Data register

Key	No	h (cm)	f (cm)	v (y/n)
Aal	1	36	25 × 20	y
Aal	2	38	35 × 30	y
Aal	3	33	20 × 20	y
Aal	4	24	20 × 17	y
Aal	5	22	18 × 25	y

Date: November 2nd, 2015

2.4 Results

The results of the factorial experiment where we measured the behavior of plant species against different concentrations of soil are:

Asclepias linaria: Although it had a loss of plants in the three mixtures of tailings to 12%, in terms of height was not have a great variation, we detected in the four mixtures was 13% maximum between the substrate 100–0 and the 0–100, so the plant can develop in soil tailings directly, in terms of coverage we not detected a great variation too, having a maximum of 30% between the substrate 100–0 and the 0–100.

Carpobrotus edulis: In all soil concentrations there was no loss of plants, so that this plant is very resistant, however variations in height were recorded up to 30% between the substrate 0–100 and the 100–0, also coverage had a variation of up to 35% between substrate 0–100 and the 100–0. The trend indicates good growth of vegetation in height and coverage in all mixtures, so the plant can be used directly on mining tailings.

Dietes vegeta: The plant did not report great variation in height in the four substrates, the largest variation was 20% between substrate 0–100 and the 100–100; the coverage registered a variation of up to 40% between substrate 0–100 and the 100–0. As for the survival of plants not recorded any loss on any substrate, so that the plant can grow in soil tailings directly.

Hedera helix: Had a similar percentage of 20% loss of plants in the three mixtures with tailings, there were no losses in natural soil. As for the height and coverage, the substrate 60–40 had less variation as in the other two, there were up to 70% variation in growth compared to the substrate 0–100. It is recommended to plant this species on mining tailings mixed with 40% of topsoil.

Sedum praealtum: There was no loss of plants on any substrate. We recorded a short variation in the other two variables with a maximum registration of 30% in coverage, and 25% in height between substrate 0–100 and the 100–0. So it is highly recommended to plant this species directly in the mining tailings.

Senna multiglandulosa: There was no loss of plants in any substrate. As for the height, there was a variation of up to 50% between substrate 0–100 and the 100–0. Regarding the coverage, there were variations of up to 55% between substrate 0–100 and the 100–0. Although no losses were recorded, we suggest to plant on mining tailings mixed with 40% of topsoil.

Trofolium repens: We recorded a loss of plants of up to 55% in the substrate 100–0. As for the height, there were variations of up to 55%, and in terms of coverage, up to 60% with respect to the substrate 100–0 and the 0–100. It is recommended to plant this species around the mining wastes, where the soil does not have many changes in physical and chemical parameters, and contamination by toxic elements is less (Puga et al. 2006).

The factorial experiment determined that the seven plant species proposals, four of them can be established on tailings, since plants have few variations, these plant species are: *Asclepias linaria*, *Carpobrotus edulis*, *Dietes vegeta*, and *Sedum praealtum*.

Senna multiglandusa and *Hedera helix* can grow on tailings if the soil is improve with 40% of topsoil. It is further recommended that *Trifolium repens* is planted in settlements around the tailings where changes in soil parameters and contamination are lower.

2.5 *Plant Palette for Green Areas in Settlement on Contaminated Soils*

With the results, we can performed in detail, a specific card like in the Fig. 1.

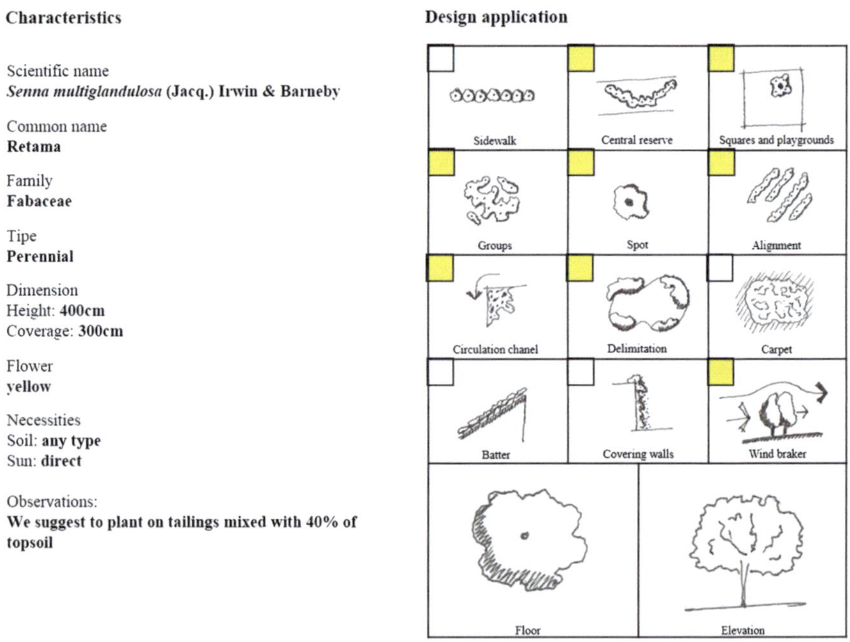

Fig. 1 Specific card of *Senna multiglandulosa*

The specific cards are for the use of plants that were tested in the factorial experiment and showed good results in terms of adaptation on the mining tailings. In these tabs is detailed information on taxonomic and biological data for each plant species. In addition, within this information, the user or designer will have technical data to help the choice of plants in the design and use of these plants in space and thereby confer habitability to green areas in urban developments on mine tailings.

3 Conclusions

According to the bioassay conducted using a factorial experiment that defines the concentration of tailings against different ornamental plant species, we find that there are plant species capable of being used in the landscape design of the spaces destined to green areas in urban developments on mine tailings, which results in improving the quality of life of users and reduce pollution by wind dispersion to which the population exposed.

As for the analysis of physical and chemical properties of soil that contains the vegetation, it is observed that the content of organic matter is very low in sample containing 100% of tailings, it is associated with a poor soil structure and poor buffering capacity to changes and contaminants, as well as an inadequate reservoir of water and nutrients, however, the content of organic matter increases when the contaminated soil is mixed with topsoil.

To the primary macroelements, nitrogen (N), phosphorus (P) and potassium (K), that plants use in g/l, we found that the tailings are deficient in nitrogen (N), even this deficiency is maintained in mixtures with topsoil, which is directly related to the difficulty of vegetation development (Allen 1989). Phosphorus (P), potassium (K) and the secondary macroelements, calcium (Ca), magnesium (Mg) and Sulphur (S), they are within acceptable ranges for plant growth.

In the case of microelements where the ranges present toxicity, it is very narrow because the requirements correspond to mg/l (ppm). However, they can cause deficiency for plants development and soil microorganisms. We found heavy metals, toxic for vegetation and any living being, like strontium (Sr), cadmium (Cd), lead (Pb) and silver (Ag), this analysis coincides with different reports made on tailings from Pachuca (Hernández Acosta et al. 2009).

Of the seven plant species tested, we found a satisfactory development in *Carpobrotus edulis* (aizosaceae), *Dietes vegeta* (iridaceae), *Sedum praealtum* (crasulaceae), and *Asclepias linaria* (asclepidaceae). We observed that *Hedera helix* (araliaceae), tailings soil and contaminants have a deleterious effect on the development of this plant. *Senna multiglandulosa* and *Trifolium repens* (leguminoseae), are plant species associated with nitrogen-fixing bacteria; it was expected that the response of this plant was satisfactory, because it not depend of nitrogen on soil, but in the sample with tailings soil the nitrogen is very low, and this can be a limiting factor to the development of vegetation. The unsatisfactory response in development to leguminoseae plants may be related to the inhibitory effect of heavy metals on the symbiotic bacteria of this plants for nitrogen fixation (Olivares 2015).

The experimental design was made ex situ, the plant species were planted in containers placed in the Laboratory of Conservation of Natural and Cultural Heritage building, the laboratory belongs to the Master and PhD Program in Architecture of UNAM. In this place the variables like irrigation were controlled, however it is recommended to make the experiment in situ to observe the behavior of plant species with the weather change of site.

It is also recommended a physical and chemical analysis of soil to determine the contribution of biomass from plants to the ground, and identify whether plants provide nutrients, so that in later plantings, the inhibitory effect will be less. Moreover, it is recommended a chemical analysis of plants species to identify if they are heavy metal accumulators and thereby begin to recovery of the contaminated soils.

Acknowledgements This project was developed thanks to the sponsorship of PAPIME PE401613, "Consolidación del laboratorio para la conservación del patrimonio natural y cultural" in the research "Apoyo al proceso de conservación de la region Zempoala—Pachuca, Hidalgo, México". And the equipment provided by laboratory to conservation of natural and cultural heritage, through the program CONACyT INFRA2014 225248.

References

Allen SE (1989) Chemical analysis of ecological materials. Blackwell Scientific Publications, Londres, Great Britain

Becerril M et al (2007) Especies nativas de suelos contaminados por metales: aspectos ecofisiológicos y su uso en fitorremediación. Ecosistemas, revista científica y técnica de ecología y medio ambiente 16(2):50–55

Comisión del Fomento Minero (1959) Planta de Tratamiento de Jales Unidad "Santa Julia" Pachuca, Hidalgo. Boletín Número 6(6):8–24

Contreras López C (2016) Oblitopías: recuperación del espacio abierto contaminado en asentamientos humanos ubicados sobre depósitos de jales mineros. Facultad de Arquitectura, UNAM, Mexico

De Haro A, Del Río M, Font R (2003) Potencial de especies cultivadas de Brassica y de especies silvestres para la fitorremediación de suelos contaminados. In: Ciencia y Restauración del Río Guadiamar. Resultados del programa de Investigación del corredor Verde del Guadiamar 1998–2002. KADMOS, Seville, Spain, pp 276–283

Flores Delgadillo L, Alcalá Martínez JR (2010) Manual de Procedimientos Analíticos. Instituto de Geología, UNAM, México

FOSS Tecator AB (2003) Application Sub Note ASN 3313, The determination of nitrogen according to Kjeldahl in soil. FOSS, Sweden

Hernández Acosta E et al (2009) Vegetación, residuos de mina y elementos potencialmente tóxicos de un jal de Pachuca, Hidalgo, Mexico. Revista Chapingo Serie Ciencias Forestales y del Ambiente 15(2):109–114

Hernández Ávila J (2012) Caracterización de la escombrera de Velazco en el Estado de Hidalgo y su uso alterno como material de construcción. V encuentro de investigación del AACTyM 6

López de Juambelz R (2015) Aspectos técnicos para el manejo de la vegetación en el diseño. En: Tecnología. Una contribución sistemática de acción, reflexión y análisis. Facultad de Arquitectura UNAM, Mexico, pp 333–353

Mendez MO, Maier RM (2008) Phytostabilization of mine tailings in arid and semiarid environments an emerging remediation technology. Environ Health Perspect 116(3):278–283

Olivares PJ (2015) tiene futuro la fijación biológica de nitrógeno. [On line] Available at: www.eez.csic.es/[olivares/ciencia/futuro2/index.html. Last access Jan 2016

Ortiz Bernard I, Sanz García J, Dorado Valiño M, Villar Fernández S (2007) Técnicas de recuperación de suelos contaminados. Universidad de Alcalá, Dirección General de Univesidades e Investigación, Madrid

Puga S et al (2006) Concentraciones de As y Zn en vegetación nativa cercana a una presa de jales. Revista Internacional de Contaminación Ambiental 22(2):75–82

Saavedra Silva EE, Sánchez Salazar MT (2007) Mineria y espacio en el distrito minero Pachuca—Real del Monte en el siglo XIX. Investigaciones geográficas, Boletín del Instituto de Geografía, UNAM, Issue 65:82–101

SEMARNAT (2003) Norma Oficial Mexicana NOM-021-SEMARNAT-2000 que establece las especificaciones de fertiliad, salinidad y clasificación de suelos, estudio, muestreo y análisis. Diario Oficial de la Federación, Mexico

SEMARNAT (2004) Norma Oficial Mexicana NOM-141-SEMARTNAT-2003 que establece el procedimiento para caracterizar los jales, así como las especificaciones y criterios para la preparación del sitio, proyecto, construcción, operación y postoperación de presas de jales. Diario Oficial de la Federación, Mexico

SEMARNAT (2012) Metales Pesados. Instituto Nacional de Ecología y Cambio Climático, Mexico

Soto Oliver N (1982) La minería, El Distrito Minero Pachuca-Real del Monte a través de la historia. Gobierno del Estado de Hidalgo, Pachuca(Hgo)

Tordoff GM, Baker AJM, Willis AJ (2000) Current approaches to revegetation and reclamation of metalliferous mine wastes. Chemosphere 41:219–228

Part VI
Energy Efficiency

Update of the Urban Heat Island of Madrid and Its Influence on the Building's Energy Simulation

Miguel Núñez Peiró, Carmen Sánchez-Guevara Sánchez and F. Javier Neila González

1 Introduction

Nowadays building energy evaluation tools play a key role in establishing the basic energy strategies in the building sector. They do not only help to improve the decision-making about building design and its inhabitants' comfort, but also to stablish the payback periods and select the best solution among all the possibilities in both new and renovated buildings.

Despite the last decades have witnessed great advances in building evaluation, there are still some disagreement between the simulation results and reality (Coakley et al. 2014). Several studies have addressed the issue, focusing on the models' calibration through the improvement of the thermodynamic algorithms, the user interaction profiles and the weather conditions (Evins 2013; Foucquier et al. 2013; Fumo 2014).

The Urban Heat Island (UHI) relates to the last one, and it consists in the temperature differences between the built (urban) and the unbuilt landscape (rural). Regardless the wide knowledge about the UHI, available climate data does not include its effects, as they undertake the regional scale. Previous studies in this research project have already anticipated the relevance of this phenomenon in the building energy evaluation (López Moreno et al. 2015), and so the significance of addressing the integration of the UHI effects in the buildings' energy simulations.

M. Núñez Peiró (✉) · C. Sánchez-Guevara Sánchez · F.J. Neila González
ABIO Research Group, Universidad Politécnica de Madrid, Madrid, Spain
e-mail: miguel.nunez@upm.es

P. Mercader-Moyano (ed.), *Sustainable Development and Renovation in Architecture, Urbanism and Engineering*, DOI 10.1007/978-3-319-51442-0_28

2 Background

Due to the sum of factors that influence the UHI formation, as well as its temporal and spatial variation, having an in-depth record of the phenomenon requires a large amount of measurements. While the first studies focused on detecting the existence of the UHI and identifying those areas with the greatest intensity (Arnfield 2003; Santamouris 2007, 2015), in recent years much of the research efforts have been linked to the development of numerical models, which allow to study the urban heat island dynamically (Mirzaei and Haghighat 2010).

The first UHI study in Spain was carried out in the city of Madrid (López Gómez et al. 1988, 1993), where the transects methodology[1] was used. It was followed by many other works related to specific questions of urban climate (Almendros and López Gómez 1995; Fernández García 2001; Sobrino et al. 2009), and even a numerical model that tried to explain the urban heat island's behaviour at both the mesoclimatic and microclimatic scale (Salamanca and Martilli 2010; Salamanca et al. 2011).

These studies, however, have not proven to be consistent with the objective of this research, given its age, their inability to generate hourly records, and their lack of accuracy at the microclimatic scale (Grimmond et al. 2010, 2011). Based on the recommendations that point towards the development of models built upon the target of the research (Mirzaei et al. 2015), this project aims to generate an empirical model from the collected data at multiple fixed urban locations (Kolokotroni et al. 2006, 2010; Mihalakakou et al. 2002).

This paper addresses the first phase of the development of the model, carrying out an update of the last surveys in the spatial distribution of the UHI, which dates back to the late 1980s.

3 Means and Methods

This research replicates the methodology used in the first study of Madrid's UHI (López Gómez et al. 1988), based on the development of proximity measurements through urban transects. In order to reckon the highest intensity of the urban heat island, the data gathering was performed under anticyclonic weather conditions with a cloudless sky, calm wind and few hours after the sunset (2100–2300 UTC). Three simultaneous transects configured a set of observations, conducting each of them twice on every set (return trip), and providing the temperature records used in the results section.

[1]The transects methodology refers to transversal itineraries through the city, commonly carried out simultaneously, and from which the air temperature at multiple points is obtained.

Needs to be mention that our study refines the process of planning, data gathering, and analysis. These improvements allowed for the reduction of the needed means, helped to automate the process, and to better the precision and the resolution of the results. In the next sections we explain its characteristics.

3.1 Itineraries and Geolocation

Since the original itineraries from the study of 1988 are not suited to the current road hierarchy of the city, new routes were created for each transect based on the original measuring points. It has been necessary to generate a digital transport network, continuous along the entire city, and able to emulate the current traffic restrictions. For its development it was used the *Network Analyst* extension of *ArcGis 10.3* as well as the *Openstreetmap* base map. Afterwards, the new routes were integrated into a mobile application to enable monitoring and geolocation at every single position of the vehicle.

3.2 Equipment and Data Gathering

Three temperature sensors and data loggers were used for the temperature record. Each sensor consisted on a NTC (N*egative Temperature Coefficient*) exposed thermistor, with an accuracy of ±0.2 °C and a response time of t_{90} = 60 s (Fig. 1). Virtually immediate response was guaranteed for the sensor at speeds above 30 km/h, and thus enabling to detect temperature differences in up to 10 s intervals.

Datalogger Testo 175 T2
Dimensions: 89x53x27 mm
Channels: 2 (int/ext)
Measuring rate: 10s – 24h
Measuring range: -35… +55ºC
Accuracy: ±0,5 ºC

Sensor NTC exposed thermistor
Dimensions: Ø 5 x 115 mm
Measuring range: -50… +125ºC
Accuracy: ±0,2 ºC
Response time: t_{90} = 60s

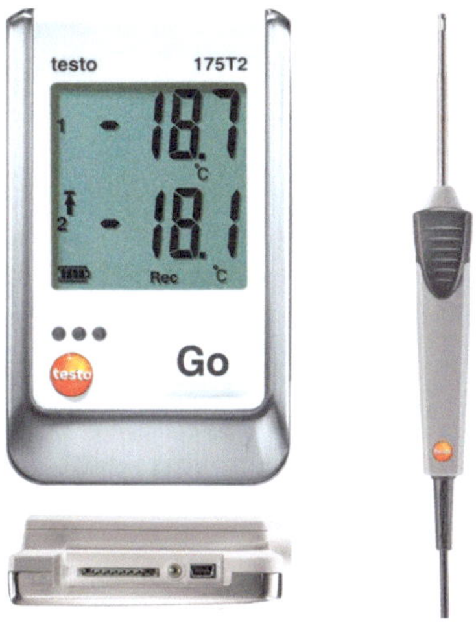

Fig. 1 Technical data of the equipment used in the measurements. *Source* Testo AG

The transects were carried out by car, with the sensor placed at the roof and distanced from the surface and the exhaust pipe. Since the measurements took place several hours after sunset, it was not necessary to protect the sensor from solar radiation.

3.3 Analysis Tools, Interpolation Techniques and Maps Generation

On the basis of the obtained data, a frequency distribution analysis of each data set was carried out in order to identify common patterns among them.

Also several isotherm maps were created for each set of observations. For that process a geostatistical analysis method called *cokriging* was used. The *kriging* is a technique that allows, by interpolating a particular batch of data, to predict the value on a given point:

$$\hat{Z}(s_0) = \sum_{\mathrm{i}=1}^{n} \lambda_{\mathrm{i}} Z(s_{\mathrm{i}}) \tag{1}$$

The *variogram*, which is the function that describes the spatial correlation in the *kriging*, was set to be exponential (2), as it has been widely used in meteorology and contamination prediction (Palomino et al. 2015).

$$\gamma(h) = c_0 + c\left(1 - \exp\left(\frac{-h}{r}\right)\right); \quad h > 0 \tag{2}$$

Previous research have also pointed out that the stream beds influence the UHI distribution (Fernández García et al. 1996). Therefore, the model was completed with a *DTM* layer (Digital Terrain Model) in order to provide information on the altimetry. All the process was fulfilled with the *Geostatistical Analyst* package from *ArcGIS 10.3*.

In addition to the 101 points from the transects, collected data from six observatories of the National Meteorological Agency (AEMET) and from nine urban stations of the air quality system (SICAM) were added to the process, giving a total of 116 measuring locations (Fig. 2).

4 Results and Discussion

Four data sets were obtained from the measurements that were carried out between July 2015 and April 2016. Each data set corresponds to one season, as the measurements took place around one month after the solstices and equinoxes. As it is

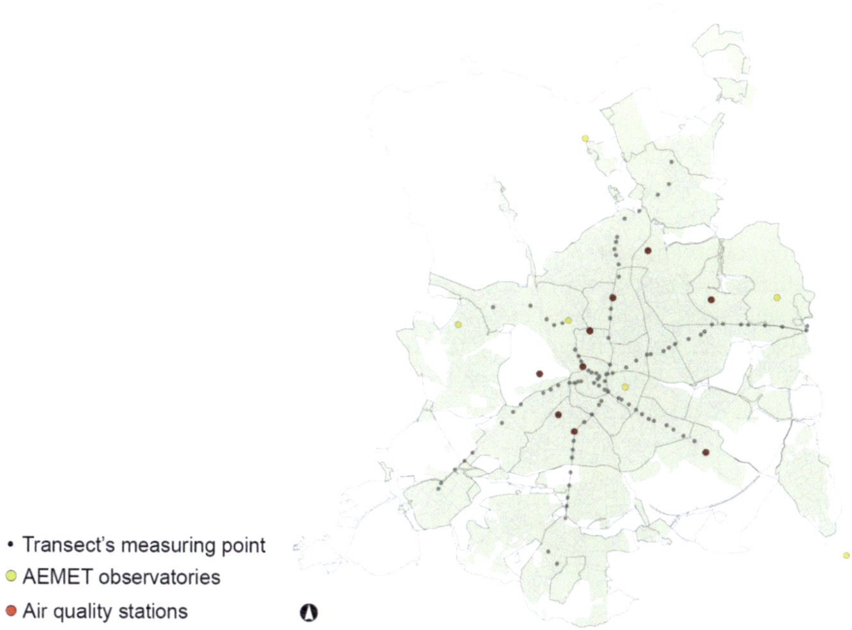

Fig. 2 Location of the observation points used in the map's generation

Table 1 Statistical parameters from each temperature data set

Parameter	July 15th, 2015	October 28th, 2015	February 16th, 2016	April 25th, 2016
Average	31.2	12.8	2.6	15.6
Minimum	27.3	11.2	−0.6	12.1
P_{10}	28.7	11.9	1.5	13.7
P_{50}	31.6	12.9	2.6	15.6
P_{90}	33.1	13.4	3.8	17.4
Maximum	33.7	13.8	5.9	18.0
Range	6.4	2.6	6.5	5.9
Standard deviation	1.6	0.6	1.0	1.3

explained in *3 Means and Methods*, all measurements were performed under the same atmospheric conditions, at the same time and following the identical itineraries.

Table 1 shows a summary of the most important statistical parameters that were used during the analysis of the frequency distribution and the isotherm maps' generation.

4.1 Frequency Distributions

The analysis of the frequency distribution (Fig. 3) shows that it can be inferred the intensity of the urban heat island from the data dispersion of each data set. Thus, it can be observed a decreasing pattern on the UHI intensity from summer to winter (*a–b–c*) and backwards (*c–d–a*), in line with the results of previous research (Núñez Peiró et al. 2016).

While confirming that the greater UHI intensity happens in the warmest month, the differences between the measurements from autumn (*b*) and spring (*d*) are remarkable, since both were performed at times of the year in which historical temperature records are very similar (Tmax = 16–20 °C; Tmin = 6–7 °C).

This can be explained by the fact that, at this latitude, the UHI intensity seems to have a high dependence upon solar radiation. It is noted, therefore, that the urban

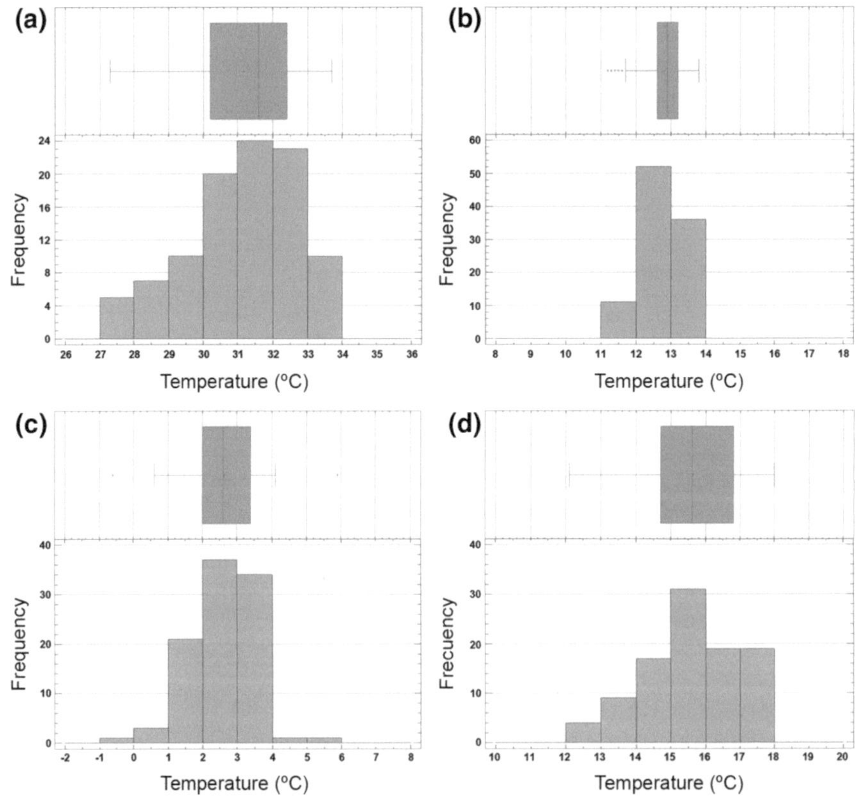

Fig. 3 Frequency distribution of temperature for each data set, presented in chronological order. **a** July 15th, 2015; **b** October 28th, 2015; **c** February 16th, 2016; **d** April 25th, 2016

heat island does not have a symmetrical behavior regarding the seasons but in relation to the solstices, intensifying or softening its effect depending on the proximity to them.

Consequently, the measurements could be presented from the highest to the lowest intensity: the data set *a* distances itself 24 days from the solstice, the *d* does 57 days, the *c* does 125 days and the *b* does 129 days. Moreover, it can be explained the resemblance between the autumn (*b*) and the winter (*c*) measurements, located symmetrically from the solstice and under the same solar radiation.

4.2 Isotherm Maps

Based on the temperature records collected during the measurements, and by following the methodology described in Sect. 3.3 *Analysis tools, interpolation techniques and maps generation*, there have been generated different maps of isotherms for each season (Fig. 4). These maps represent the temperature difference between each observation point and a reference value.

Usually the reference value is taken from a rural location placed in the surrounding of the city. It should be said that it is not an easy task to locate it, as there are consistent doubts about what can be considered a rural location, or how to ensure that the measurement of that particular point is truly representative of the surroundings of the city. Those doubts have led, among other things, to question the methodology and results from hundreds of UHI studies developed around the world (Stewart 2011).

Another approach to determine the reference temperature value consists, when developing urban transects, in taking the minimum temperature of the series. However, as it can be seen in Fig. 4c, there are cases where the minimum temperature corresponds to an atypical value. Although this does not mean that measurements are necessarily erroneous, using them as a reference value when generating a map could cause huge distortions and come to wrong conclusions. It should also be noted that in this study the coldest record was never found at the same location, moving from *Ciudad Universitaria* at the west (summer) to the exit of Madrid at the north (autumn), to the vicinity of the Jarama river at the east (winter) and to the entrance of Getafe in the south (spring).

This study, instead of setting a temperature value for the coldest point, it poses the use of the first decile of the data set as a reference value for the first isotherm. It was found that this process downplays the weight of any atypical and leads to a better and easier comparison between different maps of isotherms.

Figure 5 shows in more detail the spatial distribution of the UHI at its greatest intensity (July 15th, 2015). This has been compared with another isotherm map from the same time of the year, but using data from 30 years ago (July 11th, 1985; Fig. 6). The temperatures were obtained from the temperature profiles published in the first UHI study of Madrid (López Gómez et al. 1988).

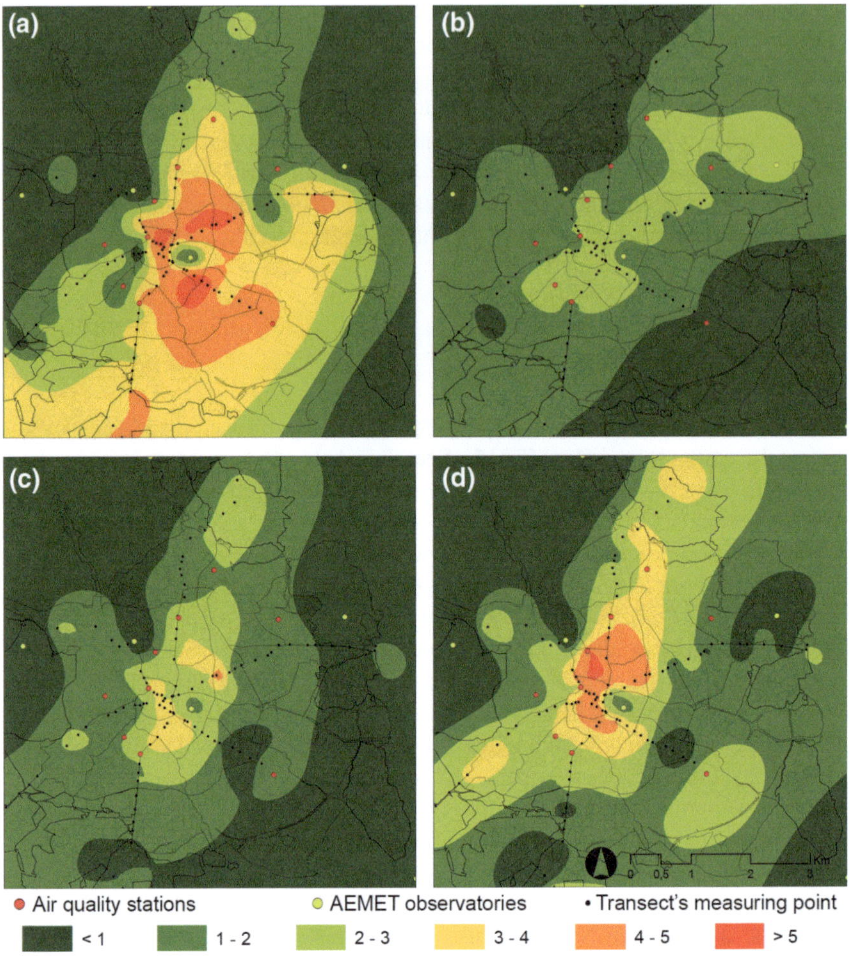

Fig. 4 Isotherm maps based on temperature variation for each data set, presented in chronological order. **a** July 15th, 2015; **b** October 28th, 2015; **c** February 16th, 2016; **d** April 25th, 2016

It is observed that the morphology of both of them is very similar. The concentric distribution of the urban heat island is tinged with a more intense development in the NE-SW axis, in line with the urbanization expansion. It is found a repeated pattern in *Ciudad Universitaria,* as it is in both cases the closest point to the city with the lowest temperature. The surroundings of the park *El Retiro,* with a cool island, show a similar behaviour too. Results confirm that the villages around the city of Madrid are under the effect of micro-heat islands, with an intensity of 1–3 °C. Confirming that there are some urban fabrics that cause either a rise or a fall in temperatures is worth of interest as well.

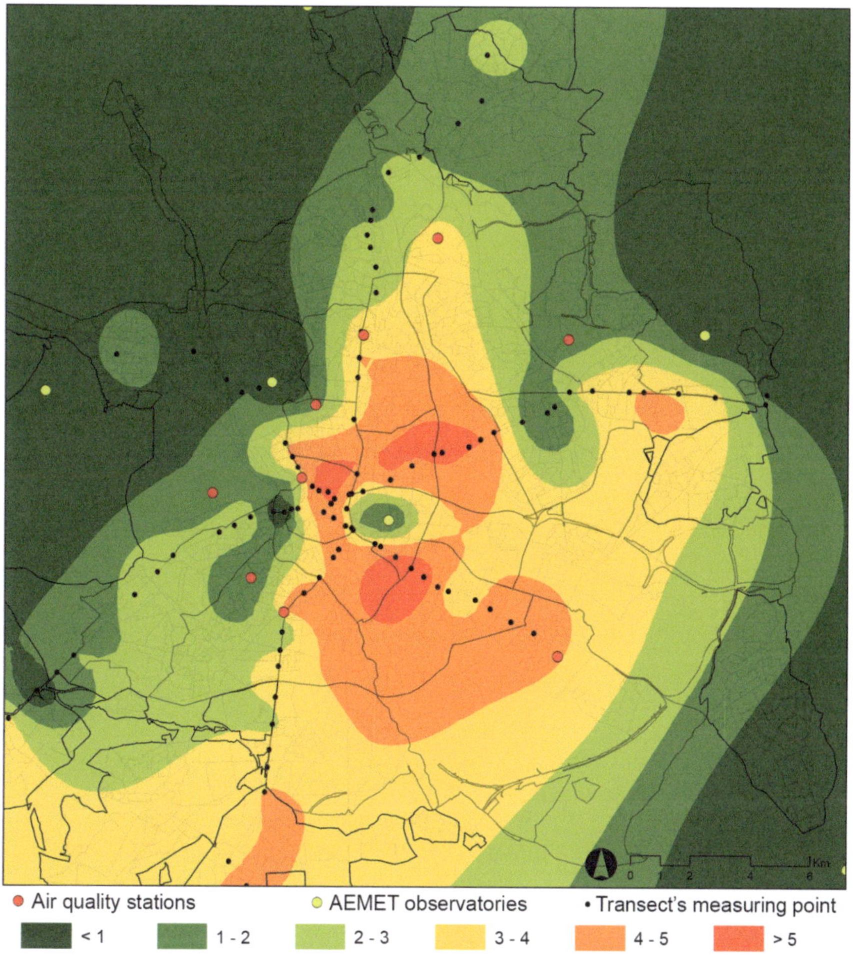

Fig. 5 Isotherm map based on temperature variation at midnight. July 15th, 2015

Regarding the evolution of the urban heat island during the last 30 years, it seems clear that the UHI has enlarged its size and increased the number of urban areas that bear the hottest temperature records (those where the intensity is over 4 °C). However, the UHI intensity does not seem to be increasing itself, as the maximum values are in both cases around 5–6 °C.

It appears that the growth of the city has effectively increased the domain of the urban heat island, spreading its higher effect over new areas. However, it does not seem enough to increase its intensity at the points where already had reached the maximum, probably because the urban fabric is consolidated and has changed little or nothing over the last three decades.

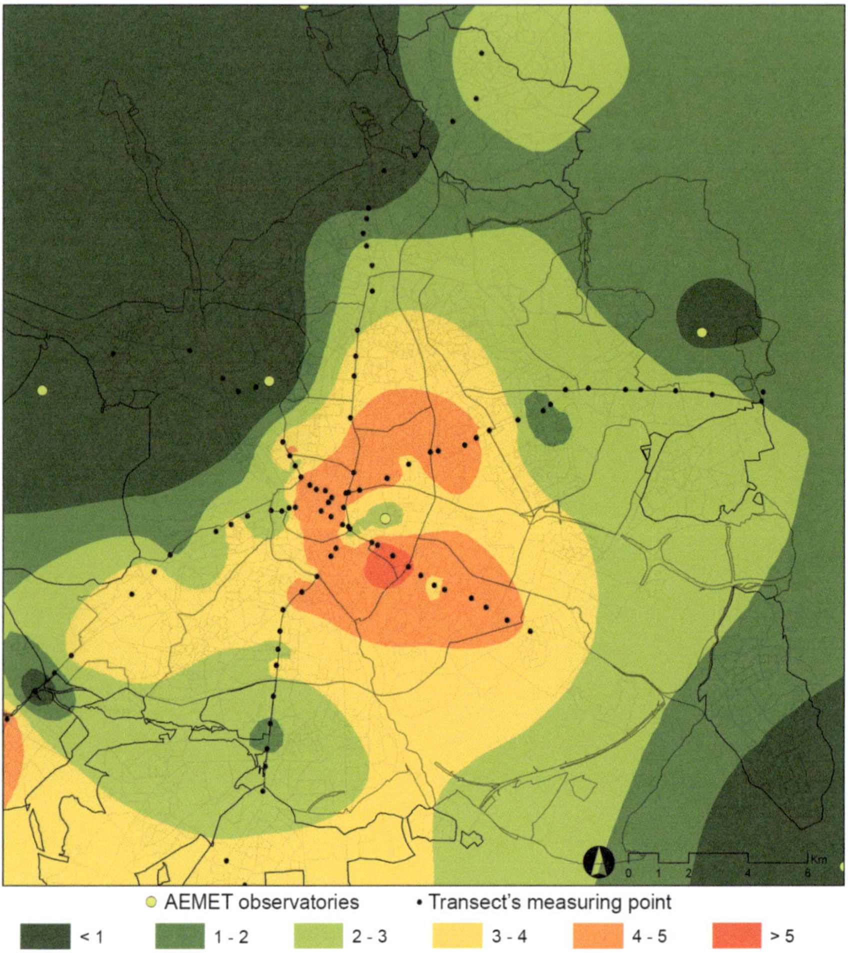

Fig. 6 Isotherm map based on temperature variation at midnight. July 11th, 1985. After López Gómez et al. (1988)

5 Conclusions

This research confirms that the urban heat island of Madrid has grown from the study that was firstly developed between 1984 and 1986 (López Gómez et al. 1988). It has been observed an increase in its area of influence, in particular on those parts with the highest temperatures.

Despite the significant human effort and the material means that were needed, the methodology that was used to carry out this updating of the UHI has turned out to be extremely useful. Due to the improvements that were implemented on the process, we expect to perform an urban climate analysis on a lower scale.

The comparison of the data sets from different times of the year suggests that the highest intensity happens around the summer solstice, mainly due to the greater amount of solar radiation that is received and accumulated by the urban fabric. Contrary to what is said in the study from 1988, the intensity of the UHI appears to reach its minimum around the winter solstice. The differences between both studies seems to be on the rural reference value.

In conclusion, the findings of this study have provided an updated base for the decision-making of UHI related aspects. These results could be used in many fields related to the urban climate, and it stands as an essential tool to select the most relevant locations of the urban microclimate. Ultimately, it completes the first stage of the dynamic model that will integrate the UHI effects into the building energy evaluation.

Acknowledgements This research is funded by *Programa de I+D+I orientada a los retos de la sociedad 'Retos In-vestigación'* of the Ministry of Economy and Competitiveness. Grant code BIA2013-41732-R, MODIFICA Project: Predictive Model For Dwellings Energy Performance Under The Urban Heat Island Effect.

Authors would also like to thank the State Meteorological Agency (AEMET) and Madrid Air Quality Integral System for weather data provided to carry out this research.

References

Almendros MÁ, López Gómez A (1995) La isla de calor en Madrid y las situaciones sinópticas. Estud Geogr 56(219):207–221

Arnfield AJ (2003) Two decades of urban climate research: a review of turbulence, exchanges of energy and water, and the urban heat island. Int J Climatol 23(1):1–26

Coakley D, Raftery P, Keane M (2014) A review of methods to match building energy simulation models to measured data. Renew Sustain Energy Rev 37:123–141

Evins R (2013) A review of computational optimisation methods applied to sustainable building design. Renew Sustain Energy Rev 22:230–245

Fernández García F (2001) El clima urbano de Madrid y su influencia sobre el confort térmico. Boletín de La Real Sociedad Geogr 137–138:12

Fernández García F, Almendros Coca MÁ, López Gómez A (1996) La influencia del relieve en la isla de calor de Madrid: las vaguadas del Manzanares y del Abroñigal. Estud Geogr 57 (224):473–494

Foucquier A, Robert S, Suard F, Stéphan L, Jay A (2013) State of the art in building modelling and energy performances prediction: a review. Renew Sustain Energy Rev 23:272–288

Fumo N (2014) A review on the basics of building energy estimation. Renew Sustain Energy Rev 31:53–60

Grimmond CSB, Blackett M, Best MJ, Baik J-J, Belcher SE, Beringer J, Zhang N (2011) Initial results from Phase 2 of the international urban energy balance model comparison. Int J Climatol 31(2):244–272

Grimmond CSB, Blackett M, Best MJ, Barlow J, Baik J-J, Belcher SE, Zhang N (2010) The international Urban energy balance models comparison project: first results from phase 1. J Appl Meteorol Climatol 49(6):1268–1292

Kolokotroni M, Davies M, Croxford B, Bhuiyan S, Mavrogianni A (2010) A validated methodology for the prediction of heating and cooling energy demand for buildings within the Urban Heat Island: case-study of London. Sol Energy 84(12):2246–2255

Kolokotroni M, Giannitsaris I, Watkins R (2006) The effect of the London urban heat island on building summer cooling demand and night ventilation strategies. Sol Energy 80(4):383–392

López Gómez A, López Gómez J, Fernández García F, Arroyo Ilera F (1988) El Clima urbano de Madrid: La isla de calor. CSIC, Madrid

López Gómez A, López Gómez J, Fernández García F, Moreno Jiménez A (1993) El Clima urbano: teledetección de la isla de calor en Madrid. MOPT, Madrid

López Moreno H, Sánchez-Guevara Sánchez C, Román López E, Neila González FJ (2015) Thermal characterization of urban heat island according to urban morphology of Madrid. In: Proceedings of the III International Congress on Construction and Building Research

Mihalakakou G, Flocas HA, Santamouris M, Helmis CG (2002) Application of neural networks to the simulation of the heat island over Athens, Greece, using synoptic types as a predictor. J Appl Meteorol 41(5):519–527

Mirzaei PA, Haghighat F (2010) Approaches to study urban heat island—abilities and limitations. Build Environ 45(10):2192–2201

Mirzaei PA, Olsthoorn D, Torjan M, Haghighat F (2015) Urban neighborhood characteristics influence on a building indoor environment. Sustain Cities Soc 19:403–413

Núñez Peiró M, Román López E, Sánchez-Guevara Sánchez C, Neila González FJ (2016) Hacia un modelo dinámico para la isla de calor urbana de Madrid = Towards a Dynamic Model for the Urban Heat Island of Madrid. Anales de Edificación 2(1):49–58

Palomino I, Vivanco MG, Theobald M, Garrido JL, Martín F (2015) Evaluación de la calidad del aire en España utilizando modelización combinada con mediciones. Preevaluación 2014. Madrid

Salamanca F, Martilli A (2010) A new building energy model coupled with an urban canopy parameterization for urban climate simulations-part II. Validation with one dimension off-line simulations. Theoret Appl Climatol 99(3–4):345–356

Salamanca F, Martilli A, Yagüe C (2011) A numerical study of the urban heat island over madrid during the DESIREX (2008) campaign with WRF and an evaluation of simple mitigation strategies. Int J Climatol 32(15):2372–2386

Santamouris M (2007) Heat island research in Europe: the state of the art. Adv Build Energy Res 1 (1):123–150

Santamouris M (2015) Analyzing the heat island magnitude and characteristics in one hundred Asian and Australian cities and regions. Sci Total Environ 512–513:582–598

Sobrino JA, Sòria G, Romaguera M, Cuenca J (2009) Desirex 2008: Estudio de la isla de calor en la Ciudad de Madrid. Revista de Teledetección 31:80–92

Stewart ID (2011) A systematic review and scientific critique of methodology in modern urban heat island literature. Int J Climatol 31(2):200–217

Thermal Energy Refurbishment of Envelope in Mass Neighbourhood Housing, Located in Semi-arid Climate of Argentina

Irene Blasco Lucas

1 Introduction

In December 2015 Argentina declared energy emergency state until the end 2017, because the whole system is near to collapse.

Residential sector is shown as the most appropriate to undertake energy efficiency policies (Evans 2010; Czajkowski and Gomez 2011), because on the one hand, represents 26% of national energy consumption, and on the other, shows the highest increase in the period 1999–2014 (59%).

In the city of San Juan is varying between 25% to humble users and 100% for those with low and medium income. In this oasis city there are a lot of neighborhoods built with low building density, intended for families of middle and low socioeconomic status.

The telluric characteristic of the area, led neglecting the hygrothermal comfort conditions of building quality, which is critical due to the semi-arid climate of the place. Faced with this problem, technological variants are formulated to enhance the thermal-energy behavior of the envelope (Blasco Lucas 2011) in a representative case of mass housing neighborhood and the performance in each one of them is evaluated, considering the associated lifetime costs.

I. Blasco Lucas (✉)
Instituto Regional de Planeamiento y Hábitat, Facultad de Arquitectura Urbanismo y Diseño, Universidad Nacional de San Juan, Av. Ignacio de La Roza y Meglioli, 5400 San Juan, Argentina
e-mail: iblasco06@gmail.com

P. Mercader-Moyano (ed.), *Sustainable Development and Renovation in Architecture, Urbanism and Engineering*, DOI 10.1007/978-3-319-51442-0_29

2 Methodology

The research was composed of three instances:

(a) *Climate analysis of a "typical year"* (*CTY*) formed by the average of 10 years, using different methods to determine the most appropriate design strategies for the region

- *CTY Elaboration*: Measurements recorded during the years 2002–2012 (Pontoriero and Hoese 2013) with the weather station type DAVIS installed on the roof of the Instituto de Energía Eléctrica de la Facultad de Ingeniería de la Universidad Nacional de San Juan (IEE-FI-UNSJ) were used. Processed climatic parameters were Temperature, Relative Humidity and Solar Irradiance. Databases were conformed in programmed MS-Excel spreadsheets.
- *Bioclimatic Strategies definition*: Time computation of inner conditioning needs through design strategies adapted to the climate is made by applying the models of Extended Bruce Novell (BNE, Mesa 2002), Architectural Bioclimatic Classification (ABC, USC 2006), Mahony-Evans Method (ME-MET, Blasco Lucas 2013) and Givoni-Watson-Szockolay—Hourly Bioclimatic Strategies (GWS-HBS; Blasco Lucas 2013).

(b) *Hygrothermal and energy dwelling's characterization and diagnose*, and also of their users, from stationary calculations, measurements and surveys in three witness-years.

- The procedures recommended by the series 11,600 of IRAM (Argentinean Institute for Rationalization of Materials) were applied in the original housing and the improved cases, using the developed informatic support "*Model for calculation of K and G*" (*KG-MOD*) (Blasco Lucas 2013).
- To evaluate the energy consumption were used records provided by gas and electricity utilities for three witness-years (1999, 2006 and 2013).
- The assessment of hygrothermal monitoring performed with dataloggers type HOBO was done with the PROMEDI-HTL (Processing of Hygrothermal and Lighting Measurements) (Blasco Lucas 2013).

- The statistical analysis of surveys of demographic and opinion type surveyed timely on large sample allowed to obtain information about family composition, dwellings equipment and how the users valuate their comfort level.

(c) *Micro-Economic Evaluation in the Life Cycle* (*MEELC*) of different selected alternatives, which contribute to greater thermal-energy efficiency of the housing.

- Method of the Present Value (PV) (Sapag Chain 2008) also called Life Cycle Cost Analysis in (LCCA) (Duffy and Beckmann 1993) is used.
- Three indexes were selected in order to compare results: the Net Present Worth (NPW), Return of Investment (ROI) and the Payback Time (PT), since the three together provide clear information for make informed decisions. The results interpretation is subject to the environmental conditions and the policies derived from them.

3 Climate Type-Year and Definition of Bioclimatic Strategies

CTY schedule monthly type days, which are shown in line graphs (Figs. 1 and 2) and surfaces graphs (Fig. 3) in its monthly statistical mean, maximum and mínimum values, and those calculated on its base, such as temperature and relative humidity amplitudes, and sunlight daily hours.

Three of the four models used to defining bioclimatic strategies based on the CTY data allow to obtain qualitative and quantitative results in different levels of disaggregation, and one of them only qualitative.

The strategies disaggregation is lower in BNE and higher in ME-MET which is a good complement to the other models, because performs constructive and building recommendations.

Fig. 1 Linear graph of average temperatures in CTY type days per month

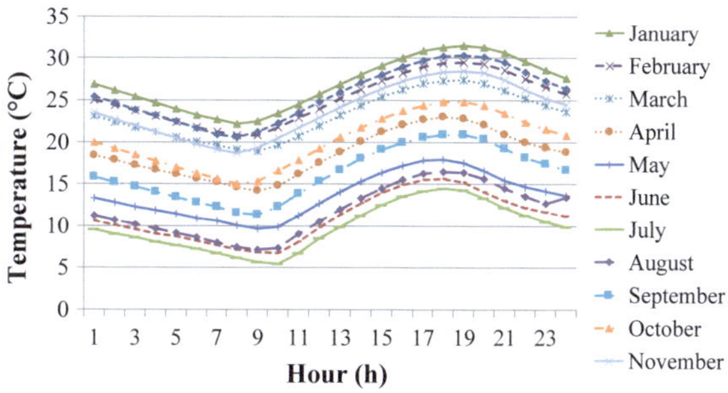

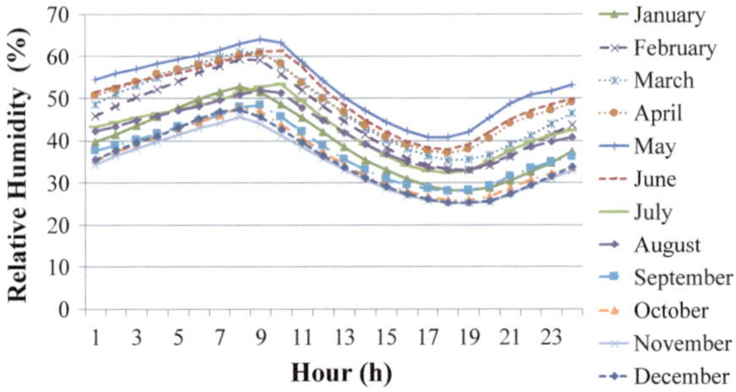

Fig. 2 Linear graph of average relative humidity in CTY type days per month

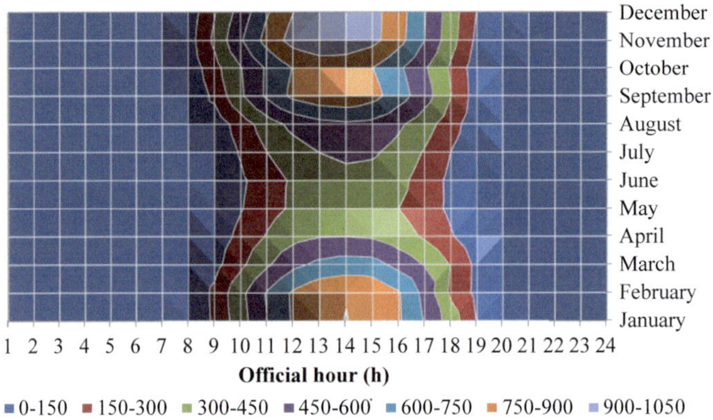

Fig. 3 Surface graph of average irradiance in CTY type days per month

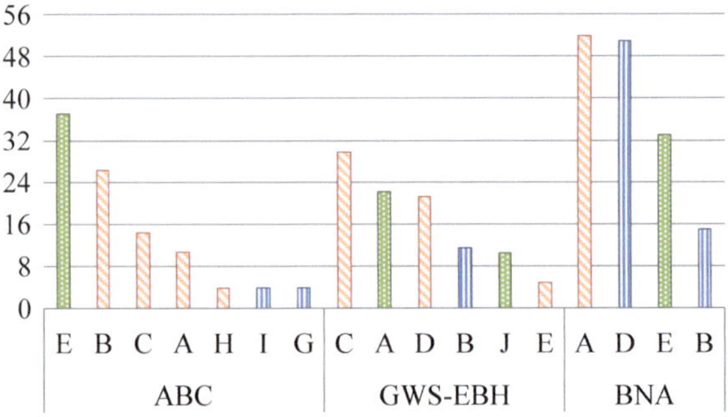

Fig. 4 Annual validity of bioclimatic strategies according to three methods, based on the CTY

In Fig. 4 and Table 1 obtained results sorted from highest to lowest for each method are summarized. All agreed needs with certain difference margin in values.

Table 1 Synthesis of bioclimatic strategies according to four models

Model	Code	Annual % or Aspect	Strategies
ABC	E	37.1	Thermal comfort
	B	26.3	Solar heating
	C	14.4	Thermal inertia (Heating)
	A	10.8	Artificial heating
	H	3.8	Thermal inertia (Cooling)
	I	3.8	Evaporative cooling
	G	3.8	Night ventilation
GWS-HBS	C	29.9	Thermal inertia (Heating) + internal gains
	A	22.2	Thermal inertia + Shadow
	D	21.2	Thermal inertia (Heating) + internal gains + passive solar
	B	11.5	Extended thermal comfort + shadow
	J	10.4	Thermal Inertia (Cooling) + evaporative cooling + selective ventilation
	E	4.9	Thermal inertia (Heating) + internal gains + passive solar + activa solar
BNE	A	52.0	Heating
	D	51.0	Shadow
	E	33.0	Comfort
	B	15.0	Evaporative cooling
ME-MET	A	Tipology	Closed organization with inner courtyard, and compact building
	B	Walls, ceilings and floors	Heavy outside and inside, with delay of more than 8 hours thermal transmission
	C	Windows	10% to 20% of surface walls, with sunscreen. Located to the North and South, at the height of the body and to windward, to promote cross ventilation
	D	Surrounding	Evaporative Cooling

Colors highlight: Green, confort; pink, heating and light blue, cooling

4 Energy-Hygrothermal Diagnosis

The case selected for the study is Lieutenant General Pedro Eugenio Aramburu neighborhood (Ar). Built between 1983 and 1985, it is placed 5 km from the center in northwestern suburb of San Juan, Argentina (Blasco Lucas 2008) (Fig. 5). Terrestrial coordinates of the city are: Latitude −31° 32′; −68° 71′ length, and HASL 620 m. Ar is composed of 103 residential building's blocks with 226 apartments and 383 single housing with 2, 3 and 4 bedrooms (Fig. 6). The analysis is focused in the most numerous isolated typology, which is those of three bedrooms (Fig. 7).

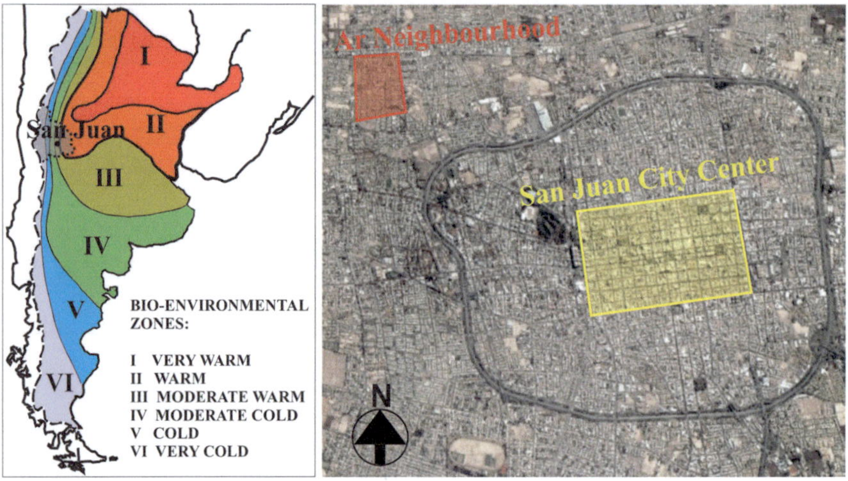

Fig. 5 Bio-environmental zones of Argentina as IRAM 11,603 (2012) with geographical location of the province and the city of San Juan (Links), and Ar neighbourhood placed in the city (*Right*)

Fig. 6 Ar neighbourhood planimetry, views of collective and individual housing

The distribution of housing in Ar is a continuous building "strip" on the block, they are paired and leave no free side spaces, but have 3 m retirement in front, and a large partially covered space, which deals as a transition filter in the income.

The typology is placed with its front facing either toward any quadrant, being 56% almost evenly in the direction N–S and S–N, which are optimal to allow cross ventilation, as concluded in the bioclimatic analysis. Built area is 71.18 m^2 with 170.83 m^3 of volume, 0.75 m^{-1} Form Factor, 19% fenestration coefficient, and 20.5% of relative volume of thermal mass (Blasco Lucas 2011).

In Table 2 is summarized the Thermal Transmittance K of building components and its deviation from the permissible values according to IRAM 11,605 (2002) for an average building quality (B) in Bioambiental Zone IIIa, where the neighbourhood is located (Fig. 5). K of the walls exceed between 22 and 133% the stipulated thresholds.

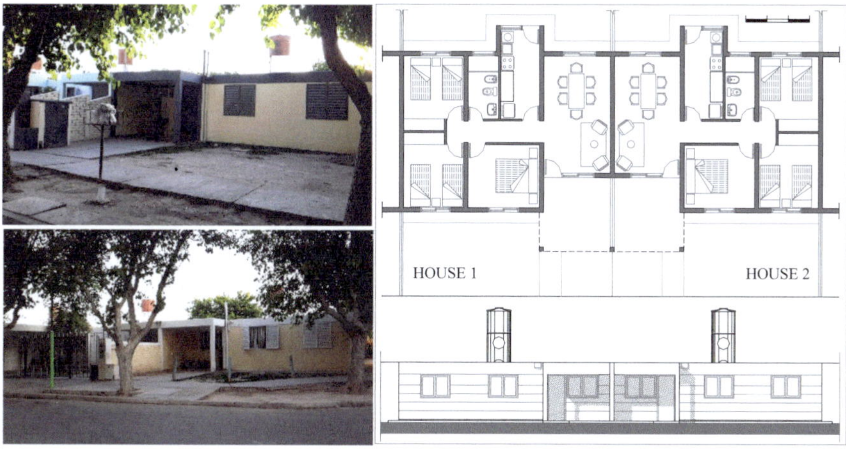

Fig. 7 Views, plant and façade of paired houses with three bedrooms

Table 2 Thermal transmittance K of components and differences with the permissible values stipulated by the Standard IRAM 11,605 (2002) for summer and winter

Component or material	Thermal transmittance K	K summer difference		K winter difference	
	(W/m^2 °C)	(W/m^2 °C)	(%)	(W/m^2 °C)	(%)
Concrete block wall	2.33	−1.08	−86	−1.33	−133
Dividing wall	1.93	−0.68	−54	−0.93	−93
Roof	0.65	−0.14	−29	0.18	22
Floor	0.40	0.05	11	0.43	51
Wooden doors	2.29	−1.04	−83	−1.29	−129
Glazing	5.75	−4.50	−360	−4.75	−475
Metal sheet	5.88	−4.63	−371	−4.88	−488

Its outer walls are of hollow concrete blocks 19 cm thick plastered on both sides, the roof is composed with 10 cm reinforced concrete and 8 cm volcanic granulate, and the carpenters are of low quality folded metallic sheet and single glazing. The total K of the envelope is 1.69 W/m^2 °C.

The Global Volumetric Heating Coefficient (Gheat) is 2.87 W/m^3 °C, and the Volumetric Cooling Coefficient (Gcool) is 40.83 W/m^3, exceeding respectively 69 and 54% thresholds established by Standards IRAM 11,604 (2001) and 11,659-2 (2007). The annual Thermal Load results in total 21.36 MWh, corresponding 14.99 MWh to heating (Qheat) and 6.98 MWh to cooling (Qcool) (Fig. 8) by comfort temperatures of 20 °C in winter and 24 °C in summer, with 1000 h of cooling needs (42 days). However, calculations made on the base of actual measurements indicate respectively 76 and 82% lower consumption, without reaching those comfort conditions, which were relieved through surveys and contrasted with thermal monitoring.

Using a procedure elaborated in Blasco Lucas et al. (2012) over the three witness years analyzed, is estimated that while in 1999 an 81% of housing lacked air conditioners (HVAC), in 2013 an 83% had HVAC equipments with normal and high consumption, which resulted in a higher electricity demand, reaching increases of 1121% between 1999 and 2013, and 242% between 2006 and 2013, gas consumption, mostly used for heating, it had an increase of 30% between 2013 and 1999, although it declined 13% between 2006 and 2013.

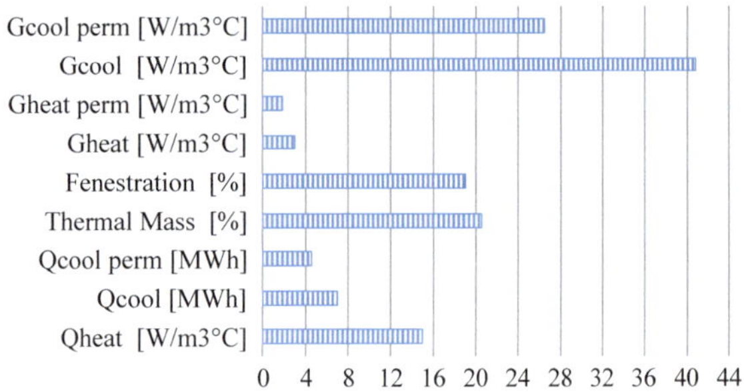

Fig. 8 Sizing-morphological indexes and thermal rates for heating and cooling

Table 3 Constructive alternatives by bioclimatic strategy and orientation of the house

House orientation	Alternative	Saving: thermal insulation in				Ventilation	Shade	
		1	2	3	4	5	6	7
		Roof	Walls	Openings (Burletes)	Windows (louvers)	Ducts	Pergolas	Vegetation
NS	A	X		X				X
	B	X		X			X	X
	C	X		X	X			X
	D	X	X	X				X
EW	E	X		X		X		X
	F	X		X		X	X	X
	G	X		X	X	X		X
	H	X	X	X		X		X

Based on the diagnosis obtained, in Table 3 simple constructive measures improvements that respond to some of the above detailed bioclimatic strategies (López Asiaín 2002). Those of annual validity contribute to energy conservation by incorporating thermal insulation and sealing infiltrations, while those appropriate only for summer, provide shade, evaporative cooling, and selective ventilation through ducts in homes facing the East or West side, which represent 44% of the total.

This will make a total of seven possibilities, which combine to give rise to eight alternatives, four of orientation to the North and four in the East, which are identified by letters. To facilitate the metric calculation details of each variant are drawn, as represented by Figs. 9 and 10, for the case F.

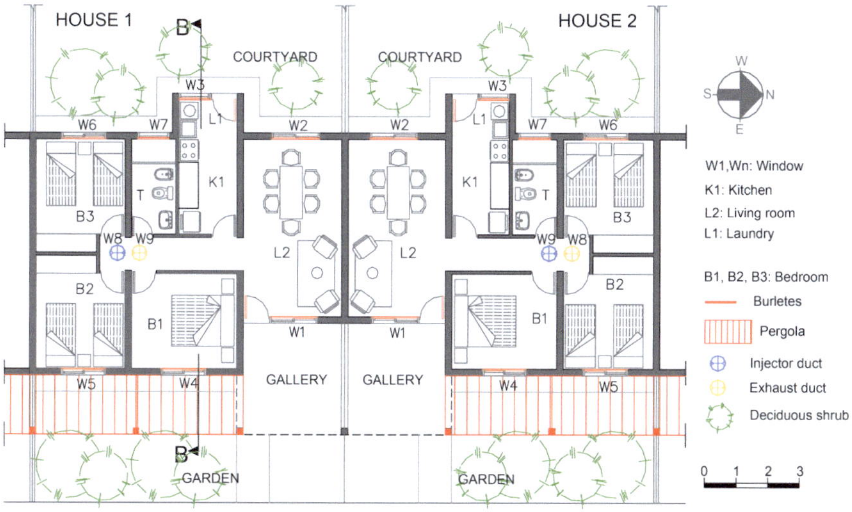

Fig. 9 Plant of the houses with proposed improvements of alternative F

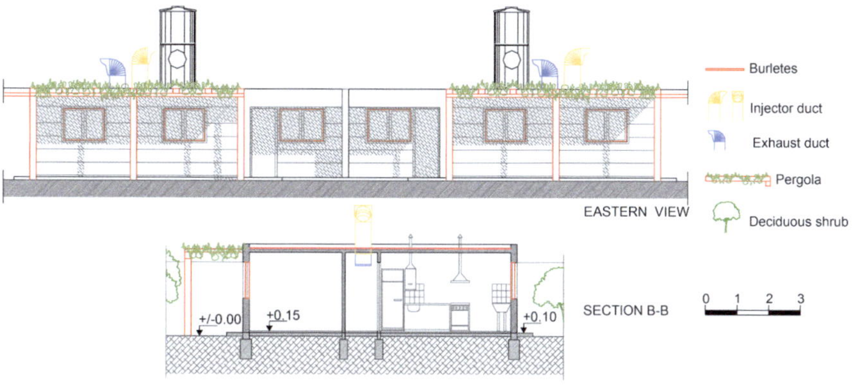

Fig. 10 Façade and cross section of the houses with proposed improvements of alternative F

Thermal-energy calculation for each alternative is made and the percentage of potential savings compared to the original housing is determined. These relative values ranging between 33 and 56% (Table 4) are applied to real consumption increased by 30% for gas and 200% for electricity, partially addressing the growing trend described above, and they are multiplied by the unit cost of each energy fluid.

5 Microeconomic Assessment in Life Cycle

The calculation of costs per alternative is obtained by adding the market prices of the composing strategies—for August 2016—reducing the values to the surface unit, and multiplying by the respective area. Table 4 shows the results of cost and energy savings for each case.

Annual operation and maintenance costs are considered as a percentage of Ii for each strategy and they addition allow to obtain the corresponding by alternative. The expected life cycle for each system is 25 years, thus housing is 30 years old. The PW method is applied because is the most appropriate of financial-mathematics to evaluate projects using natural energy, as they higher initial investment is offset by saving conventional energy throughout the useful life of the asset, which in this case is housing (Samuelson and Nordhaus 2004).

This annual benefit in monetary values must be subtracted in the equation of PW cash flow, which is calculated multiplying savings by the unit price of each energy fluid, considering increases of 200% for gas and 100% for electricity, although the Government has announced that will decree much larger augments. To estimate the NPW a low annual interest rate (2%) is applied, which partially offset the large price difference between those inherent in the construction and energy, being much higher the first.

Table 4 Results of costs (€ = $ AR 16.18) and energy saving (%) by alternative

Alternative	Initial investment (€)	Annual maintenance (€)	Annual Oper. (€)	Electrical saving (%)	Gas saving (%)	Total saving (%)
A	4428	21	398	4	29	33
B	5160	24	470	6	29	35
C	5839	28	536	15	29	44
D	5839	28	536	13	41	54
E	6187	29	570	7	29	36
F	7178	34	667	7	29	36
G	6187	29	570	23	29	52
H	6187	29	570	15	41	56

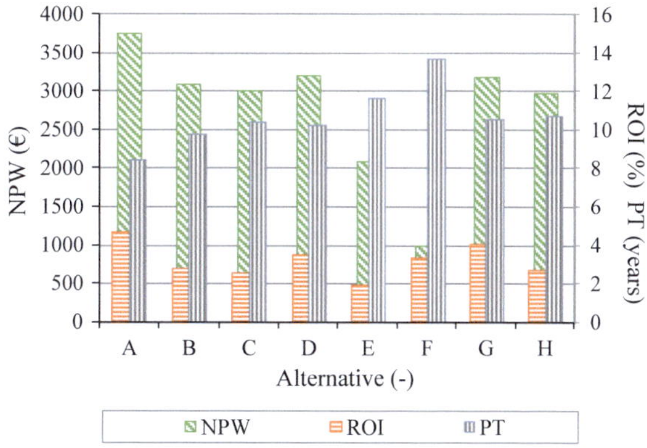

Fig. 11 Indexes of microeconomic profitability (NPW, ROI and PT) by alternative

Table 5 Priority of alternatives according each IR and all of them together

RI	1	2	3	4	5	6	7	8
ROI	A	G	D	F	B	H	C	E
PT	A	B	D	C	G	H	E	F
NPW	A	D	G	B	C	H	E	F
All	A	D	B	G	C	H	E	F

Also the Residual Value (RV) of assets at the end of life is included in the equation, having adopted for them 15% of current price of housing market, plus-valued 10% higher for having an alternative. Therefore Ar VR is €81,582. In Fig. 11 results for each alternative are represented.

The convenience of each proposal is determined when the NPW is positive and the ROI exceeds the interest rate used for the calculation, both being better the higher the amount. In turn, a lower PT makes a project more attractive. The nearest investment more profitable are those that meet the requirements better way for all three indexes. Table 5 shows the alternatives listed in order of priority as each RI and considering them all together.

However, depending on the investors interest the valuation can be weighted in accordance with only one of them. So to get the most energy savings over the housing life cycle should be chosen the H that reach 56% in oriented E–W or W–E cases, or the D with 54% for the cases N–S or S–N, as well permit good economic performance based on the three indexes.

6 Conclusions

Making global figures, if an average investment of €6000 would be done on each home of Ar, this would amortize in 11 years, with 3% ROI, allowing during 25 years an annual average energy savings of 45%, and reporting a net profit of €3000 at the end of the period. Under the assumed conditions the total amount of the gain achieved with all the 383 individual houses in the neighbourhood Ar is then it raised millions €1149, leading to a decrease in the order of 34 GWh in energy consumption and the associated environmental pollution, enabling also improvements in hygrothermal comfort levels and quality of construction.

The evaluation carried out on the issue both at initial diagnosis and proposals, and its projection in scale, it shows that urgent action is widely justified in San Juan residential sector through strong policies that encourage investment in technological improvements, while allowing a proper refurbishment, before housing become older. The criteria, procedures and tools developed and applied in this investigation, are practical tools for this purpose, whose stationary calculations have the limitations inherent to the 11,600 series (1996–2012) of the IRAM Standards.

References

Blasco Lucas I (2011) Evaluation of thermal-energy behavior of improvements with bioclimatic alternatives on FONAVI typologies. Adv Renew Energy Environ (AVERMA) 15(05):09–18

Blasco Lucas I (2013) Sustainable architecture in arid rural habitat of seismic zone: theoretical and methodological contributions. Doctoral Thesis in Architecture. Mendoza University

Blasco Lucas I, Rosés R, Facchini M, Hoese L (2012) Procedure for estimating energy consumption housing intended for air conditioning. Renew Energy Environ (ERMA) 27:1–10

Czajkowski J, Gomez A (2011) Journal of Sustainable Architecture. Faculty of Architecture and Urban Development, National University of La Plata, Buenos Aires

Duffy J, Beckmann W (1993) Solar engineering of thermal processes. Wiley Interscience, NY

Evans J (2010) Sustainability in architecture. Professional Association of Architects Planners, Buenos Aires

Lopez Asiaín J (2002). Architecture, city and environment. Editorial University of Sevilla, Publications Secretariat

Mesa A (2002) BNA: bruce-novell expanded method. Laboratory of Human Environment and Housing, Scientific and Technological Center Mendoza

Pontoriero D, Hoese L (2013) Bank of meteorological data of San Juan city. Electric Power Institute, National University of San Juan

Samuelson P, Nordhaus W (2004) Microeconomics. McGraw-Hill Ed. SA, NY

Sapag Chain N, Sapag Chain R (2008) Preparation and evaluation of projects. McGraw-Hill

Standards IRAM 11,600 Series (1996–2012) Thermal conditioning of buildings. Argentinian Standardization Institute, Buenos Aires

USC (2006) ABC: architectural bioclimatic classification software 1.3. University of San Carlos, State of Sao Paulo, Brazil

Method for the Implementation of Active Solar Systems in Hospitals, in the Hospitalization Unit of the Hospital Clínico del Sur, Concepción, Chile

Alberto Nope Bernal, Rodrigo García Alvarado and Ariel Bobadilla Moreno

1 Introduction

Design in hospitals is usually based on the program of needs according to the number of people that must be met, services and staff operations flow, patients and visitors (Cisneros 1999). Involving construction concepts, operations and premises seeking to ensure the functioning of user care and internal activities, with a design process that starts from generating the layout, to the detailed definition of areas such as clinic, technical support, administration and general services among others (Horacio and González 2006), usually in different stages. Located in central urban sites that generate complex morphologies characterized by variability in heights, producing high percentages of annual shading, which weakens the active or passive solar collection (Fig. 1).

In Chile, health facilities represent approximately 15% of the total energy consumption in the public sector, which involves around 240 GWh year (Walter et al. 2014). This type of buildings generate high energy demands due to its continued operation and concentrated construction, often without technical or efficient facilities, for which several actions of passive improvement have been recommended (Arquiambiente Alguasol 2010), although without addressing specific capabilities for active solar generation.

A. Nope Bernal (✉) · R. García Alvarado · A. Bobadilla Moreno
Universidad del Bío-Bío, Concepción, Chile
e-mail: albertonope@gmail.com

R. García Alvarado
e-mail: rgarcia@ubiobio.cl

A. Bobadilla Moreno
e-mail: abobadil@ubiobio.cl

P. Mercader-Moyano (ed.), *Sustainable Development and Renovation in Architecture, Urbanism and Engineering*, DOI 10.1007/978-3-319-51442-0_30

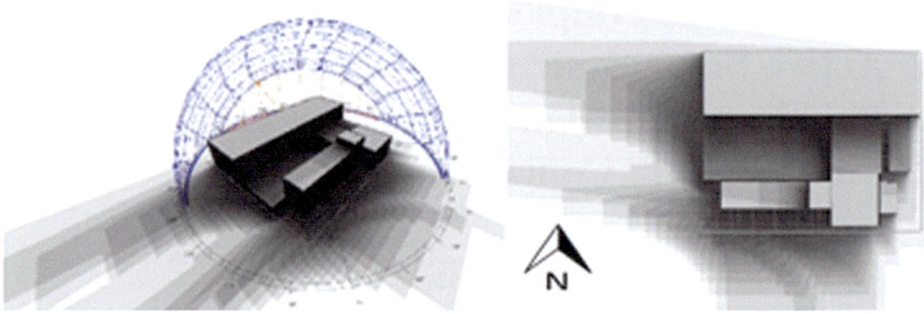

Fig. 1 Maximum shading annual at the hospital clinic of the south. Developing own

Studies of active solar potential integrated in buildings, propose a high capacity in central locations in Chile, which would allow establish a balance between the energy consumed and generated energy (Wegertseder 2016).

Suggested studies for the installation of thermal or photovoltaic technologies in buildings usually address only technical aspects in a bounded way, omitting volumetric conditions, services and the financial analysis which would allow the comparison of different alternatives, identifying initial costs and residual value, that contribute substantially to the establishment.

The objective of this work is to review the existing methods, abstracting important attributes and considering missing actions, to formulate a comprehensive methodology that allows the implementation of solar technologies. Providing an effective strategy for architects and/consultants in existing health facilities or those who are in early stages of design.

2 Existing Methods

Method, is defined as the organized and systematic way to proceed in order to arrive to a result. In this case, technical, energy and architectural recommendations described in different sources oriented to implement thermal or photovoltaic technologies in the building have been explored from the architect and/or consultants perspective.

An analysis of the procedures set out in guides and manuals issued by public and professional institutions (Table 1), as the Ministry of housing and urbanism of Chile (Minvu 2014), the Corporation of technological development (CDT 2013), the Chilean Chamber of construction (CDT—CChC 2007), the National Ministry of energy of Spain (CNE 2013) and the Foundation of the energy of the community of Madrid of Spain (DGIEM, Dirección General de Industri 2010), was carried out to identify from each one of these, the raised technical, energy, architectural and financial analysis, and propose missing considerations.

Table 1 Actions identified in guides and manuals

	Method				
	DGIEM (2010)	Minvu (2014)	CNE (2013)	CDT (2013)	CChC (2007)
Integration with architecture	x				x
Adaptation to architecture	x	x	x	x	x
Shading	x	x	x	x	x
Installation	x	x	x	x	x
Modulation	x		x	x	x
Structure of support	x	x	x	x	x
Place of installation	x	x	x	x	x
Accessories and add-ons		x	x	x	x
Performance	x	x	x	x	x
To measure	x	x	x	x	x
Analysis of investment and residual value			x		
Evaluation of the solar potential					
Simulation of demand					
Quantification of consumption			x		x
Maintenance	x	x	x	x	x
Technical specifications of the panels		x	x	x	x
Base weather		x	x	x	x
Cases of study		x	x		x

Developing own

On the other hand, also is the review of digital tools that emerged for the implementation of systems of solar energy in buildings (Table 2), integrated in recognized design software by Autodesk and Graphisoft, among others. Which consider implementation variables, allowing to evaluate surfaces, but in the majority of cases they do not bind types of technologies, equipment data, and demand contribution to architectural technical feasibility, demand and return on investment, which is a limitation for the average local, because of its international reach.

All these resources are characterized by independently involve technical and/or energy analysis of formulas of calculation between teams, but do not have a linked review of volumetric aspects and projections in services to ensure an effective implementation.

Revised attributes and methods are described below.

For new and existing projects were recognized in general, activities such as (Fig. 2) the consumption quantification (1), evaluation and selection of technologies (4), and the supply capacity calculation (5).

Table 2 Digital tools and attributes identified

	Digital tools employed to implement solar systems				
	Autodesk Ecotect	Autodesk Revit	Graphisoft Archicat	Autodesk Vasari	Valentin T- Sol Pro 5.0
Development of conceptual model	x	x	x	x	
Study solar year, month, day, hour	x	x	x	x	
Calculation of solar radiation, considering location.	x	x	x	x	x
Estimation of the energy produced		x		x	
Estimation of the energy demand for domestic hot water	x	x	x	x	x
Estimation of the energy demand for electricity	x	x	x	x	x
Estimation of the energy demand for heating	x	x	x	x	x
Estimating economic					x
Estimating economic					x
Quantification of technologies					x
Projection of shadows	x	x	x	x	

Developing own

Fig. 2 Attributes identified for the implementation of solar panels on buildings. Developing own

However, these calculations do not seem sufficient to fully determine the contributions of solar systems in a building. Some activities to complement these studies might be considered (Fig. 3), as the energy demand simulation (new or existing projects) (2), envelope (3) solar potential assessment, the assessment of the technical feasibility of architectural (6), and financial assessment of the different alternatives of investment (7), allowing to propose a revision more general possibilities for efficient installation.

Fig. 3 Suggested attributes to complement the implementation of solar panels in buildings. Developing own

3 Case of Study

The Clinical Hospital of the South, is located in the centre of the city of Concepción Chile, which is located in the South—36.81 latitude and longitude—73.05. Located in a temperate maritime climate with a Mediterranean influence, with temperatures in summer between 11 and 22 °C and in winter between 6 and 13 °C. Annual average accumulated radiation on flat surface is 1.629 kWh (DTI 2012).

A survey of the health services was conducted in the central municipality of the metropolitan area of Concepción, Chile, identifying hospitals and evaluating their morphological characteristics and location that affect the possibilities of solar collection. Then, an illustrative case was selected with increased incidence of shadow, to experiment the proposed methodology and assess an integrated implementation, and thus, suggest design strategies that allow to optimize the area of catchment in roofs and facades, allowing higher contributions.

Figure 4 shows the location of health services present in the commune of Concepcion, Chile, obtaining in the Table 3 an estimate of the number of services by types, surface of useful cover, built surface, consumption, area of useful roof and initial solar potential, calculated according to the area of general cover and annual radiation received.

Hospital Clínico South (Fig. 5), is one of the most important health canters in the Region of Bio Bio. It has 117 beds and 7 surgical pavilions of high technology; with an infrastructure of first level, including a new emergency room, with 8 attention boxes, observation room, resuscitation box and procedures and plaster box. Also counts with UCI and UTI which have been refurbished, counting currently with 16 beds for critical patient and 4 beds of UTI neonatology.

This facility is located in an area of 9794 m^2, with a total area of 11,300 m^2, on a main building construction and one secondary. The main building consists of volumes A-A1-A2-B (Fig. 5) that are part of the hospitalization, surgical wards and maternity unit. While the secondary D-C-E is aimed principally to administrative functions of the institution, emergency room, attention box, observation room, resuscitation box and procedures box.

Hospitalization unit was selected as an example of higher concentrated demand (volume A-A1), which includes a total 3000 m^2 of construction, housing 58 rooms with 117 beds, in a rectangular Pavilion facing the North-western, which presents 20 m in height, 14.80 m wide and 56.22 m long, with a total roof cover of 1385 m^2 surface. Main energy consumptions of this unit are hot water, electricity and heating.

Fig. 4 Health services present in the commune of Concepcion, Chile, in *red*. In *blue* hospital clinic of south case study. Extracted from (Nope 2015)

Table 3 Consumption estimation and energy generation potential for different health facilities present in the commune of Concepcion Chile

Destiny	Type of establishment	Roof (m^2)	Surface built (m^2)	Consumption (kWh/año)	Area useful in roof	Potential of energy (kWh/año)
Health	Hospital	27,174	93,351	56,980	10,870	16,370
	Clinics	14,817	32,176	19,852	20,032	8926
	Office medical	50,080	140,307	86,569	20,032	30,168
Total		92,071	265,834	163,402	50,934	55,464

Extracted from (Nope 2015)

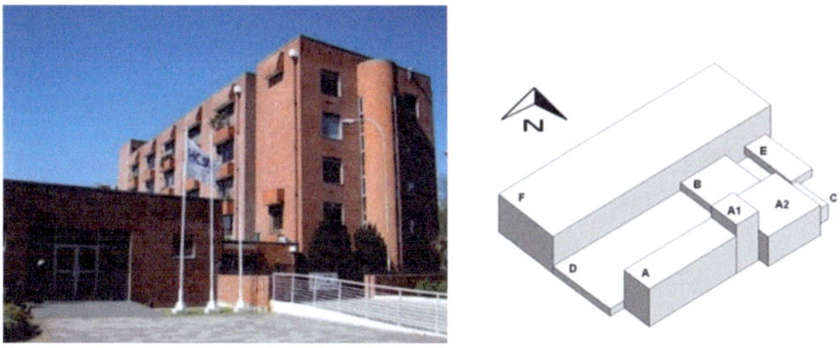

Fig. 5 Hospital clinical of the south Concepcion Chile. Extracted from (Revista nos 2010; Nope 2015)

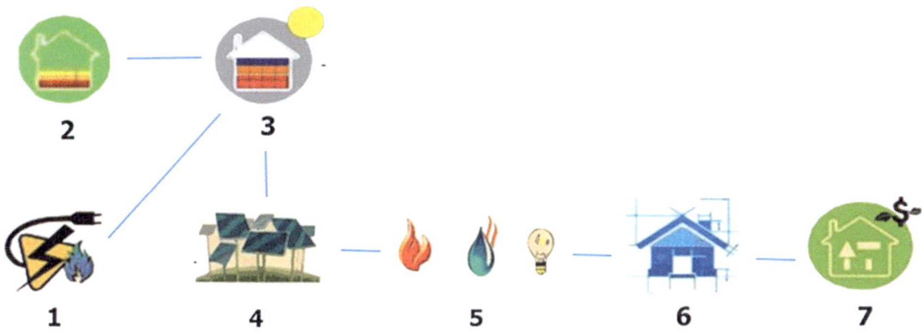

Fig. 6 Attributes suggested for the formulation of the method integral. Developing own

For analysis purposes and application of the method, it was decided to work with hot water and lighting requirements since low conditions of air infiltration and thermal transmittance of the envelope does not guarantee the adequate conservation of the heating in the enclosures of the inpatient unit, and they can be reduced with envelope or equipment improvements.

4 Proposed Method

Once recognized the detected attributes and suggested in the methods and reviewed tools, a process was formulated, allowing architects and/or energy consultants to implement an integrated solar technology in the architecture of new or existing hospitals, considering 7 methodological attributes (Fig. 6) that involve technical, architectural and economic conditions.

For existing buildings, the method should be developed starting from the attribute 1 followed of the three 3-4-5-6 and 7, and running the attribute 2 as a comparison between consumption and energy demands.

For projects that are in early stages of design method will be developed from the attribute 2 followed by the 3-4-5-6-7.

1. Quantification of gas and electricity consumption.
2. Simulation of HW, heating and lighting energy demands.
3. Evaluation of the solar potential of the different areas that comprise the building envelope.
4. Evaluation and selection of technologies to implement.
5. Supply capacity calculation.
6. Evaluation of constructive-architectonic feasibility.
7. Financial evaluation of the different alternatives of investment.

5 Application of the Method in the Study Case

5.1 *Quantification of Gas and Electricity Consumption*

For existing health facilities, this analysis consists in the review of annual and monthly energy consumption generated by the building, which commonly are generated by gas and electricity consumption, calculated in m^3 and kWh respectively.

In this case, with expenditure information provided by the Hospital Authority, was quantified the annual gas and electricity consumption of used to meet the needs of domestic hot water (DHW), heating and lighting of the 11,300 m^2. This way, is noted that, for the year 2013, the total gas consumption was 322,849 m^3, equivalent near 3,489,999 kWh; and the electricity consumption for the same year were of 1,844,100 kWh; its previous is equivalent to an approximated annual cost of USD 457,852,080, which represents, in environmental and financial terms, a relevant expenditure that could reduce the implementing of technologies alternatives.

5.1.1 Results—Quantification of Gas and Electricity Consumption

The patient unit, has a built area of 3000 m^2 distributed over 4 levels. The monthly energy consumption (Figs. 1 and 2) intended for domestic hot water (DHW) for 2013 was 123,780 kWh/m^2 (371,335 kWh), for heating of 185,670 kWh/m^2 (557,003 kWh) and lighting 139,370 kWh/m^2 (489,420 KWh).

5.2 *Energy Demand Simulation*

It was formulated as a tool to determine the energy demand according to General conditions of domestic hot water (DWH), heating and lighting in the early stages of design. For this case, it was used as comparison measure of energy consumption of the evaluated case, identifying percentages that exceed an acceptable consumption.

To assess the energy demand of DWH, heating and lighting of the Hospital Clínico del Sur hospitalization pavilion, Design Builder software was used, developing a three-dimensional model (Fig. 7), consisting of 44 thermal zones for which are assigned occupation, climate, latent loads, sensitive loads, and envelope material properties, according to the actual state of the building.

5.2.1 Results—Energy Demand Simulation for DWH, Heating and Lighting

The energy simulation allowed to identify that, for DWH consumption, it exceeds the demand in a 20%, for heating a 41% and for lighting a 39% (Table 4).

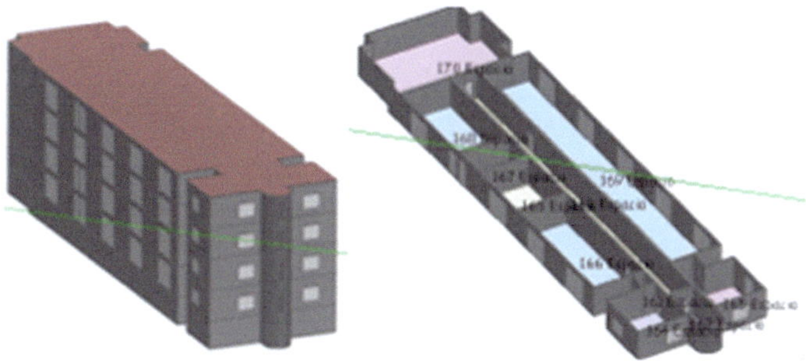

Fig. 7 Energy simulation model for the hospital clinical of the south patient unit in Concepcion, Chile. Developing own

Table 4 Energy demand for lighting, heating and DWH in patient unit

Month	Lighting (kWh/m$^{2)}$	Heat generation (Gas) (kWh/m^2)	DHW (Gas) (kWh/m^2)
January	20,291.38	1766.59	8475.92
February	17,644.67	2150.49	7655.67
March	18,526.91	4097.27	8475.92
April	19,409.14	8066.68	8202.51
May	20,291.38	16,781.80	8475.92
June	17,644.67	21,493.73	8202.51
July	20,291.38	24,019.70	8475.92
August	19,409.14	19,847.23	8475.92
September	18,526.91	15,249.69	8202.51
October	20,291.38	10,220.57	8475.92
November	18,526.91	5327.98	8202.51
December	19,409.14	2561.56	8475.92
Total	230,263.01	131,583.29	99,797.15

Developing own

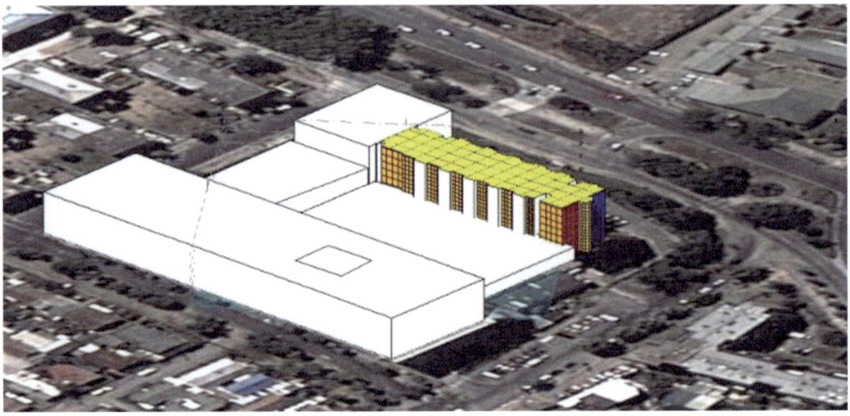

Fig. 8 Solar potential evaluation for the hospitalization unit of the hospital Clínico de Chile Concepción South. Developing own

5.3 Evaluation of the Solar Potential of Different Surfaces that Comprise the Building Envelope

The solar potential evaluation in the different areas that comprise the envelope of hospitalization unit of the Clinical Hospital of South, was VASARI, a software from Autodesk, which is developed based on the construction of conceptual masses who shared the same morphological case study selected (Fig. 8). This program includes a meteorological data type TMY2 which is based on averages of data collected in periods of 30 years (often in airports) for this case corresponded to the 8682290 Carriel Sur airport, in Concepcion. The assessment type is cumulative considering values in kWh/m^2 year on deck, and facade surfaces considering the volumetric relationship between the Patient Unit and the rest of the hospital complex, projecting shading on roofs and facades.

5.3.1 Results—Solar Potential Evaluation over Different Surfaces that Comprise the Building Envelope

The greater potential solar is located in the roof deck, comprising 1,124,170 kWh/m^2 año. Whereas a total of 1385 m^2 useful area total solar energy is 1,556,975 kWh year (Table 5).

5.4 Evaluation and Selection of Technologies to Implement

This activity addressed the review of 15 thermal panels (flat plate and vacuum tubes) and 15 photovoltaic (monochrystalline and polychrystaline) present in the

Table 5 Potential of solar radiation for roofs and facades of the hospitalization unit

Facade	Surface	m^2	KWh/m^2 año	Total
Roof	1	1385.00	1124.17	1,556,975.45
Facade North-eastern	1	411	513.33	35,932.86
Eastern South facade	1	196	1,849,762	15,538.00
Western South facade	1	280	194,379	5442.61

Developing own

market, whereas for its selection aspects as: local availability, installed power, thermal capacity, dimensions, weight by unit, efficiency, unit cost, integration in the architecture, connection systems, product life, generated energy and finally, the demand priority to meet. Three thermal panels and three photovoltaic were selected, in order to propose at least three alternatives of installation combining these technologies (Table 6).

5.4.1 Results—Evaluation and Selection of Technologies to Implement

See Table 6.

5.5 *Supply Capacity Calculation*

Final supply estimation with solar thermal and PV for each of the alternatives of installation, was performed using spreadsheets, which quantified the supply capacity for DHW and lighting through the incorporation of technical data.

To determine the supply capacity of DHW using thermal panels, were considered data as: manifold performance, inlet and outlet temperature out of the manifold, ambient temperature for each month, surface in m^2 for each collector, number of collectors, capacity in lt/m^2 in the accumulation pond, radiation data in kWh/m^2 day, kWh/m^2, kWh/month for the city of location.

For the case of lighting capacity calculation through the use of photovoltaic panels the supplied entry data was: power (w), number of panels, panel surface in m^2, inclined surface radiation ($/m^2$ day kWh) and sun hours a day for Concepcion. The results are set out in Tables 7 and 8.

5.5.1 Results—Supply Ability Calculation

See Table 7.

Table 6 Selection of thermal and photovoltaic alternative 1, 2 and 3 of installation technologies

Technology	Unit value all cost	Produced kWh generated energy	Weight/unit (kg)	Lifespan (year)
Alternative 1				
Solar thermal (vacuum tubes) Ref: TB solar collector	USD 6,054,959	9.172 Kcal day	114	25
Solar photovoltaic (polycrystalline) Ref solar module Kyocera KD19GH-4FU-Artikel Nr 0101544	USD 2,646,000	1 kWp (1.500 kWh/year)	16	25
Alternative 2				
Solar thermal (plate flat) Ref: CU solar collector	UDS 2,776,139	6.646 kcal day	37.2	25
Solar photovoltaic (Mono crystalline) Ref 300 W monocrystalline solar panel	USD 1,814,965	1 kWp (1.500 kWh/year)	14.5	25
Alternative 3				
Solar thermal (plate flat) Ref: C-CORE 88	USD 2,296,316	6.646 kcal day	38	25
Solar photovoltaic (Polycrystalline) Ref stand-alone (off-Grid) Kit OFF—GRID 0.1 kWp	USD 2,352,000	1 kWp (1.500 kWh/year)	16	25

Developing own

5.6 Architectonic Feasibility Assessment

The technical feasibility of alternatives 1, 2 and 3, (Fig. 9) is carried out using the RAM Elements program, simulating the load capacity of the evaluated building roof structure, whereas living loads, own weight and wind pressures; resulting in weight in kg per m^2 admissible for this structure and establishing the type for structural reinforcement (if required) for each alternative installation.

Both the thermal case for the photovoltaic water accumulators and power inverters were placed in rooms for this purpose in each of the levels of the building reducing the weight on deck.

5.6.1 Results—Constructive-Architectural Feasibility Assessment

The analysis to the structure of trusses and Rafter (Fig. 10) considered live loads, wind pressure, seismic loads, weight in kg/m^2 technologies and weight of the structure, obtaining the following results for the alternatives 1, 2 and 3.

Table 7 Supply capacity calculation for DHW and lighting alternative 1, 2 and 3 of installation

	Alternative 1	**Alternative 2**	**Alternative 3**
Thermal panel	Solar thermal (vacuum tubes) Ref: TB solar collector	Solar thermal (flat plate) Ref: CU solar collector	Solar thermal (flat plate) Ref: CU solar collector
Panel surface in m^2	5.53	2.43	2
Number of panels	58	127	131
Total installation área in m^2	310.3	308.33	262
% occupancy in roof deck (%)	33	36	37
DHW consumption kWh/year	916.11	916.11	916.11
Supplementing % kWh/year (%)	70	70	70
	Alternative 1	**Alternative 2**	**Alternative 3**
Photovoltaic panel	Solar photovoltaic (polycristaline) Ref Solarmodule Kyocera KD19GH-4FU-Artikel Nr 0101544	Solar photovoltaic (Mono crystalline) Ref 300 W monocrystalline solar Panel	Solar photovoltaic (Polycristaline) Ref stand-alone (off-Grid) Kit OFF—GRID 0.1 kWp
Panel surface	1.2	1.26	1.31
Number of panels	126	166	196
Total installation area	151.2	209.16	256.76
Performance (%)	14	16	10
% occupancy in roof deck (%)	58	36	37
Supplementing % kWh/year (%)	58	79	91

Developing own

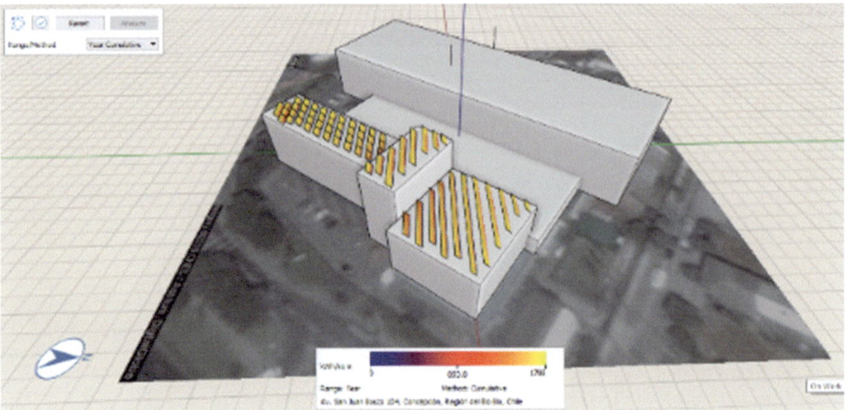

Fig. 9 Location of thermal panels and photovoltaic in roofs decks to-A1-A2 hospitalization unit of the hospital clinical of the south Concepcion Chile. Developing own

Fig. 10 Tension in Metallica truss and beam Kg/cm^2. Developing own

For required Truss:
br to 46% of their ability to take efforts, with a charge per m^2 of 7 kg whereas afferent area of 3.5 m.
Warp: L/1129 < L/700

For required belt:
br to 62% of their ability to take efforts with a charge per m^2 of 7 kg whereas afferent area of 3.5 m.
Warp: L/1436 < L/700.
Of the results obtained is concludes that not is necessary to reinforce the structure.

The architectural feasibility depends on the concept under which is intended to install the technology. This can be in an integrated and/or adapted way considering the most appropriated disposition within the hospital layout in relation to the heights of each building, shading and solar potential areas. Varying parameters of technologies such as height, density and orientation, in order to determine which surfaces, at a ensemble level, show higher and lower potential to implement solar energy systems on roof deck and a lower impact in the architectural image.

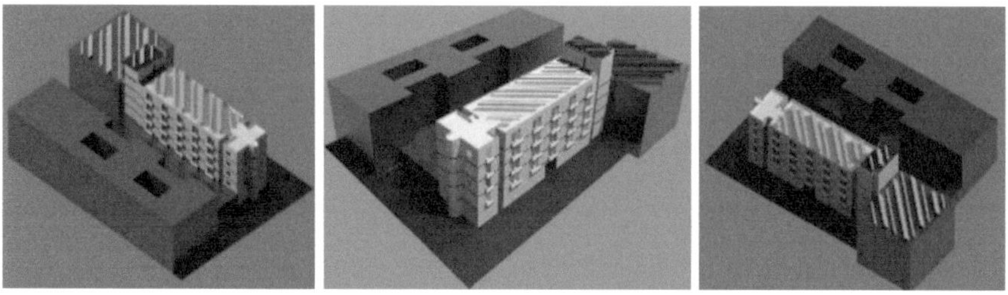

Fig. 11 Target image for implementation of thermal and photovoltaic technologies adapted on roof. Alternative 3. Developing own

In this case, the architectural layout offered high possibilities to implement solar systems preferably in cover (Fig. 11), since the greater useful area without obstruction is present on these surfaces. Contrary case happens with facades presenting major shadow areas reducing the usable area and, therefore, any solar energy system efficiency due to a configuration of balconies and morphology of the building itself.

5.7 *Financial Evaluation of the Different Investment Alternatives*

A financial analysis was performed to determine the most profitable installation using the format of analysis developed by SMART SPP (ICLEI and European 2009), which considered data as: nominal discount, end-of-life costs or elimination, inflation rate, life cycle rate costs (CCV), time horizon of planning, remnant value, and periodic costs during a time of 25 years.

This review allowed to identify the costs associated with each alternative and return on the initial investment of each time (Table 9).

5.7.1 Results—Financial Evaluation of Different Alternatives of Investment

See Table 8.

5.8 *Revision of the Method*

The application of this sequence of activities to a hospital building with concentrated energy and morphological features, allowed to determine the installation of photovoltaic and thermal technologies on roof, which obtains, in the best of cases, a

Table 8 Evaluation of investment and residual value for alternative 1, 2 and 3 of installation

Alternatives of investment			
	Alternative 1	**Alternative 2**	**Alternative 3**
Number of implemented thermal panels	58	127	131
Number of implemented panels photovoltaic	126	166	196
% occupancy in roof deck	33	32	33
% of housing demand (DHW-heating)	70	70	70
% of covered demand (lighting)	58	79	91
Investment initial Cost $	$326,434,138	$324,331,434	$281,489,151
Investment initial Cost USD	USD 528,638.28	USD 525,233.09	USD 455,852.88
Annual savings $	$30,826,073.00	$34,186,501.00	$36,236,460.00
Annual savings USD	USD 49,920.77	USD 55,362.75	USD 58,682.53
Payback in years	10.6	9.5	7.8

Developing own

contribution to the consumption of DHW and lighting of a 70 and 91% respectively, with a payback time of 7.8 years. While the revised procedures, with partial activities, usually suggests more limited facilities and smaller contributions, or are unaware of its implications and General possibilities. Raising of this mode, a strategy integral and effective to determine equipment solar, in stages early of design and allotments of buildings existing.

The described method, can be used also for buildings with other services, regardless of consumption and energy demand, showing morphological characteristics and similar location.

6 Conclusions

The analysis of various methods such as guides, manuals and used digital tools for the implementation of solar technologies in buildings, showed lack of integral architectural and functional aspects.

Three technical aspects were selected, such as the consumption quantification, calculating the delivery capacity, evaluation and selection of technologies to implement, by formulating four actions such as: energy demand simulation for DHW, heating and lighting, evaluation of the solar potential of the different areas that comprise the building envelope, the constructive-architectural feasibility assessment, and financial evaluation of different alternatives of intervention. Enabling the design of a comprehensive method that will involve technical, constructive and architectural conditions.

The development of a strategy involving technical, architectural and economic aspects in a case of high consumption and high predominance of shadow incident on its envelope (hospitalization wing of the Clinical Hospital of the South), allowed to identify by means of the combination of 6 different thermal and photovoltaic technologies for three alternatives of installation, the most convenient option for the studied unit. Suggesting that the proposed method can be used in early stages of design of projects or on retrofitting projects to improve its environmental performance and mitigate their energy consumption, with a higher adaptation to the demand and architectural integration.

An organized analysis of morphological conditions and techniques in health facilities allows to define appropriated systems for the use of the solar energy, that contribute substantially to their energy consumption. Enabling from the architecture to establish methods that integrate solar energy systems in the health building envelope, varying volumetric features, considering functional and technical aspects of this type of buildings.

Acknowledgements This research work was developed thanks to the project with * end Conicyt-AKA NCRE-007: evaluation of the Solar potential in urban Area.

References

Alfredo Plazola C (1999) Enciclopedia de Arquitectura Plazola V6 Hospitales. In: Arquifutura

Arquiambiente Alguasol (2010) *Guia de Eficiencia Energetica Para Establecimientos de Salud*. ed. Agencia Chilena de eficiencia energetica AChEE. Santiago de Chile

CDT, Coporación de desarrollo tecnologíco—Cámara Chilena de la construcción (2013) Diseño Y Dimensionamiento de Sistemas Solares Fotovoltaicos Conectados a Red. In: Coporación de Desarrollo Tecnológico

CDT—CChC, Corporación de desarrollo tecnologico (2007) Sistemas Solares Térmicos, Manual de Diseño Para El Calentamiento de Agua. In: Coporación de Desarrollo Tecnológico

CNE, Consejo Nacional de Energía (2013) Instalación de Sistemas Solares Sobre Techos. In: Consejo Nacional de Energía

DGIEM, Dirección General de Industri, Energía y Minas de la Consejeria y Hacienda y la Fundación de la Energía de la Comunidad de Madrid (2010) Guía de Integración Solar Fotovoltaica. In: Fundación de la energía de la comunidad de Madrid FENERCOM

Horacio M, García G (2006). Propuesta Arquitectónica Hospital General de Enfermedades Del Instituto Guatemalteco de Seguridad Social. Universidad de San Carlos de Guatemala

Minvu, Ministerio de Vivienda y Urbanismo de Chile (2014) Manual de Sistemas Solares Térmicos. In: Ministerio de vivienda y Urbanismo

Nope, A (2015) Método Para La Integración Arquitectónica de Energía Solar En Establecimientos de Salud. Universidad del Bío Bío

Revista nos (2010) Hospital Clínico Del Sur Duplicó Su Capacidad En Servicio de. Revista nos. http://www.revistanos.cl/2010/11/hospital-clinico-del-sur-duplico-su-capacidad-en-servicio-de-urgencia/

Walter L, Karin G, N S, Zamorano Alicia (2014) Igarss 2014 Eficiencia Energética En Hospitales, Públicos edn. GTZ - Dalkia, Santiago de Chile

Wegertseder P (2016) Potencial Solar En La Arquitectura Y La Ciudad. In: Universidad del Bío Bío

Validation of a Dynamic Simulation of a Classroom HVAC System by Comparison with a Real Model

Miguel Ángel Campano, Armando Pinto, Ignacio Acosta and Juan J. Sendra

1 Introduction

When a HVAC system is selected for the hygrothermal and air quality conditioning of buildings, usually some uncertainty arises as how the energy diffusion/emission system will actually behave in the rooms under study, as well as how energy-efficient it will be in comparison with other alternative systems with similar characteristics.

Despite there are many predictive methods for determining the energy balance in generic rooms in order to size their air conditioning systems (HVAC), either by measurements or nodal simulation, there is also a certain lack of knowledge as to whether the use of the building's energy is correct, as well as its transmission efficiency in habitable venues considered as tridimensional spaces, with occupants, furniture, equipment, and other heat sources that actively intervene in the system.

In order to predict the behaviour of these exchange systems, it is necessary to resort to Computational Fluid Dynamics (CFD), previously validating the computer tool by comparison with measurements in existing buildings.

M.Á. Campano (✉) · I. Acosta · J.J. Sendra
Instituto Universitario de Arquitectura y Ciencias de la Construcción,
Escuela Técnica Superior de Arquitectura de la Universidad de Sevilla, Seville, Spain
e-mail: mcampano@us.es

I. Acosta
e-mail: iacosta@us.es

J.J. Sendra
e-mail: jsendra@us.es

A. Pinto
Laboratório Nacional de Engenharia Civil, Núcleo de Acústica, Iluminação,
Componentes e Instalações, Lisbon, Portugal
e-mail: apinto@lnec.pt

P. Mercader-Moyano (ed.), *Sustainable Development and Renovation in Architecture, Urbanism and Engineering*, DOI 10.1007/978-3-319-51442-0_31

Thus, it is proposed to apply this validation protocol by comparison to a secondary school standard classroom, given that is one of the most common room types with high occupation and due to the sensitivity of their occupants.

1.1 State of the Art

When a dynamic energy simulation process is performed for architectonical enclosures, it is possible to solve it just by CFD techniques. However, it can be an extremely long and complex process, since it requires a dynamically characterization of the energy balance with the outside and all the rest of rooms of the building through both the envelope and the exchanged air flows. Thus, this process is usually performed by coupling (Negrão 1995), being separated in two consecutive phases, either steady-state or dynamic-state simulations (Zhai et al. 2002; Zhai and Chen 2004):

- Nodal calculation, in which a global thermal balance is established for obtaining the boundary conditions.
- Calculation through Computational Fluid Dynamics (CFD). Using previous boundary conditions obtained from nodal calculation, a detailed and exhaustive characterization of the movement and thermal variation of the air volume in the interior of a room is performed, as well as its dynamic interaction with other existing elements in such room.

In general, an energy simulation process is not reliable until it has been validated by comparison. In this way, and as a basis for the present study, research by Karimipanah et al. (2007) is worthy of mention; it is focused on the analysis of the thermal distribution efficiency both of displacement systems and confluent jets, through CFD calculation, being previously validated by comparing with a series of measurements in a classroom model, built inside a climatic chamber. In addition, there are also relevant the studies by Yang et al. (2014) for German classrooms, both for natural cross ventilation and mechanical ventilation systems, either conventional or by displacement, as well as research by Conceição and Lúcio (2011) for Portuguese classrooms with radiant and cooling surface systems.

1.2 Objectives

The aim of this study is the validation of the Design Builder tool as a CFD simulation software, applied to the characterization of the thermal behaviour in medium-sized spaces with high internal loads and a high occupation density.

To that effect, a validation protocol is developed, in which the results of the predictive simulation models are compared with the measured air temperature values inside an existing classroom, which is used as a reference.

2 Methodology

For establishing the work methodology of this software validation study (Campano 2015), it is necessary to define a study sample for being used as a comparison element, also selecting the objective parameters that will be measured and used as reference. After selecting the calculation tool, the comparative analysis methodology is defined, as well as its regulatory framework.

2.1 Selection of the Study Sample

The room model selected for this validation study is the standard classroom 24 of the "Eça de Queirós" secondary school, which can be considered as a representative example of an educational facility classroom of the Iberian Peninsula. It is located in the freguesia de Olivais, within the municipal area of Lisbon (Portugal), a city with Csa conditions according to the Koppen climate classification (Kottek et al. 2006). This building is owned by the public entity "Parque Escolar" of the Government of Portugal and was designed by the architect Jorge Martins.

The classroom is located on the second floor of the secondary school, on its south cornet. It is in contact with the outside both by its southeast-facing and southwest-facing walls, as well as by its roof. Its northwest-facing inner partition is in contact with the classroom 23 and its northeast-facing inner partition is adjacent to both the corridor and the classroom 25. Finally, it is located over the teacher's room of the first floor.

Dimensions are 7.00 m × 7.00 m with a height of 3 m, accommodating 31 tables plus the teacher desk (Fig. 1). Its only access is located on the partition in contact with the corridor with a door 0.9 m × 2.10 m and a ventilation opening above with a free area of 0.15 m × 0.80 m.

The classroom is equipped with one desktop PC placed on the teacher's desk, which has an associated projector in the centre of the ceiling of the room. The room has a suspended lighting system of nine fluorescent lamps of 48 W each, arranged in three rows parallel to the window and levelled with the suspended ceiling.

The composition of the classroom envelope is shown in Table 1.

The HVAC system of this educational facility consists on a four-pipe hydronic system with one makeup air unit (MAU) for each four classrooms, being located on the roof. These units provide outside air to their classrooms through 300 mm diameter individual galvanized sheet metal ducts arranged over the window. Three double deflection grilles of 0.10 × 0.05 m are located on this, disposed with an

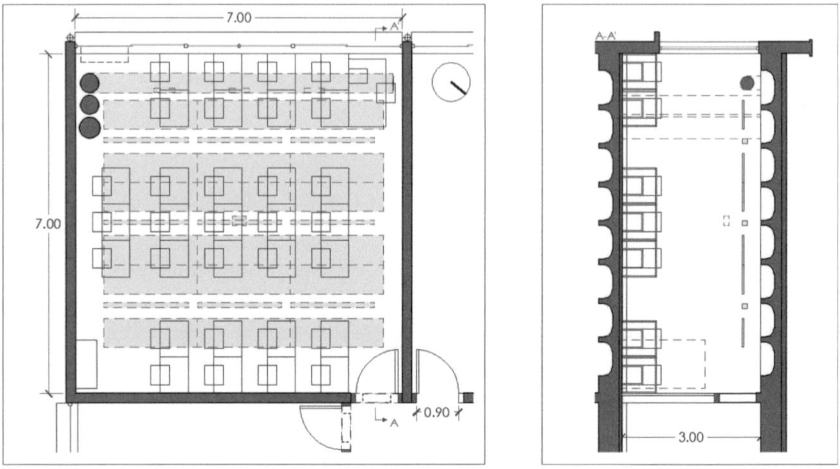

Fig. 1 Plant and section of the standard classroom 24 of the "Eça de Queirós" secondary school

Table 1 Envelope composition of the "Eça de Queirós" secondary school (Lisbon)

Element	Composition	U (W/(m²K)
Façade	Single-layer cement mortar (1 cm); thermal clay block (14 cm); air chamber without ventilation (1 cm); mineral wool (0.031 W/m K, 2 cm); 2× gypsum plasterboard (1 cm)	0.69
Vertical partitions	2× gypsum plasterboard (1 cm); air chamber without ventilation (1 cm); thermal clay block (14 cm); air chamber without ventilation (1 cm); 2× gypsum plasterboard (1 cm)	0.96
Horizontal partitions	Linoleum (1 cm); cement mortar (2 cm); sand layer (4 cm); grid slab (35 cm)	2.4
Roof	Gravel layer (5 cm); XPS extruded polystyrene (0.034 W/m K, 5 cm); cement mortar (1 cm); bitumen sheet (0.5 cm); cement mortar (1 cm); lightweight aggregate concrete, formation of slope (mean 10 cm); grid slab (35 cm)	0.55
Windows	6 mm single glass pane Aluminium frame (2 cm) with no thermal break Solar protection with light gray opaque inner sunblind	5.8 5.9

angle of 45°. Air extraction is developed through an air passage over the door. Thus, the ventilation efficiency in cooling mode is 1.0, while in heating is 0.8 (CR 1752:1998).

The value of the infiltrations measured in this classroom is 0.51 ACH, but they are considered as zero when the HVAC system is operating, since it overpressures the room.

The climate conditions during the measurements are shown in Table 2, which were obtained from the weather station of the Portela Airport, located a kilometre and a half to the west of the educational facility.

Table 2 Climate data summary table (weather station of Portela Airport)

Location: Lisbon (Portugal) Longitude/latitude: −9.11° (W)/38.76° (N) Time zone: GTM +0:00	Elevation above the sea level: 114.0 m Wind exposure: Regular Mean wind speed: 5.0 m/s
Climate data template: Portela_2012.swec Ground temperature in winter: 8.8 °C	Summer AT/RH (0.4%): 34.1 °C/29.5% Winter AT/RH (99.6%): 4.1 °C/85%

2.2 Measurement of Objective Parameters

To perform this CFD validation process, the classroom under study was monitored during 5 days of November, in which the indoor thermal parameters were recorded, both in a continuous and in a sporadic way, in order to obtain both the initial conditions (boundary conditions) for calculation models and the reference values for the validation process. All this information was obtained using the following equipment, disposed as shown in Figs. 2 and 3:

- Continuous measurement (contrast for CFD validation):
  - 80 type J thermocouple sensors, distributed throughout four columns with 20 sensors per column (C1, C2, C3 and C4).
  - 3 hot wire sensors (TA5 Thermal Anemometer) linked to columns C1 and C2.

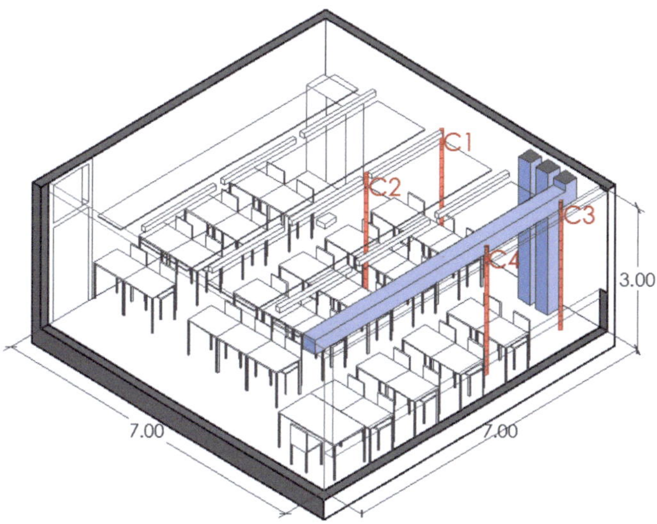

Fig. 2 3D view of the classroom 24 with the location of the four thermocouple sensor columns

Fig. 3 Inner view of the classroom 24 with the location of the four thermocouple sensor columns

Table 3 Measured indoor conditions, internal loads and HVAC system operation during the measurement instant

Internal loads			
Occupation	22 students (16 years old) and 1 teacher (metabolic rate 1.20 met, 130 W)		
Lighting	3 suspended lines of 3 48 W fluorescent lights (40% convective)		
Equipment	EPSON EB85 projector: 275 W (45% convective) Desktop PC: 140 W (70% convective) Laptop PC 140 W (70% convective)		
Measured indoor conditions			
Mean indoor air temperature (°C)	19.6 °C	Mean radiant temperature (°C)	19.2 °C
Mean outdoor air temperature (°C)	16.2 °C	RT frontal interior wall (°C)	20.1 °C
RT lateral interior wall (°C)	19.4 °C	RT lateral exterior wall (°C)	19.0 °C
RT rear exterior wall (°C)	19.1 °C	RT floor (°C)	19.3 °C
RT ceiling (°C)	19.6 °C	RT window (°C)	17.8 °C
HVAC			
System	Four-pipe hydronic system. Thermal production by boiler and chiller Thermal treatment by a multi zone makeup air unit (22,000 m^3/h, 125 kW for cooling) with heat recovery (60 kW) Air distribution to the classrooms by individual ducts		
Air supply	1100 m^3/h (3 × 366.67 m^3/h) at 16.2 °C (7.48 ACH) Three double deflection grilles of 0.10 × 0.05 m located on a sheet metal duct arranged over the window, disposed with an angle of 45°		
Extraction	Air outlet over the door, dimensions of 0.15 × 0.80 m		

  - 2 Vane Sensor Anemometer AIRFLOW AV6 (35 mm for HVAC system air supply and 100 mm for exterior conditions).
  - Data Loggers DT85 DataTaker and DT800 DataTaker.

- Punctual measurement (boundary conditions):
  - Testo 810 sensor for surface temperature of the walls.
  - Testo 0635.1535 thermal velocity sensor for air temperature and air velocity on the air outlet over the door.

These type J thermocouple sensors have a bias error of ±1.138 °C, as well as an average error of 0.069 °C and a maximum error of 0.207 °C regarding to the used Testo 0635.1535 reference sensor, type RTD PT-100 (Campano et al. 2015).

The selected measurement instant in these five days is at 9:25 of the day 21.11.2012, since the classroom was at full capacity with the HVAC in operation. In Table 3 is possible to see all data about occupation, lighting, computer equipment, indoor conditions and operating parameters of the HVAC system that are associated to this measurement instant.

2.3 *Calculation Tools*

The software chosen to solve the Navier-Stokes equations and obtain the indoor air temperatures, indoor mean radiant temperature and air velocity was the Design Builder 2.42.026 package, which develops a coupling process using a steady-state Type CFD module. This module has low computational needs since calculates snap-shots of the model studied using nodal simulation data from *Energy Plus* as boundary conditions. For the purposes of this study, which is focused on the validation of the dynamic calculation module, these boundary conditions were directly obtained from the measurements performed in the existing classroom (Table 3). These measurements mainly corresponded to the air temperature and air velocity values inside the room, the radiant temperature of its inner surfaces, the thermal emission of the existing elements and the air inlet/outlet flows.

This tool was previously validated both by School of Built and Natural Environment of the Northumbria University (2011) and by Campano et al. (2015).

A two-equation (Standard $k - \varepsilon$) turbulence model was chosen for CFD calculation conditions, since it is the most complete model included in this software, despite its assumption of a fully turbulent flow. A Renormalization Group $k - \varepsilon$ model could solve laminar flow with greater accuracy, but the relative deviation between the results of both models is acceptable for this type of indoor environment (Stamou and Katsiris 2006; Srebric et al. 2008). In addition, 'Upwind' was chosen as a second-order discretization method as, under non-extreme conditions, it is simpler to calculate cases with air as the sole working fluid and without significant losses in the expected results.

Seated occupants are modelled on this tool through four prisms, corresponding to the head, torso (including arms) and bent legs. Its total area is 1.85 m^2, which corresponds to the mean skin area that has a human being (ISO 7730:2005) and young students. Its thermal emission is 1.2 met, sedentary activity, equivalent to 130 W (ISO 8996:2004; ISO 7730:2005). Convective fraction for this activity is 30% (Novoselac et al. 2006).

The domain discretization was performed using a hexahedral structure with straight uniform sides and a variable density. The mesh has a maximum spacing of 0.05 m, which is progressively reduced near surfaces and objects until it reaches 0.0125 m. The used junction tolerance is of 0.0065 m and the maximum ratio between the edges of the resulting cells is established in 1–10, with a minimum residual error of 10–5.

Given that a finer mesh requires higher number of cells, and therefore, a greater computing cost, it is necessary to achieve a balance using a mesh that suits to the model conditions without a significant loss of accuracy with:

- A higher cell density in regions where the fluid under study abruptly changes its variables (significant variation of the Reynolds number).
- A lower cell density in those regions where the fluid under study maintains significantly homogeneous its physical variables.

Two parallel discretization hypotheses have been developed for the calculation model, in order to guarantee the grid independence. To that effect, these two meshing hypotheses, which have different grid densities, cannot significantly vary in their overall results.

- Trial model A: mesh of 8,028,426 cells.
- Trial model B: mesh of 5,451,444 cells.

2.4 Analysis Methodology

The validation of the tool is performed by comparing the air temperature values obtained from both simulations with the air temperature measurements performed with the 80 thermocouple sensors, which were distributed throughout columns C1 to C4 (Fig. 2). To that effect, both the mean values and their standard deviation (95% confidence interval) are used.

For the comparison of such air temperature values it is necessary to resort to ISO 7726:2002 standard on “Ergonomics of the Thermal Environment, instruments for Measuring Physical Quantities”, which requires a maximum standard deviation of ±0.5 °C for obtaining a minimum category C in the thermal comfort measurement.

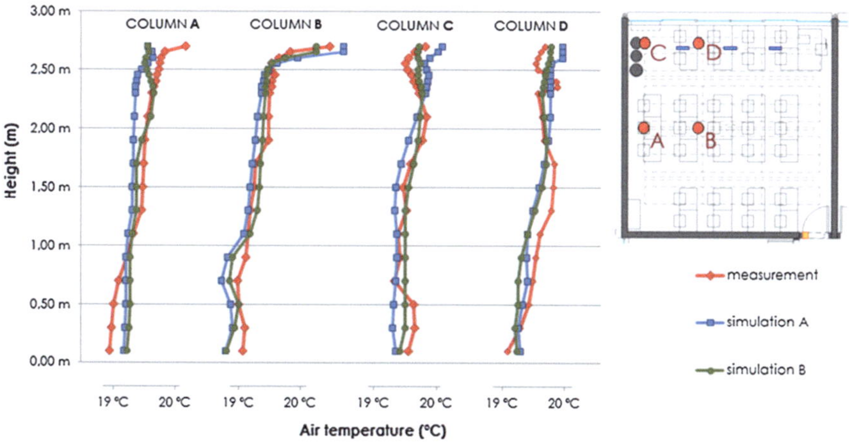

Fig. 4 Air temperature vertical variation curves for measurements and simulation models A and B

3 Results

The obtained air temperature values, both from dynamic simulation of models A (8,028,426 cells) and B (5,451,444 cells) and measurements are shown in Fig. 4 and in Table 4.

Simulation model B shows slightly closer air temperature values to measurements (Table 4), with a mean bias error (MBE) of 0.14 °C and a standard deviation of 0.33 °C. When it is compared to simulation model A (denser grid), the difference between the mean values of both simulations is not significant, being less than 0.07 °C, but it is bigger between their standard deviations, 0.14 °C, although both trials comply with standard ISO 7726; thus, both grid densities are appropriate to meshing these type of simulation models.

4 Conclusions

In this work, a validation protocol has been proposed for dynamic simulation tools in medium-sized spaces with high internal loads, by comparing the obtained air temperature values according to ISO 7726:2002. As an example of application, Design Builder simulation tool has been validated, comparing its results with the measurements in an existing room, a standard classroom of the "Eça de Queirós" secondary school of Lisbon (Portugal). To that effect, a series of air and radiant temperature values has been measured in an array of evaluation points inside the classroom through the use of 80 thermocouple sensors, obtaining a mean bias error (MBE) of 0.21 °C and a maximum standard deviation of 0.47 °C, which is under the maximum limit of ±0.5 °C established by this standard. The development and

Table 4 Air temperature average values at evaluation points (Eva) from the measurements (m) and the model simulations (A and B) with their bias errors (95% confidence interval, in bold)

Eva (m)	Column A (°C)			Column B (°C)			Column C (°C)			Column D (°C)		
	m	A	B	m	A	B	m	A	B	m	A	B
2.70	20.18	19.56	19.56	20.47	20.70	20.26	19.83	20.09	19.72	19.69	19.98	19.79
2.65	19.85	19.62	19.57	19.84	20.70	20.26	19.66	20.02	19.71	19.63	19.98	19.78
2.60	19.78	19.64	19.53	19.66	19.96	19.75	19.55	19.89	19.71	19.57	19.98	19.77
2.55	19.76	19.56	19.51	19.55	19.63	19.55	19.52	19.81	19.73	19.54	19.78	19.73
2.50	19.75	19.46	19.54	19.52	19.49	19.48	19.54	19.83	19.72	19.57	19.77	19.71
2.45	19.72	19.40	19.57	19.59	19.42	19.46	19.60	19.88	19.71	19.73	19.77	19.69
2.40	19.69	19.38	19.61	19.55	19.41	19.45	19.65	19.87	19.73	19.88	19.77	19.67
2.35	19.66	19.36	19.64	19.54	19.38	19.45	19.69	19.83	19.75	19.89	19.78	19.65
2.30	19.63	19.36	19.66	19.50	19.37	19.44	19.73	19.82	19.78	19.58	19.78	19.64
2.10	19.56	19.34	19.60	19.50	19.31	19.42	19.84	19.70	19.74	19.64	19.77	19.67
1.90	19.51	19.33	19.45	19.49	19.27	19.39	19.77	19.56	19.71	19.68	19.74	19.71
1.70	19.50	19.32	19.38	19.27	19.23	19.36	19.59	19.44	19.65	19.84	19.68	19.71
1.50	19.47	19.30	19.37	19.26	19.19	19.34	19.48	19.36	19.56	19.83	19.59	19.64
1.30	19.45	19.31	19.37	19.19	19.16	19.31	19.52	19.35	19.51	19.79	19.49	19.52
1.10	19.32	19.24	19.31	19.16	19.09	19.19	19.39	19.37	19.50	19.61	19.41	19.42
0.90	19.23	19.21	19.28	19.12	18.83	18.91	19.42	19.37	19.50	19.55	19.39	19.31
0.70	19.08	19.21	19.27	19.00	18.73	18.86	19.34	19.36	19.50	19.49	19.41	19.27
0.50	19.00	19.20	19.27	19.01	18.87	19.00	19.64	19.33	19.51	19.43	19.32	19.26
0.30	18.98	19.19	19.27	19.11	18.92	18.94	19.66	19.31	19.51	19.29	19.26	19.21
0.10	18.94	19.17	19.22	19.07	18.80	18.81	19.55	19.35	19.43	19.09	19.29	19.23
2σ	–	**0.40**	**044**	–	**0.48**	**0.30**	–	**0.58**	**0.22**	–	**0.40**	**0.34**
Measured mean air temperature (°C)							m: 19.53	A: 19.50		B: 19.51		
Measured standard deviation (95%, °C)							–	A: 0.47		B: 0.33		

application of this validation methodology by comparison allows to prove the reliability of the tool for the thermal behaviour prediction in such type of venues, as well as the viability of the room discretization hypotheses performed.

Acknowledgements This work has been partially funded by the IV Plan Propio de Investigación de la Universidad de Sevilla. The authors wish to express their gratitude to the "Laboratório Nacional de Engenharia Civil" and the Public Entity "Parque Escolar" from Portugal.

References

Campano MA (2015) Confort térmico and eficiencia energética en espacios con alta carga interna climatizados: Aplicación a espacios docentes no universitarios en Andalucía. PhD Dissertation, Universidad de Sevilla, Sevilla

Campano MA, Acosta I, Fernández-Agüera J, Sendra JJ (2015) Towards finding the optimal location of a ventilation inlet in a roof monitor skylight, using visual and thermal performance criteria, for dwellings in a Mediterranean climate. J Build Perform Simul 8(4):226–238. doi:10.1080/19401493.2014.913683

Conceição EZE, Lúcio MMJR (2011) Evaluation of thermal comfort conditions in a classroom equipped with radiant cooling systems and subjected to uniform convective environment. Appl Math Model 35:1292–1305. doi:10.1016/j.apm.2010.09.006

European Committee for Standarisation (1998) Ventilation for buildings. Design criteria for the indoor environments, CR 1752:1998. European Committee for Standarisation, Bruxelles

International Organization for Standardization (2002) Ergonomics of the Thermal Environment. Instruments for Measuring Physical Quantities, ISO 7726:2002. International Organization for Standardization, Geneva

International Organization for Standardization (2004) Ergonomics of the thermal environment—Determination of metabolic rate, ISO 8996:2004. International Organization for Standardization, Geneva

International Organization for Standardization (2005) Ergonomics of the Thermal Environment. Analytical determination and interpretation of thermal comfort using calculation of the PMV and PPD indices and local thermal comfort criteria, ISO 7730:2005. International Organization for Standardization, Geneva

Karimipanah T, Awbi HB, Sandberg M, Blomqvist C (2007) Investigation of air quality, comfort parameters and effectiveness for two floor-level air supply systems in classrooms. Build Environ 42(2):647–655. doi:10.1016/j.buildenv.2005.10.016

Kottek M, Grieser J, Beck C, Rudolf B, Rubel F (2006) World map of the Köppen-Geiger climate classification updated. Meteorol Z 15:259–263

Negrao COR (1995) Conflation of computational fluid dynamics and building thermal simulation. PhD Dissertation, University of Strathclyde, Glasgow

Novoselac A, Burley BJ, Srebric J (2006) Development of new and validation of existing convection correlations for rooms with displacement ventilation systems. Energy Build 38:163–173. doi:10.1016/j.enbuild.2005.04.005

School of Built and Natural Environment of Northumbria University. (2011) An inter-program analysis of computational dynamics based on PHOENICS and design builder. Available: http://www.designbuilder.co.uk/component/option,com_docman/task,doc_download/gid,39/Itemid,30/. Accessed 10 Mar 2016

Srebric J, Vukovic V, Guoqing H, Yang X (2008) CFD boundary conditions for contaminant dispersion, heat transfer and airflow simulations around human occupants in indoor environments. Build Environ 43(3):294–303. doi:10.1016/j.buildenv.2006.03.023

Stamou A, Katsiris I (2006) Verification of a CFD model for indoor airflow and heat transfer. Build Environ 41(9):1171–1181. doi:10.1016/j.buildenv.2005.06.029

Yang W, Fu-Jun Z, Kuckelkorn J, Di L, Jun L, Jun-Liang Z (2014) Classroom energy efficiency and air environment with displacement natural ventilation in a passive public school building. Energy Build 70:258–270. doi:10.1016/j.enbuild.2013.11.071

Zhai ZJ, Chen QY (2004) Sensitivity analysis and application guides for integrated building energy and CFD simulation. Energy Build 38(9):1060–1068. doi:10.1016/j.enbuild.2005.12.003

Zhai ZJ, Chen Q, Haves P, Klems JH (2002) On approaches to couple energy simulation and computational fluid dynamics programs. Build Environ 37:857–864. doi:10.1016/S0360-1323(02)00054-9

Study on Envelope in Office Buildings Under the Influence of Climate Change in Santiago, Chile

Carlos Rubio-Bellido, Alexis Pérez-Fargallo and Jesús A. Pulido-Arcas

1 Introduction

Predictions of future climate scenarios and its influence on energy demand in buildings represents a challenge, regarding research and development in the field of building science. Several prediction models have been generated for various climate scenarios (IPCC 2014), being the majority of them aimed at the conditions for the United Kingdom (Jentsch et al. 2008; Mylona 2012). However, they have increasingly extended along the international framework (Guan 2009; INN 2008).

Currently, IPCC, supported by United Nations Environment Programme (UNEP) and World Meteorological Organization (WMO) envisages multiple emission scenarios for the near future (years 2020, 2050 and 2080). Depending on them, climate will vary accordingly and the energy demand of buildings will do accordingly.

Nowadays, the building industry accounts, according to diverse sources, for roughly 40% of the total energy consumption of human activities (Pérez-Lombard et al. 2008), and several studies have been conducted in order to assess the impact of an eventual variation of climate over the energy demand of this economic sector (Sorrell 2015).

C. Rubio-Bellido (✉)
Department of Building Construction II, Higher Technical School of Building Engineering, University of Seville, Seville, Spain
e-mail: carlosrubio@us.es

A. Pérez-Fargallo · J.A. Pulido-Arcas
Department of Building Science, Faculty of Architecture, Construction and Design, Center for Research in Construction Technologies (CITEC), University of Bío-Bío, Concepción, Chile
e-mail: aperezf@ubiobio.cl

J.A. Pulido-Arcas
e-mail: jpulido@ubiobio.cl

P. Mercader-Moyano (ed.), *Sustainable Development and Renovation in Architecture, Urbanism and Engineering*, DOI 10.1007/978-3-319-51442-0_32

In this context, due to the diversity of climates around the globe, each country represents a different case-study. Besides, the legal framework and the method followed to determine the energy demand differs by each location. Directly related to the aims and objectives of the present study, researches by Wang and Chen (2014), Kalvelage et al. (2014) were proved to be effective when calculating the future trend of energy demands of buildings in different climates across the USA, considering a probable change in their conditions. Another study by Gangolells and Casals (2012) also forecasted the variation in both cooling and heating demand of several building located in Spain, with several scenarios for climate change. Speaking about the convenience of using morphed weather data to undertake the simulations, Jentsch et al. (2013) concluded that morphing actual weather files in format EnergyPlus/ESP-r Weather (EPW) with previsions from IPCC using HadCM3 files for the scenario A2 gives a reliable starting point to forecast variations in energy demand. They have also made significant contributions on the technical aspects of the morphing process of weather files (Jentsch et al. 2008), giving a technical base to undertake the present research.

So that, taking into account the preceding literature, the present study, focused on the city of Santiago (Chile) poses a compelling case-study for two main reasons. First, it visualizes the necessity of adapting a general procedure (IPCC HadCM3 weather morph model) to the local conditions of Santiago, with its own legal framework and several specifications regarding the construction techniques. Secondly, as proved by former researches, future scenarios of climate change may have controversial affects over the energy demand of buildings; that is, that demand may increase, decrease or even stay the same in the future.

This study is particularly focused on the basic design parameters that have influence over the building design on the first stages of the project: Size, orientation, shape and basic information about constructive systems and conditioning systems. Test models have been generated using a large number of combinations of these design parameters. The outcomes of the research are divided in three main groups.

- Clarify how the climate conditions will change according to the A2 'medium-high' emissions Greenhouse Gases GHG scenario for the city of Santiago (Chile).
- Assess the effect that these changes in climate may have on the energy demand for office buildings in this city. This repercussion shall be evaluated with regard to different building shapes and facades types.
- Clarify which is the most optimal combination of these parameters that would provide with the lowest possible energy demand under the influence of climate change.

2 Methodology

This study relies on a parametric simulation of a wide number of cases that aims at finding statistical trends for a set of output data. In order to perform as much as calculations as possible in a reduced amount of time, the procedure uses a routine developed in MS Excel Visual Basic.

In the first stage, the input data for the simulation process are set up, making a distinction between two groups: Climate data and what has been called "test models".

Files containing the current climate data in EPW format have been compiled, giving the base for calculating the energy demand of the test models. These files have been "morphed" according to the IPCC HadCM3 model, forecasting the climate scenarios for 2020, 2050 and 2080.

In the second stage, test models are defined by all the input data, which are implemented into the simulation routine, following the procedure stablished by ISO-13790:2008 (2008). This methodology has been duly tested and applied within the various legal framework of the EU member countries (EC 2012/C 115/01 2012; EU 2010). In order to adapt the model to the Chilean contexts, variables regarding climate zoning (INN 2008) and constructive elements (MOP 2011) has been extracted from the local legislation. Test models are defined by their location, the geometry of the model (orientation, Form Ratio (FR), Window-to-Wall Ratio (WWR)), constructive systems (U-value of the envelope and thermal inertia of the construction), internal heat loads (set up as constant by the regime of use) and external heat loads (solar gains). Orientation, FR and WWR are set as free, producing 8.200 iterations of the test models. Energy demand is analyzed for each one of them, both for the present time and the forecasted scenarios.

In a third stage, statistical analysis of the results is performed in order to find the tendency concerning the present and future situation and finally the most optimal combination of variables for each scenario is extracted from the dataset.

2.1 *Test Models*

Test models are defined as a series of prototype buildings, which do not represent any particular case study, but standard models, statistically representative of a large number of buildings. They are defined by a series of variables, according to the specifications of ISO-13790:2008 and classified into 5 main categories: Location, building geometry and enclosure, constructive systems, internal heat loads and external heat loads.

Location: All test models are located in the city of Santiago (Chile). Climate variables have been modelled accordingly in the respective EPW file and their climate zone according to Chilean legislative framework.

Geometry: The variables in this group are let free and analyzed in the present research. Test models are considered parallelepiped volumes with rectangular plans, perfectly oriented towards a North-South axis. Gross area is fixed at 1.500 m^2 and distributed evenly across 3 floors of 500 m^2 each. Each façade ranges from 10 to 50 m in length (Fig. 1). The quotient between the length of the East-West façade and the North-South façade is denoted as Form Ratio (FR), varying from 0.1 to 5. These parameters set the total area of the thermal envelope. At last, the Window-to-Wall-Ratio (WWR) is expressed as the percentage of window are with respect to the whole area of the external walls, varying from 10 to 60% in steps of 1%; this variable is also set free.

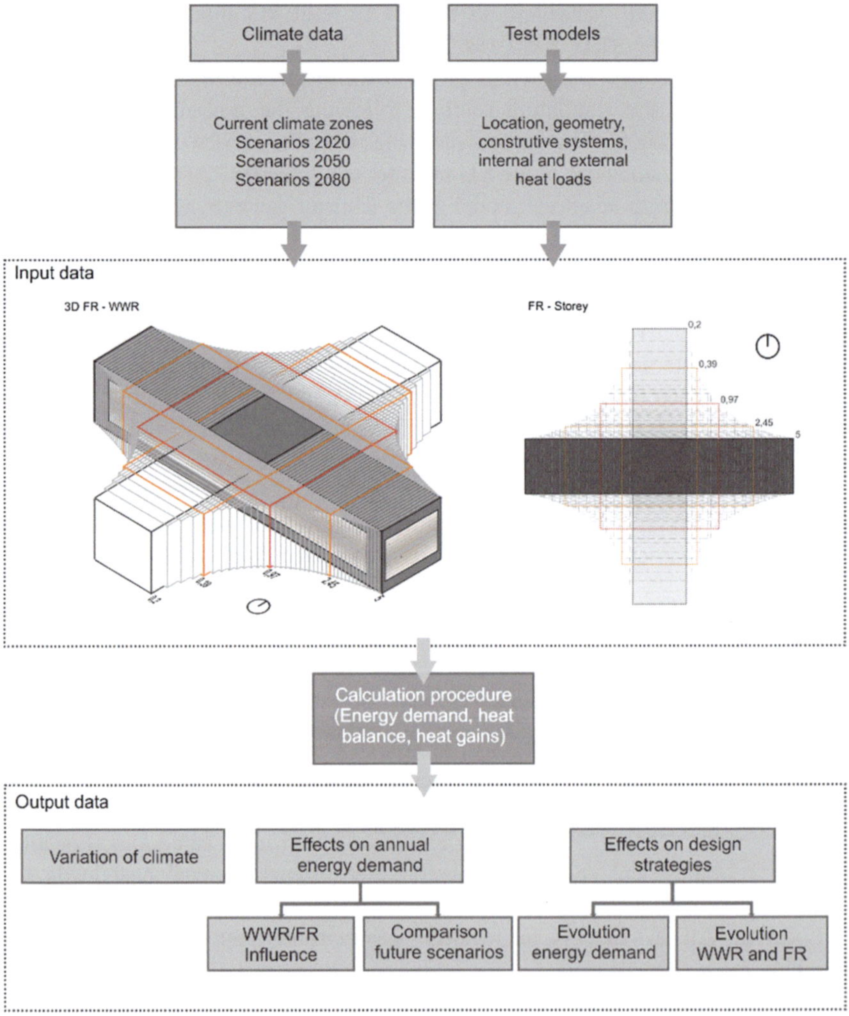

Fig. 1 Methodology for annual energy demand optimization under the impact of climate change

Constructive systems: These parameters are tabulated in accordance to the Reference Standard TDRe (MOP 2011). Limit values have been set for thermal transmittance (U), solar factor (SF), both varying in relation to WWR. Thermal inertia is fixed at 400 kJ/m^2 K (heavy construction), which is the common trend for Santiago.

Internal and external heat loads: Internal gains are established according to the regime of use, which is considered static during 9 h of office activity, with a lighting load of 16 W/m^2, a occupancy load of 6 W/m^2 and an equipment load of 4.5 W/m^2. Infiltration rate has been set at 0.3 l/s. External heat gains depend on the orientation and shape of the building, and levels of Solar Radiation (SR) have been extracted from the EPW files.

3 Effects of Global Warming on Climate

Figure 2 shows the predicted effect of the climate change in the city of Santiago, taking as a base the EPW file for the current climate and morphing it for the three scenarios (2020, 2050 and 2080). A remarkable rise in the average Dry Bulb Temperature (DBT) is observed, with steady increments of +1 °C in 2020, +2.11 °C in 2050 and +3.67 °C in 2080. Relative Humidity (RH) will see a decrease in its levels of −1.59% in 2020, −4.41% in 2050 and −5.92% in 2080. Solar levels stay quite constant, with only a slight increase of +1.02 Wh/m^2 in 2020, +3.64 Wh/m^2 in 2050 and +9.62 Wh/m^2 in 2080. Therefore, climate in Santiago is expected to be remarkably hotter, slightly dryer and with the same amount of solar radiation as nowadays.

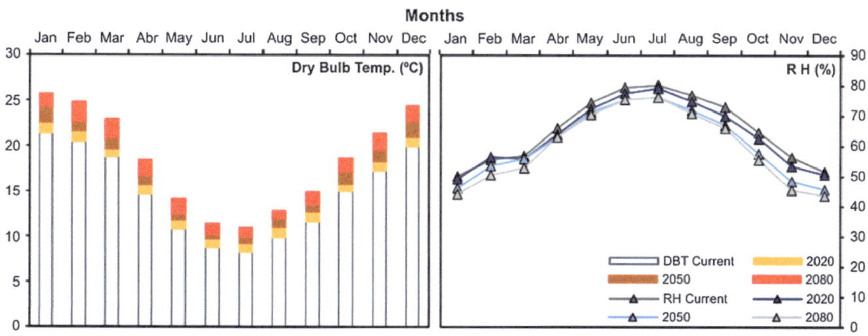

Fig. 2 Predicted change in climate variables for the city of Santiago

4 Evolution of Energy Demand

4.1 Annual Energy Demand

Taking as a base each possible value of FR (X axis), combined with the values of WWR, grouped in trenches of 10%, the curves of energy demand have been obtained for each of the given scenarios (Fig. 3). Reading the curves in horizontal, if the temporal span is fixed, energy demand is very independent from the variations in FR, as the curves stay nearly flat. Reading in vertical, for a given FR, variations in WWR have a remarkable effect on the energy demand. For the current scenario, demand ranges from 16.6 to 17.2 kWh/m^2; in 2020, it will increase, reaching a range of 16.9–17.4 kWh/m^2; in 2050, these figures will show values between 17.4 and 18 kWh/m^2; finally, the scenario for 2080 envisages energy demands in the range of 18.3–18.9 kWh/m^2. For all temporal spans it is observed that lower amounts of glass (WWR = 10%) results in lower energy demands and higher WWR (60%) gives the highest demands.

4.2 Heating and Cooling Demand

Energy annual demand for both heating and cooling are depicted in Fig. 4. For each temporal scenario, the parameters have been combined to produce 8.200 variations

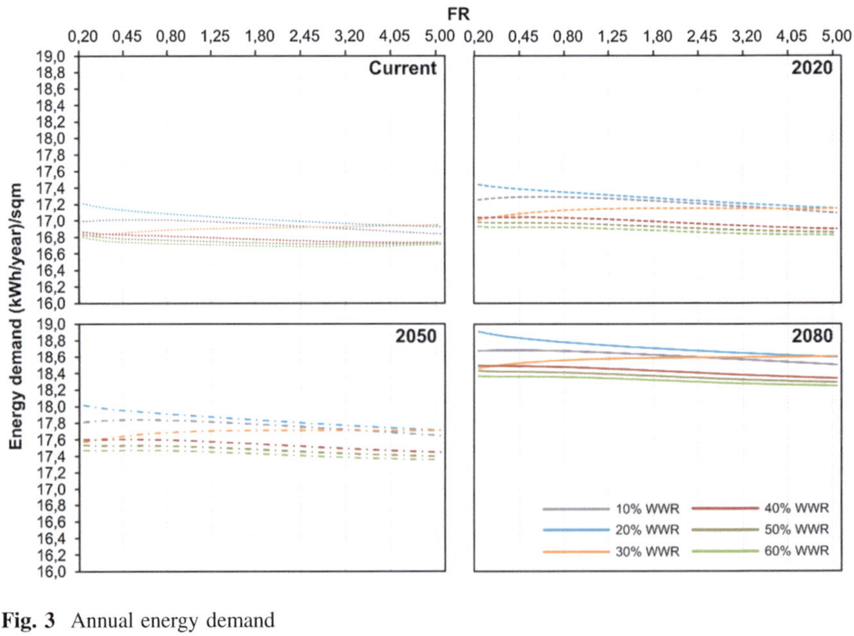

Fig. 3 Annual energy demand

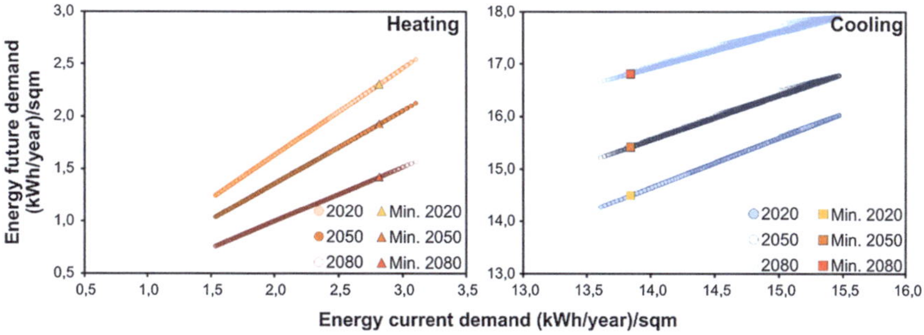

Fig. 4 Heating and cooling energy demand

of the test model and then, the optimal combination of parameters, that is, the one that results in a lower energy demand, has been highlighted.

Regarding the energy demand for heating, the gradient of the lines varies for each temporal span. In this case, values will decrease in 2020, 2050 and 2080, and the optimized values will decrease accordingly. In 2020, the best possible design will result in an energy demand for heating of 16.6 kWh/year, in 2050 15.9 kWh/year and in 2080 14.9 kWh/year.

The data for the energy demand associated with cooling is distributed according to a straight line that maintains its grade for the three temporal spans. Optimal values are 1.25 kWh/year in 2020, 1.7 kWh/year in 2050 and 2.4 kWh/year in 2080. Energy demand for cooling will rise in all cases, even for the best optimized test model.

5 Optimized Parameters

With the objective of clarifying and summing up all the data, Fig. 5 depicts the evolution of the optimized test models at present time, in 2020, 2050 and 2080.

The calculation routine extracts the combination of parameters that give the lowest cooling and heating demands for the present time, and then compares this value with their corresponding for 2020, 2050 and 2080. Optimizing the form of the building, the energy needed for cooling can be reduced to 13.8 kWh/year per m^2 (upper horizontal axis). This value will rise to 14.5 in 2020, 15.4 in 2050 and 16.9 in 2080 (right vertical axis). Regarding heating, the lowest possible demand at present time is established at 2.7 kWh/year per m^2 (lower horizontal axis). A steady decrease in this optimized value is expected in 2020 (2.4 kWh/year per m^2), 2050 (1.9 kWh/year per m^2) and 2080 (1.4 kWh/year per m^2). Combining both of them the total energy demand needed for conditioning the building can be obtained; at present time this demand is fixed at 16.5 kWh/m^2 year, it will stay nearly the same in 2020 (16.9) and rise to 17.3 in 2050 and 18.3 in 2080.

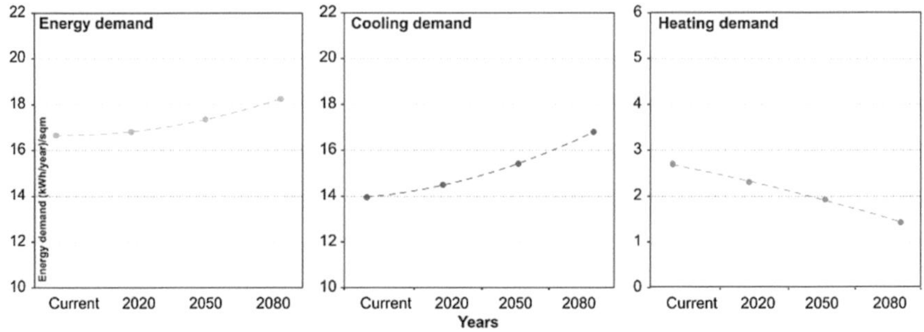

Fig. 5 WWR, FR, heating, cooling and annual energy demand

6 Conclusion

This study has clarified how climate change will affect the energy demand for office buildings located in the city of Santiago. The main scientific advances provided by the present research are as follows:

Climate in the city of Santiago will get hotter and dryer in the forthcoming years and the amount of sunshine will stay nearly the same. New EPW files have been produced, taking as a base the actual file and morphing them with the IPCC HadCM3 scenario, producing a new set of EPW files that will be useful for predicting the energy demand of buildings in the near future.

Taking as a base the calculation procedure from ISO-13790:2008 and the technical guidance of the Chilean building code TDRe a simulation routine has been compiled and duly tested for 8.200 test models of office buildings. Designers can make use of this tool in the early stages of design in order to test the influence of FR and WWR in the final energy demand of their buildings.

When trying to counteract the effects of climate change over the final energy demand of buildings, modifying the FR has very little effect on it, which means that the shape of the building is not a decisive factor. On the contrary, WWR plays a significant role, being the lower WWR associated with the lower energy demands.

With the prospect of a hotter climate, even for the best optimized model of office buildings in Santiago, with the best combination of FR and WWR, energy demand associated with cooling will inevitably rise, whereas the demand for heating will decrease. However, the former will not be able to balance the latter, and consequently the total energy demand will rise, which means that in a context of a changing climate, the formal design of the building itself (shape, orientation and percentage of glass) will not be sufficient to counteract this effect, despite it can be of help.

The results of this study suggest that further research should be made regarding other parameters of the building that were included in this study, such as the type and efficiency of conditioning devices for heating and cooling or the thermal transmittance and the thermal inertia of the building envelope.

Acknowledgements The authors extend sincere gratitude to the project, 150203/EF "Grupo de investigación en formación. Grupo de Arquitectura y Construcción Sustentable" of the University of Bío-Bío for support in this research.

References

EC 2012/C 115/01 (2012) Guidelines accompanying Commission Delegated Regulation (EU) No 244/2012 of 16 January 2012 supplementing Directive 2010/31/EU of the European Parliament and of the Council on the energy performance of buildings by establishing a comparative methodology f'. Official Journal of the European Union, p 28. doi:10.3000/1977091X.C_2012.115.eng

EU (2010) Directive 2010/31/EU of the European Parliament and of the Council of 19 May 2010 on the energy performance of buildings (recast). Official Journal of the European Union, pp 13–35. doi:10.3000/17252555.L_2010.153.eng

Gangolells M, Casals M (2012) Resilience to increasing temperatures: residential building stock adaptation through codes and standards. Build Res Inform 40(June):1–20. doi:10.1080/09613218.2012.698069

Guan L (2009) Preparation of future weather data to study the impact of climate change on buildings. Build Environ 44(4):793–800. doi:10.1016/j.buildenv.2008.05.021

INN (2008) NCh 1079. Of 2008 Arquitectura y Construcción-Zonificación climático habitacional para Chile

IPCC (2014) Climate change 2014: synthesis report. Contribution of working groups I, II and III to the fifth assessment report of the intergovernmental panel on climate change. Ipcc

Jentsch MF, Bahaj AS, James PAB (2008) Climate change future proofing of buildings—generation and assessment of building simulation weather files. Energy Build 40(12):2148–2168. doi:10.1016/j.enbuild.2008.06.005

Jentsch MF, James PAB, Bourikas L, Bahaj AS (2013) Transforming existing weather data for worldwide locations to enable energy and building performance simulation under future climates. Renew Energy 55:514–524. doi:10.1016/j.renene.2012.12.049

Kalvelage K, Passe U, Rabideau S, Takle ES (2014) Changing climate: the effects on energy demand and human comfort. Energy Build 76:373–380. doi:10.1016/j.enbuild.2014.03.009

MOP (2011) TDRe: Términos de Referencia Estandarizados con Parámetros de Eficiencia Energética y Confort Ambiental, para Licitaciones de Diseño y Obra de la Dirección de Arquitetura, Según Zonas Geográficas del País y Según Tipología de Edificios, Términos de Referencia Estandarizados con Parámetros de Eficiencia Energética y Confort Ambiental, para Licitaciones de Diseño y Obra de la Dirección de Arquitetura, Según Zonas Geográficas del País y Según Tipología de Edificios. Santiago, Chile. Available at: http://arquitectura.mop.cl/eficienciaenergetica/Paginas/default.aspx

Mylona A (2012) The use of UKCP09 to produce weather files for building simulation. Build Serv Eng Res Technol 33(1):51–62. doi:10.1177/0143624411428951

Pérez-Lombard L, Ortiz J, Pout C (2008) Review on buildings energy consumption information. Energy Build 40(3):394–398. doi:10.1016/j.enbuild.2007.03.007

Sorrell S (2015) Reducing energy demand: a review of issues, challenges and approaches. Renew Sustain Energy Rev 47:74–82. doi:10.1016/j.rser.2015.03.002

Standard EN-ISO 13790, Energy Performance of Buildings, Calculation of Energy Use for Space Heating and Cooling' (2008)

Wang H, Chen Q (2014) Impact of climate change heating and cooling energy use in buildings in the United States. Energy Build 82:428–436. doi:10.1016/j.enbuild.2014.07.034

Methodology for the Optimisation of Thermal Performance and Daylight Access to the Retrofit of Hospital Rooms in Mediterranean Climate

Juan Diego Pérez Téllez, Rafael Suárez Medina and Ángel Luis León Rodríguez

1 Introduction

The compliance with the targets of energy performance improvement of existing buildings, laid down by European directives for 2020, will be largely based on the decisions made by the Governments, concerning strategies and energy retrofit plans of housing and service building stock. Given the wide development of research for housing and office buildings, there is a considerable gap regarding hospital buildings. Hospitals are complex building with multiple functional areas (inpatient areas, administration department, outpatient services, laboratories, operating departments, etc.) with different comfort and controlled environmental requirements, with the result of an added complexity when facing the retrofit process.

In order to propose a strategic analysis to tackle the issue of the energy retrofit two questions arise: firstly, what is the current energy performance of existing hospital buildings? Secondly, to what extent can be retrofitted the existing hospitals to current requirements within feasible technological and economic criteria? Hence, to solve these questions, two consecutive and complementary lines of inquiry are presented: the energy consumption quantification of Andalusian hospitals and the evaluation of different energy retrofit alternatives.

Among the active hospitals of the Andalusian area, approximately 46% of them have been built prior the entry into force of the *Norma Básica de la Edificación sobre Condiciones Térmicas en los Edificios* (NBE CT-79), of which only about

J.D. Pérez Téllez (✉) · R. Suárez Medina · Á.L. León Rodríguez
Instituto Universitario de Arquitectura y Ciencias de la Construcción, Escuela Técnica Superior de Arquitectura, Universidad de Sevilla, Seville, Spain
e-mail: jd.pertel@gmail.com

R. Suárez Medina
e-mail: rsuarez@us.es

Á.L. León Rodríguez
e-mail: leonr@us.es

P. Mercader-Moyano (ed.), *Sustainable Development and Renovation in Architecture, Urbanism and Engineering*, DOI 10.1007/978-3-319-51442-0_33

23% have been totally refurbished afterwards (Fig. 1). Therefore, there is room for major improvement of energy performance and comfort because these requirements are far from current standards.

The published research about implementation of Energy Conservation Measures (ECM) in existing buildings generally follows a methodological approach based on three steps: energy quantification, proposal of ECMs, and thermal assessment of different alternatives. Some singular studies of energy characterisation and assessment of ECMs options have been performed on hospitals in the Mediterranean area (Buonomano et al. 2014a, b, 2016; Zorita et al. 2016) from University of Valladolid, used energy models and actual energy consumption data to quantify energy consumption in fourteen public hospitals in Castilla y León, for its potential use the energy management of the buildings. It is also worth mentioning the general study performed in the United Kingdom about the energy characterisation of public hospital buildings, in order to determine its resiliency against climate change (Short et al. 2012).

Energy models are the most powerful and widely used tool for the optimisation and energy performance improvement of existing buildings. However, the detailed representation of the actual operational conditions of the buildings through energy models is difficult due to its complexity and the large number of independent variables involved (Coakley et al. 2014; Tulsyan et al. 2013). Thence, a global energy analysis approach of the buildings entails important limitations regarding time and computational costs.

The quantification of the energy consumption of buildings is generally made by energy models simulations, onsite measurements and monitoring, and a hybrid approach of both methods in a parallel fashion (Wang and Yan 2012). Energy models of existing buildings are widely used in the research optimisation and energy performance improvement, in order to support the decision-making process, as a tool to evaluate the suitability of different retrofit alternatives against its economic and energy performance (Royapoor and Roskilly 2015; Coakley et al. 2014; Tulsyan et al. 2013; Heo et al. 2015; Roberti et al. 2015). Its wide use its due to its capability of predicting the behavior of the building given prior unobserved

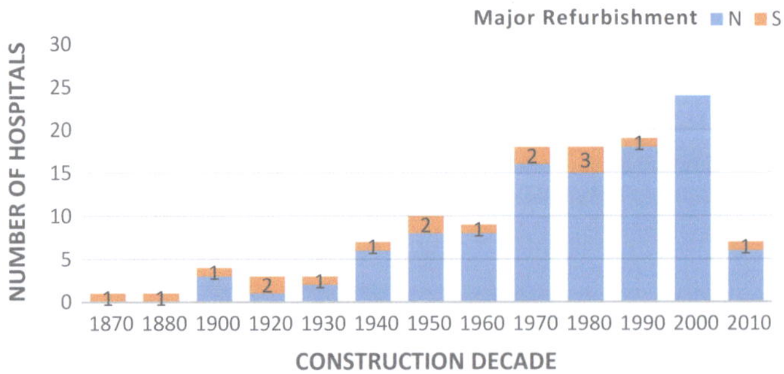

Fig. 1 Number of Andalusian hospitals built per decade, with and without major refurbishments

conditions, therefore, it allows making alterations in the configuration of the building and screening its impact on its energy performance (Coakley et al. 2014). Nonetheless, these increasingly require a greater level of accuracy in order to accomplish more meaningful studies (Royapoor and Roskilly 2015), supported by the increasing availability of detailed energy data.

Even so, due to the complexity and large number of independent variables involved in energy models, there is a considerable amount of inaccuracy and uncertainty in the outputs. This issue is partially solved by the calibration of the models, performed by reconciling simulation outputs with measured (Coakley et al. 2014; Tulsyan et al. 2013; Royapoor and Roskilly 2015). Furthermore, it allows the quantification the accuracy of the model, which despite of the great development in the field still suffers from under-determination of the input parameters (Raftery et al. 2011).

Many authors have demonstrated recently the importance of calibration, particularly in the evaluation of energy retrofit alternatives (Heo et al. 2015; Roberti et al. 2015; Coakley et al. 2014). However, the calibration of energy models is an over-specified and under-determined problem, which results multiple solutions, that is, multiple energy models can generate compatible outputs with the measured data but whose structures are not compatible to each other (Coakley et al. 2014), and not necessarily accurately represents the materiality and its thermal characteristics of the real building. The probability of obtaining a calibrated model with parameters far from actual values is especially high in existing buildings with a high level of uncertainty in the definition of the constructions components of the envelope (Roberti et al. 2015).

The issue of the uncertainty of the energy models is important to globally analyse the calibration of energy models, which is generally omitted in recent studies and is not considered at all in the validation criteria of current standards (Coakley et al. 2014). Various studies recommend the identification of the magnitude and sources of uncertainty of input parameters in order to perform a reliable evaluation of energy retrofits (Coakley et al. 2014; Heo et al. 2015; Royapoor and Roskilly 2015). For existing buildings, it largely depends on building typology, occupation, use patterns, construction and measured data availability.

In any case, the quantification of uncertainty is largely based on the resolution of data collection, which largely influences in the reliability of the model predictions (Heo et al. 2014). In this process of quantification is fundamental to identify the parameters with higher impact on the outputs (Heo et al. 2015), which can be performed by a sensitivity analysis. There are several methods available to perform this analysis. However, the selection of a suitable method depends on the purpose of the research, computational costs of energy models, number of variables of input parameters, etc. The recommended method for the energy analysis of existing buildings is the regression or Bayesian method, and, in the case of obtaining a big proportion of variance, the meta-model method can be applied without running extra simulation (Tian 2013).

Various recent studies include these methodological advances in their research in residential and office buildings, adapted to its particular case study and functional characteristics (Roberti et al. 2015; Vesterberg et al. 2014; Heo et al. 2015; Royapoor and Roskilly 2015). For example, Roberti et al. (2015) proposed a calibration methodology for historic buildings, where the definition of the envelope material properties is uncertain, in order to quantify its energy consumption and to evaluate different alternatives of energy retrofit. Therefore, their methodology is aimed to help defining the input parameters of the envelope properties, calibrating the most relevant parameters according to a sensitivity analysis targeting indoor measured temperatures.

Therefore, to make use of the full potential offered by the latest tools and techniques in building simulation, it is necessary to design a specific methodology depending of the research purpose and application, building typology, occupation and use patterns, construction and monitored data availability, climatic zone, etc.

The main objective of this work is the development of a methodology to assess the thermal performance of the standard hospital room in the climatic zones of Southern Spain. Thence, to explore possible passive energy retrofit strategies and envelope optimisation, to reduce energy demand and consumption, as well as improving comfort, within a feasible technical and economic context.

2 Methodology

To achieve the objective of this study we propose a methodological approach based on an experimental analysis from the monitoring of environmental variables and energy consumption, calibration and prediction of energy performance by means of building energy and natural lighting models. This methodology is currently being

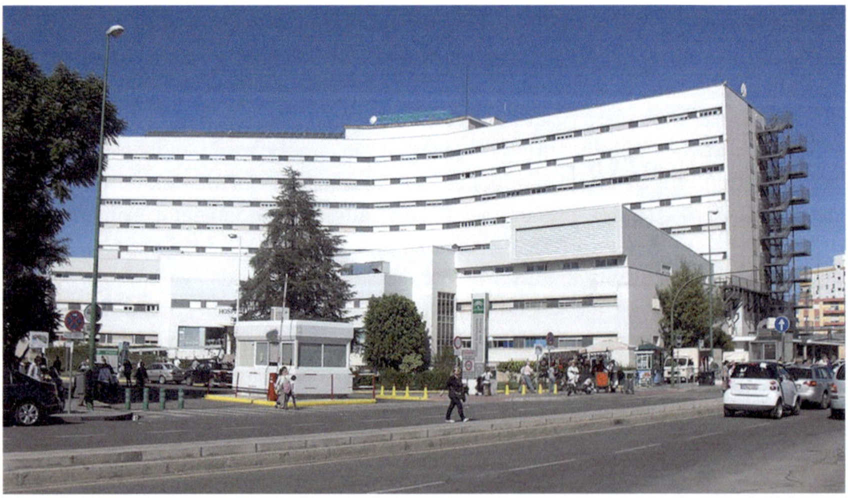

Fig. 2 Hospital universitario Virgen Macarena (www.geodruid.com, s.f.)

applied to a case study: a standard hospital room of the Virgen Macarena Hospital in Seville (Fig. 2). The methodology proposed is organised in three stages:

2.1 Identification, Classification and Selection of Representative Samples

In this first stage, the list of active hospitals and its functional and assistance characteristics is gathered from the *Catálogo Nacional de Hospitales 2015* (Ministerio de Sanidad, Servicios Sociales e Igualdad 2014). This information is completed with data about materiality, morphology, date of construction, etc., from the land register data base and based on visual inspection, along with the definition of its climatic zone according to the *Código Técnico de la Edificación* (CTE), in order to elaborate an Andalusia's hospital buildings stock data base (Figs. 1 and 3). From this data base, extended according to the purpose of this research, a classification of the most representative typologies of hospitals will be drawn, since most public hospitals built before the NBE CT-79 follow certain common design patterns and form (Isasi and Pieltain 2000).

Hospitals combine different functional zones, therefore the analysis is simplified by limiting the study to impatient areas, since they occupy the majority of the built area and they are continuously in use. Hence, a major impact on energy efficiency is presumed. Therefore, we propose a series of detailed analysis of representative case

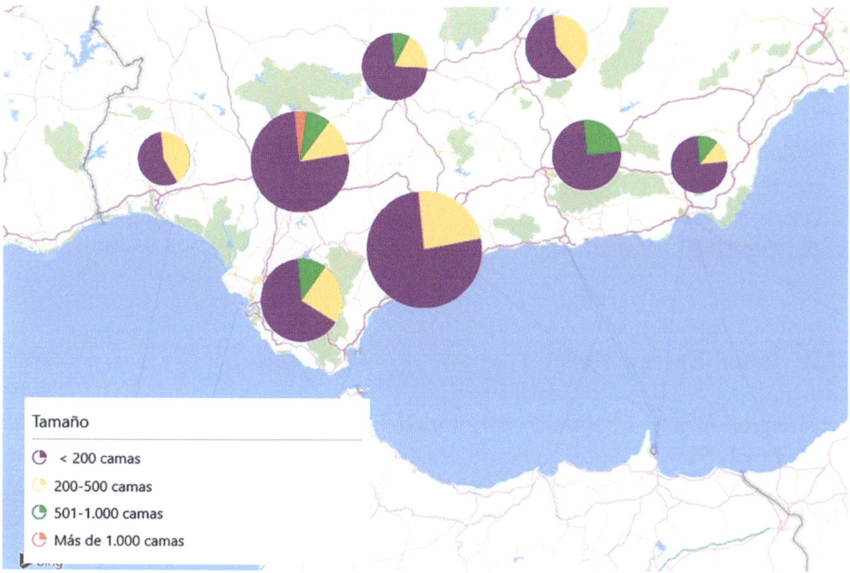

Fig. 3 Distribution of hospitals per province and size, according to its number of beds

studies to extrapolate results to other hospitals of similar characteristics, rather than a collective study of less resolution. Therefore, the data collection process is hugely simplified. Finally, the optimisation is limited to the thermal envelope, in order to limit the purpose of the energy models and, therefore, to reduce the parameters involved.

After the identification, a classification of typologies of Andalusia's hospital building stock can be made through a statistical analysis of the main characteristics of study: materiality, functionality, assistance, etc.

2.2 Energy Assessment and Analysis

This stage is developed in the following tasks:

2.2.1 Onsite Measurements

Characterisation of the selected samples through measurements, walk-through audits and monitoring, in real use conditions during a period of nine months, including summer and winter, of indoor and outdoor environmental variables and energy consumption, comfort and operating conditions of the standard room type:

- Monitoring of environmental conditions (indoor and outdoor air temperature and relative humidity, surface temperature of the envelope and partitions, CO_2 and luminance levels), managed by a multiple logger system which registers the measured data by the sensor on an hourly basis (Fig. 4).
- Detailed monitoring of energy consumption in the room, by means of an electricity and thermal energy sub-meter.
- Infrared thermography, to check thermal losses through the envelope, and air infiltration test in order to check its air tightness, by means of a BlowerDoor equipment.
- User's surveys, in order to obtain information about use and operation patterns, and perceived thermal comfort.

2.2.2 Initial Energy Model

Construction of initial building energy model in EnergyPlus, and sensitivity analysis of involved variables, in order to obtain a classification of the most influential parameters and to quantify its uncertainty level. In this particular case, we consider that the regression method is the most suitable, since we presume a major source of uncertainty in the parameters that describe the thermal envelope (Fig. 5).

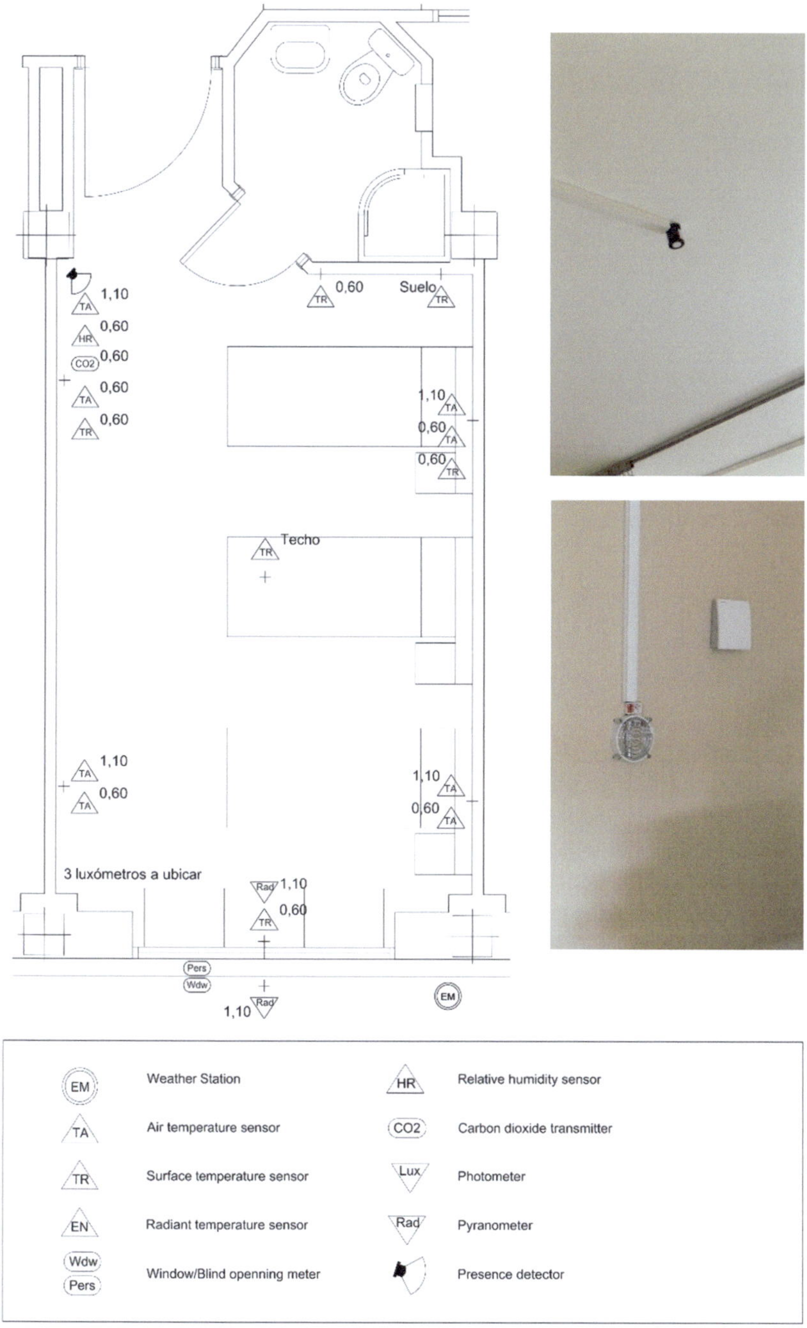

Fig. 4 Floor plant of the standard room of Virgen Macarena Hospital and monitoring equipment

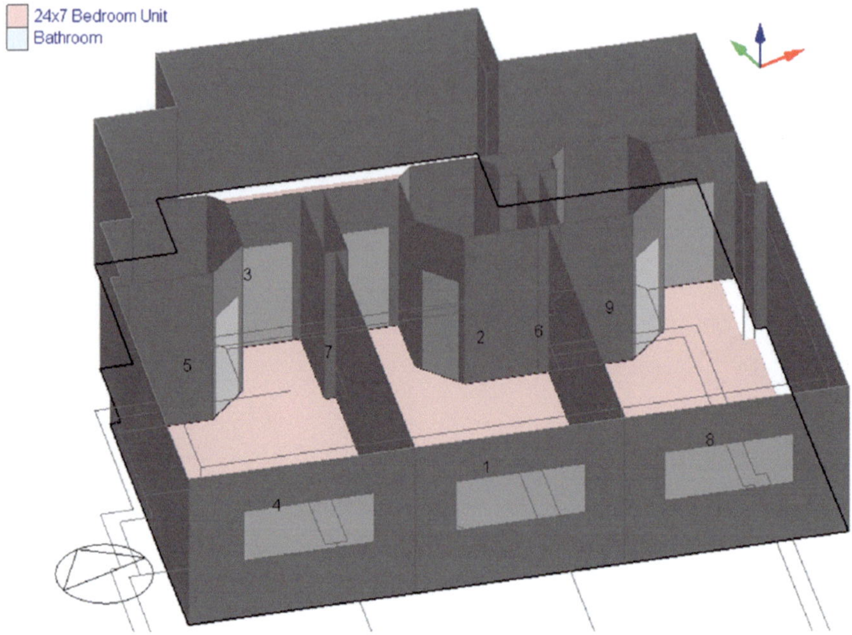

Fig. 5 DesignBuilder energy model of the standard room type and adjacent Macarena Hospital

2.2.3 Calibration and Validation of the Energy Model

In a first step, the parameters with higher level of uncertainty will be calibrated and, afterwards, the model will be calibrated against measured indoor temperatures and energy consumption, until an acceptable error is achieved. This approach is supported by the fact that the purpose of the research is to evaluate different alternatives of energy retrofit against its thermo-economic feasibility and the adaptative thermal comfort improvement achieved.

2.2.4 Evaluation of Energy and Thermal Performance

Evaluation of the energy and indoor thermal performance, in its current state, by obtaining simulated predictions of its performance: energy demand, indoor air temperature, and natural lighting levels of each simple, annually and hourly, for each day and season, for later comparison against normative adaptive comfort (UNE-EN 15251:2008/PNE-prEN 16798-1) and energy demand benchmarks.

2.3 *Optimisation Strategies*

In this last stage, a set of passive strategies will be proposed in order to improve indoor environmental conditions. This stage will be developed in two steps:

2.3.1 Evaluation of Energy and Thermal Performance

Based on the environmental and seasonal energy consumption evaluation, an optimisation analysis will be performed in order to describe a set of improvement passive strategies of the envelope, according to the ranking of the most influential parameters obtained in the sensitivity analysis. The different retrofit alternatives will be described by its material and thermal properties, combining simple and multiple hypothesis (insulation level, materiality and size of the windows, shading systems, etc.), for later implementation in the calibrated model. From the simulation of these models the energy demand, indoor air temperatures and lighting levels will be obtained for each retrofit alternative, annually and hourly for days with extreme and average climatic conditions, in each season.

2.3.2 Energy and Environmental Assessment of the Retrofit Proposals

Thermal and energy evaluation, and cost estimation of the proposed retrofit options, by comparing the energy demand and consumption outputs, indoor air temperatures and lighting levels against indicators and/or normative benchmarks. Furthermore, an analysis of the seasonal performance of each alternative and the improvement against initial conditions will be made, in order to assess its thermo-economic feasibility against its energy efficiency, by means of simple payback periods.

3 Conclusions

The necessary improvement of the energy efficiency of the Andalusia's hospital building stock should be based on the energy retrofit, which is a complex and uncertain process. In order to optimise this process it is necessary to apply a specific methodology according to its functional and material typology, proposed in this work and applied to a standard room type of the Virgen Macarena Hospital in Seville.

This paper presents a methodology for the energy consumption characterisation of a standard hospital room and its potential for improvement, as a prior step for the application of energy retrofit strategies on the Andalusian hospital building stock. The proposed methodology is organised in three stages, based on the identification, classification and selection of representative sample hospitals of Andalusia; the evaluation and energy analysis of the selected case studies and the development of a set of improvement strategies of the passive systems.

Acknowledgements This study is developed linked to the research project SUB-UMBRA: "Energy building rehabilitation in Mediterranean climate commercial use by optimisation of solar systems protection", (Ref. BIA2014-53949-R) within the I+D+i projects of the state research programs, development and innovation focused on the Challenges of society, 2014 call (*Ministerio de Economía y Competitividad*).

References

Buonomano A, Calise F, Ferruzzi G, Palombo A (2014a) Dynamic energy performance analysis: case study for energy efficiency retrofits of hospital buildings. Energy, Issue 78:555–572

Buonomano A, Calise F, Ferruzzi G, Vanoli L (2014b) A novel renewable polygeneration system for hospital buildings: design, simulation and thermo-economic optimization. Appl Therm Eng 1–2(67):43–60

Buonomano A, Montanaro U, Palombo A, Santini S (2016) Dynamic building energy performance analysis: a new adaptive control strategy for stringent thermohygrometric indoor air requirements. Appl Energy, Issue 163:361–386

Coakley D, Raftery P, Keane M (2014) A review of methods to match building energy simulation models to measured data. Renew Sustain Energy Rev, Issue 37:123–141

Heo Y, Augenbroe G, Graziano D, Muehleisen RT, Guzowski L (2015) Scalable methodology for large scale building energy improvement: relevance of calibration in model-based retrofit analysis. Build Environ, Issue 87:342–350

Heo Y, Graziano D, Guzowski L, Muehleisen R (2014) Evaluation of calibration efficacy under different levels of uncertainty. J Build Perform Simul 8(3):135–144

Isasi J, Pieltain A (2000) Hospitales: la arquitectura del Insalud, 1986-2000. INSALUD, Madrid

Ministerio de Sanidad, Servicios Sociales e Igualdad (2014) Catálogo Nacional de Hospitales. Secretaría General Técnica, Madrid

Raftery P, Keane M, O'Donnell J (2011) Calibrating whole building energy models: an evidence-based methodology. Energy Build, Issue 43:2356–2364

Roberti F, Oberegger UF, Gasparella A (2015) Calibrating historic building energy models to hourly indoor air and surface temperatures: methodology and case study. Energy Build, Issue 108:236–243

Royapoor M, Roskilly T (2015) Building model calibration using energy and environmental data. Energy and Buildings, Issue 94:109–120

Short CA, Lomas KJ, Giridharan R, Fair AJ (2012) Building resilience to overheating into 1960's UK hospital buildings within the constraint of the national carbon reduction target: adaptive strategies. Build Environ, Issue 55:73–95

Tian W (2013) A review of sensitivity analysis methods in building energy analysis. Renew Sustain Energy Rev, Issue 20:411–419

Tulsyan A, Dhaka S, Mathur J, Yadav JV (2013) Potential of energy savings through implementation of energy conservation building code in Jaipur city, India. Energy Build, Issue 58:123–130

Vesterberg J, Andersson S, Olofsson T (2014) Robustness of a regression approach, aimed for calibration of whole building energy simulation tools. Energy Build, Issue 81:430–434

Wang S, Yan CXF (2012) Quantitative energy performance assessment methods for existing buildings. Energy Build, Issue 55:873–888
www.geodruid.com, s.f. www.geodruid.com. [En línea] [Último acceso: septiembre 2016]
Zorita A, Fernández-Temprano MA, García-Escudero LA, Duque-Perez O (2016) A statistical modeling approach to detect anomalies in energetic efficiency of buildings. Energy Build, Issue 110:377–386

Urban Heat Island of Madrid and Its Influence over Urban Thermal Comfort

Emilia Román, Gloria Gómez and Margarita de Luxán

1 Introduction

This communication is part of the MODIFICA Project: Predictive model of the energy performance of residential buildings under conditions of urban heat island. BIA2013-41732-R, financed by the Ministry of Economy and Competitiveness through R&D + i 2013 program.

This part of the research aims to quantify the evolution of urban heat island in Madrid in the last 30 years. Urban heat island is a phenomenon by which the built-up areas have a higher temperature than less urbanized surrounding areas due to the progressive replacement of natural and vegetal surface by the urban area. Thus, the surfaces absorb more solar radiation and this, coupled with other anthropogenic factors, increase the air temperature and cause therefore a rise in the local temperature.

The result is a modification of the urban microclimate that affects to the comfort conditions in urban space and to the energy performance of buildings and therefore, to the quality of life of the inhabitants. The growth of the city and features of the new urban area determine the evolution of the heat island and the change in comfort conditions in different areas of the city (López Moreno 2015).

E. Román (✉)
PhD Architect Cc60 Estudio de Arquitectura,
Research Group Sustainable Architecture and Urban Planning (GIAU+S),
Universidad Politécnica de Madrid. Professor at the Urban and Regional Planning Department (ETSAM , UPM), Madrid, Spain
e-mail: emilia.roman@upm.es

G. Gómez
PhD Architect Cc60 Estudio de Arquitectura, Madrid, Spain

M. de Luxán
PhD Architect Emerita Professor at the Upper Technical School of Architecture, Universidad Politécnica de Madrid, Madrid, Spain

P. Mercader-Moyano (ed.), *Sustainable Development and Renovation in Architecture, Urbanism and Engineering*, DOI 10.1007/978-3-319-51442-0_34

2 Method

In this phase of the research we have started from the temperature data obtained during field work MODIFICA project in the period 2015–2016 for results and conclusions. Data collection was performed using urban, corresponding to three directions crossing Madrid and the towns of the surroundings, with different densities and types transects (Fig. 1). The measurements were performed simultaneously on the three routes. In addition to obtaining current values, the intention was to make a comparison with the data obtained in 1985 (López Gómez et al. 1988) on the same routes, as there are areas that remain in the same state edified and others that have changed over this period. In addition, measurements were carried out in similar summer conditions (11 July 1985 and 15 July 2015), at the same hours (from 21:00 to 23:00 solar time) and time stable in both cases. It should be noted that the temperatures recorded in July 2015 were higher than the average of the historical record of the weather station Barajas (about 4 more) due to a heat wave that happened during those days in the center of the peninsula. Along with this we have to consider the general trend of rising temperatures caused by climate change.

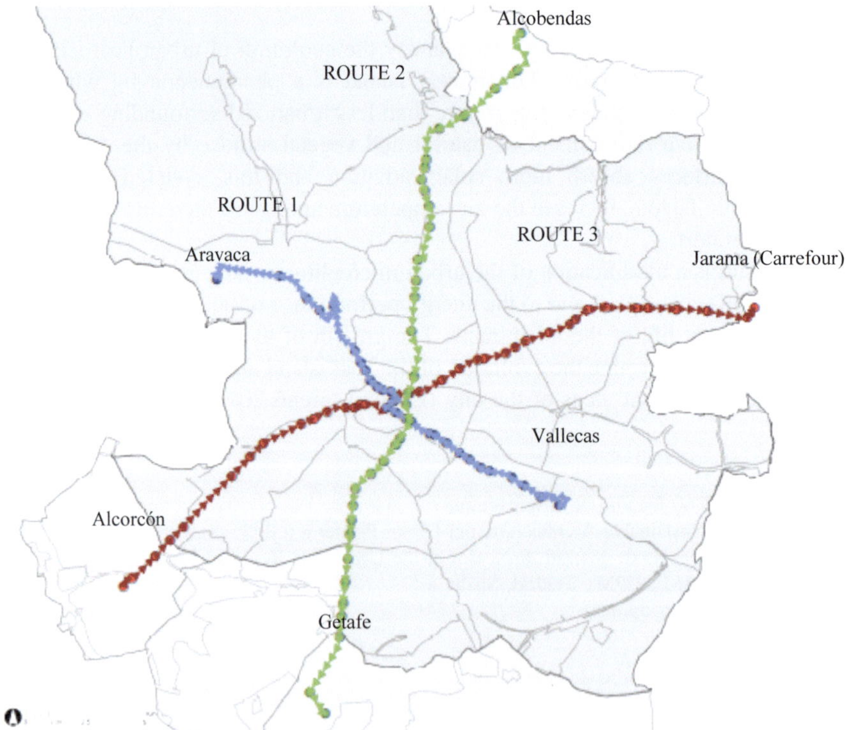

Fig. 1 Routes conducted in 1985 and 2015 to obtain data direct thermal island of Madrid. *Source* Elaboration by the authors from Núñez 2016

The ability to relate measurements at each point chosen routes with the building density and dominant type in the nearby area let see if variations in the temperatures are guidelines based on the urban area studied.

For the preparation of corresponding to the urban heat island in Madrid in July 1985 and 2015 have, in addition to the data obtained in fieldwork, used Map provided by the Stations of Quality Municipal Air (ECA) and by the stations of the State Meteorological Agency (AEMET). All data have been incorporated into a Geographic Information System (GIS).

3 Evolution of Land Use and Heat Island in Madrid During the Period 1985–2015

It has been a continuous increase in per capita land use and housing in recent decades in the region of Madrid. The evolution and expansion of the heat island in Madrid has not occurred evenly throughout the territory. Measurements taken in 1985 allowed obtaining first information about setting it in the study area. The data collected during the years 2015–2016 show a direct link between recent urban developments, driven mainly in the first decade of s. XXI, and temperature variations observed in different parts of the city, compared to the base year (1985). The question that arises is to define urban factors that have had the greatest influence in this modification of the urban climate of Madrid during this period.

3.1 Land Use in the Region of Madrid Between 1985 and 2015

According to data from Naredo and Garcia, land use rose from 112 m^2/inhabitant in 1956, to 196 m^2/inhabitant in 1980 and 269 m^2/inhabitant in 2005 (Comunidad de Madrid 2016). In other words, during the period considered land use increased 140% compared to the initial state considered, implying a major transformation of the territory and a high consumption of natural resources (Gómez 2014). In addition, housing occupancy went from 448 m^2/dwelling in 1956, to 551 m^2/dwelling in 1980 and to 580 m^2/dwelling in 2005 (Naredo and García 2008). One of the consequences of this great transformation that has suffered this territory in just 50 years, is the modification of the urban climate, particularly by increasing the heat island effect, which joins rising temperatures from climate change in the region, as numerous studies have pointed out (Tapia et al. 2015). In the last twenty years (1990–2010) the area devoted to urban uses in the metropolitan area of Madrid has increased by 45%, approximately 80,000 Ha (Núñez 2015).

3.2 Evolution of the Urban Heat Island from 1985 to 2015

The measurement of the urban heat island in Madrid in July 1985 (López et al. 1988) allowed to observe a decreasing temperature gradient that ran from the city center (records temperatures 28.6 °C) to the peripheral areas (temperature records 24.6 °C), with differences in temperature to 4 °C between the area during data collection. The cause is the thermal inertia of the city, generated in part by the transformation of natural and permeable soils to urban land, paved mostly and therefore more waterproof and with a lower coefficient of albedo (more capacity to capture energy as heat). This fact is also produced by the presence of buildings and human activities, such as vehicular traffic, which increase the effect described.

It is observed in the temperature record and in the graphical representation of the heat island that the influence of the presence of a large mass of vegetation appears in Fig. 2, in the northwest of the city of Madrid (Monte de El Pardo), whose effects are channeled through the valley of the Manzanares. It also should be noted the presence of El Retiro park, located in the city center, and whose cooling effects conducive to a temperature difference with the neighboring districts of more than 3 °C.

In the records taken 30 years later it shows how the heat island thermally accentuates and extends to areas where urban growth has been developed with more

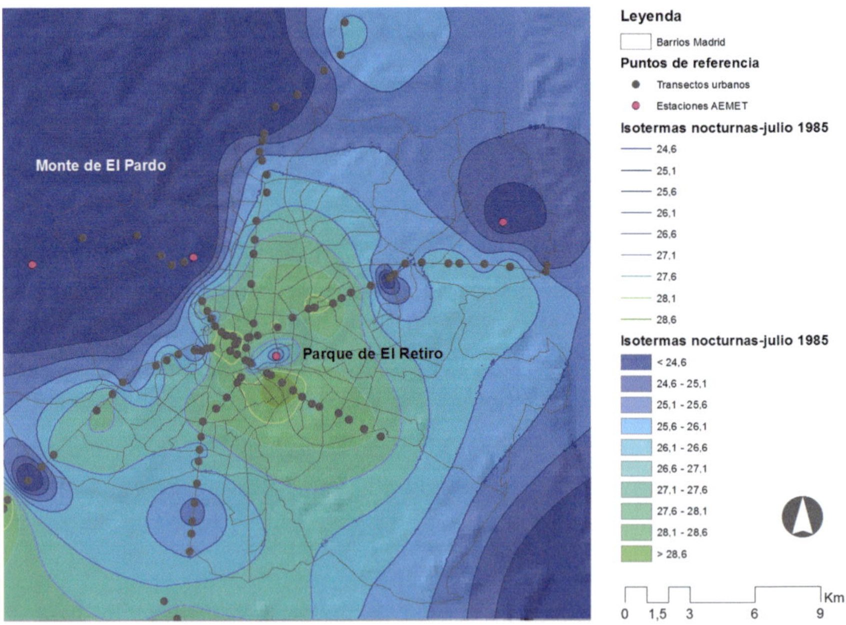

Fig. 2 Night isotherms in Madrid, July 11, 1985 (°C). *Source* Elaboration by the authors from López Gómez et al. (1988)

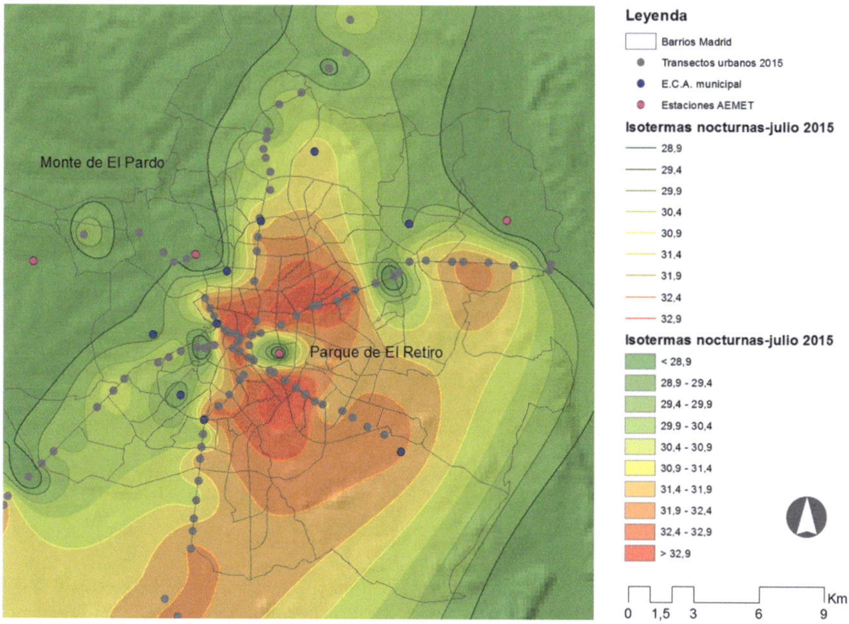

Fig. 3 Night isotherms in Madrid, July 15, 2015 (°C). *Source* Elaboration by the authors from temperature data obtained in the MODIFICA project

intensity, particularly in the northeast suburbs, southwest and southeast of Madrid (Fig. 3). In urban temperatures is also evident the influence of the intervention for "Madrid Rio", with the burial of the M-30 highway and the creation of a linear park that runs parallel to the Manzanares river and extending from El Pardo to Getafe. This action has led to a drop in temperatures and increases in extent, for 1985, the effect brought about by the Monte de El Pardo.

Notably, in the city center they have increased heat sources in intensity and surface relative to the reference year, although the influence of El Retiro Park as a thermal regulator great importance for this area is maintained.

4 Relationship Between Urban Grid and Local Evolution of Heat Island: Two Case Studies

From these general data evolution of the urban heat island in Madrid, they have identified several areas where there has been greater urban transformation over the last 30 years and therefore would be expected to have also produced more local modification of the heat island when comparing 1985 data with 2015.

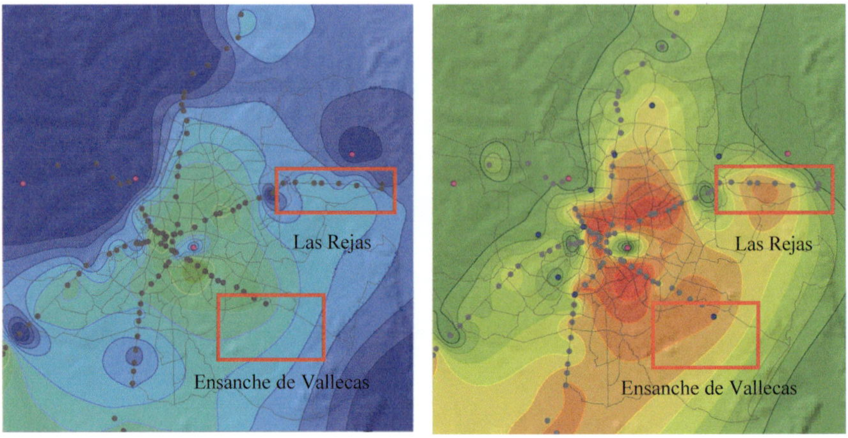

Fig. 4 Selection of case studies from the comparison of temperatures in the two periods considered, 1985 and 2015, respectively. *Source* Elaboration by the authors

For the selection of these case studies have used the comparison of temperature profiles obtained for the different transects the two dates of measurements (summer of 1985 and 2015), since the effect of transforming the microclimate effect and urban heat island climate change, can have a major impact on the health of people (Tapia 2015). In addition, we considered the maps generated by GIS, corresponding to the heat island in both periods (exposed in previous points) technology. With this information has made a selection of two areas where temperatures have increased considerably compared to 1985 measurements (>5 °C) (Fig. 4) for the local study of the evolution of urban heat island in Madrid:

- **Barrio de las Rejas**, in the district of San Blas. Increased temperatures between 5 and 6 °C compared to 1985 (from 26.1 to 26.6 °C isotherm night in 1985 to 31.9 to 32.4 ° C isotherm night in 2015).
- **Ensanche de Vallecas**, in the district of Villa de Vallecas. Variation in temperature between 5 and 6 °C compared to 1985 (from 26.6 to 27.1 °C isotherm night in 1985 to 31.9 to 32.4 °C isotherm night in 2015).

4.1 Case Study 1. Barrio de Las Rejas

This suburb is located in the district of San Blas, northeast of Madrid. Its geographical location, south of the airport of Barajas, between the A-2, M-40 and M-21 railways, as well as the abundance of industrial estates, industrial and give it an isolated character. Historically the residential population is concentrated in the western part of the neighborhood, specifically in Ciudad Pegaso. Urban

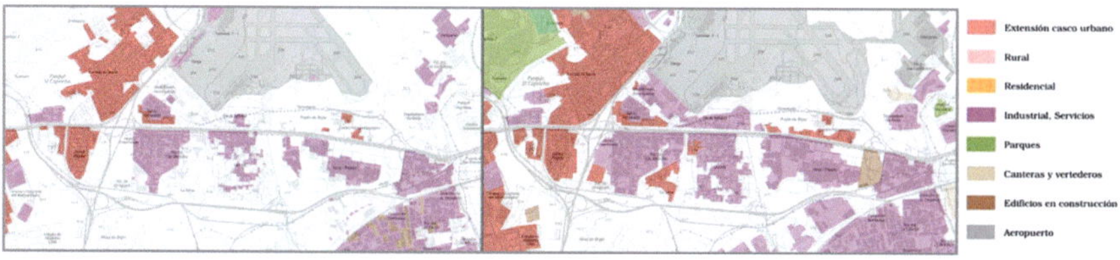

Fig. 5 Comparison of land use for the area of Las Rejas, 1980–2005. *Source* Visor Comparativo-Planea, Dirección General de Urbanismo, Comunidad de Madrid

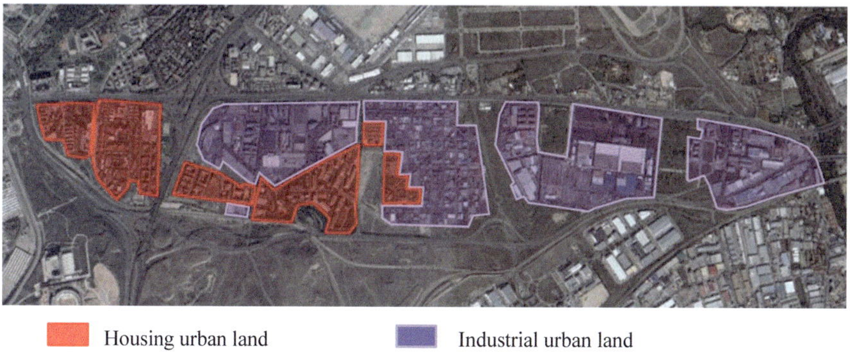

Fig. 6 Land use in the area of Las Rejas, 2015. *Source* Elaboration by the authors from Google Earth

development considered by the Urban Plan of 1997 led to the construction of over 1600 homes, greatly increased its population and occupation in recent years.[1]

As can be seen in Fig. 5, the increase in land use in 2005 is considerable compared to the year 1980, where practically the only core residential population was concentrated in Ciudad Pegaso and the rest of the area was occupied by developments industrial and undeveloped areas.

Nowadays it can be seen how the occupation of the territory continued to rise to transform a large percentage of natural soil in artificial soil, mainly asphalt and residential, services and industrial buildings (Fig. 6).

By analyzing the temperature increase of the area can be seen how the main focus of refractivity heat with the densest area in which there are several shopping centers (Centro Comercial Plenilunio, Makro Barajas, Media Markt, etc.) with large paved parking, which they are impervious areas with a low coefficient of albedo, and with sparsely vegetation (Fig. 7). However in the border east and west of the area of higher temperature areas is recorded a decrease of this variable. It could be

[1]Source: http://www.ciudadpegaso.com/cplocalizacionentorno.html [Consulted on: August 24, 2016].

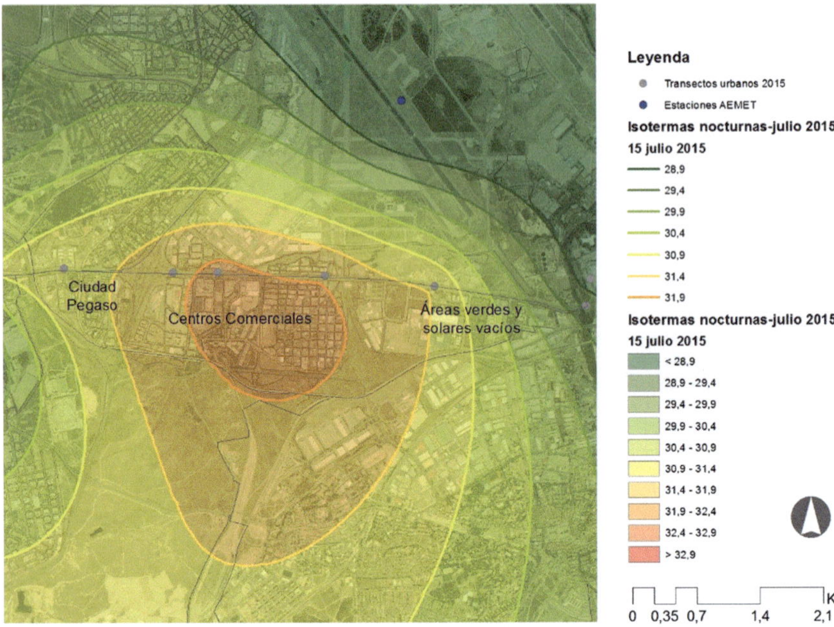

Fig. 7 Night isotherms in the neighborhood of Las Rejas, 2015. *Source* Elaboration by the authors

due to the presence in the west of the Pegaso City, residential blocks consisting of semi-detached houses and detached houses, with wide green spaces and great presence of trees in public space. In the limit of this suburb is appreciated a large vacant lots and green areas.

4.2 Case Study 2. Ensanche de Vallecas

Ensanche de Vallecas is located southeast of Madrid, between highways M-40 and M-50. In this area is also seen a considerable land occupation. In 1980, there was the old town of Villa de Vallecas and part of Santa Eugenia, which joined the west industrial estate South Vallecas, as shown in Fig. 8. However already in 2005, virtually the industrial areas of western and residential area between Vallecas and Santa Eugenia have silted. It has also begun to develop the Urban Action Plan of Vallecas, designed in the early 90s. This plan has provided for the execution of 28,000 dwellings with services, to an occupation of more than 7 million m^2.

In 2015 it shows how the occupation of the territory has continued to increase, by running large number of residential buildings and equipment covered by urban development program (Fig. 9). To this the effects of urbanization of much of the

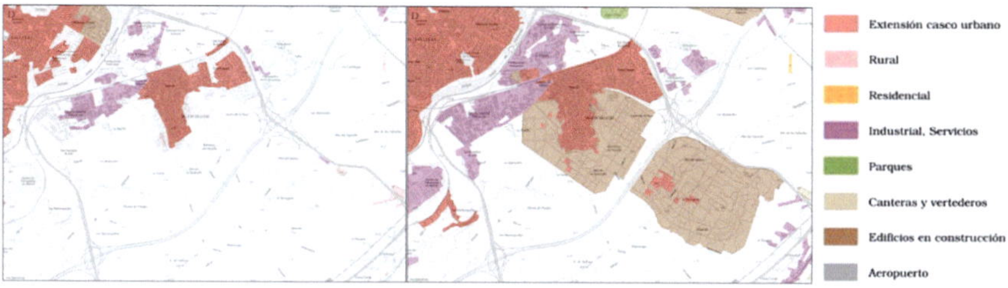

Fig. 8 Comparativa de ocupación del suelo para el Ensanche de Vallecas, años 1980–2005. Fuente: Visor Comparativo-Planea, Dirección General de Urbanismo, Comunidad de Madrid

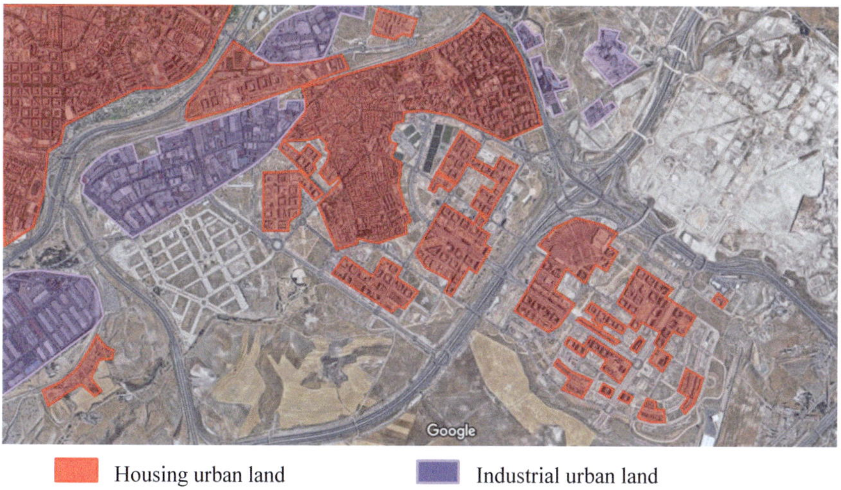

Fig. 9 Land use in the area of Ensanche de Vallecas, 2015. *Source* Elaboration by the authors from Google Earth

area, which has transformed the natural soil in asphalt and impermeable soil are added. Also the extension of these urbanized areas, still unbuilt, south of the industrial estate and southwest of the old town.

During data collection in 2015 a significant increase in temperatures over the year 1985 and an increase in the extension of the urban heat island, favored by urban developments that have been mentioned previously observed. The extension of existing heat focus in 1985 in the historical district of Villa de Vallecas increased significantly, an effect brought about by the densification of the corresponding industrial and residential development of this urban area urban fabric.

However, the temperature is reduced according begins to decrease urban density due to voids solar still predominate in the southeast area this area (Fig. 10).

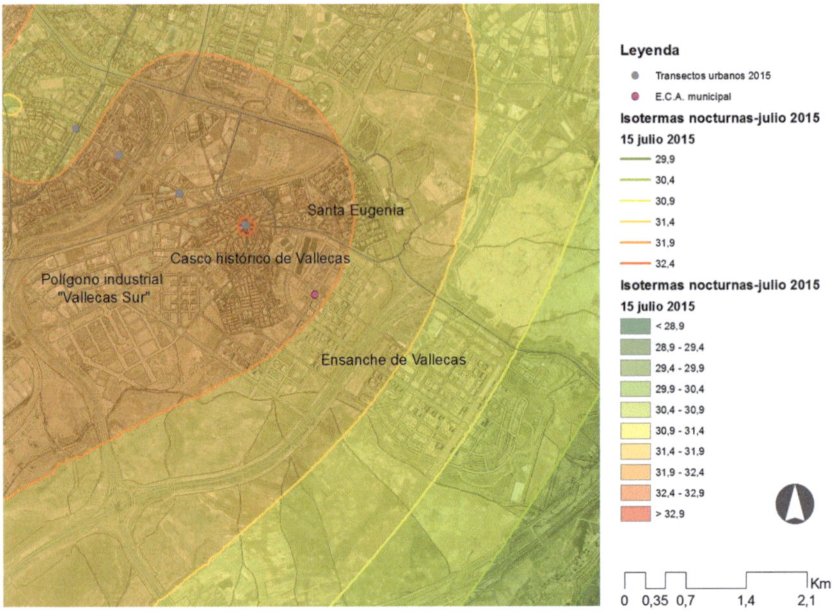

Fig. 10 Night isotherms in the neighborhood of Ensanche de Vallecas, 2015 (°C). *Source* Elaboration by the authors

5 Conclusions

This research aims to show the relevance of the local influence of heat island and its relation to the formation of the city. According to data obtained in the two case studies, the most important variables in the temperature increase in these neighborhoods are urban surface finish and building density. A major transformation of natural surface with significant presence of buildings, the greater has been the increase in temperatures in the period studied, finding differences of up to 6 °C. Despite the presence of a heat wave in 2015, it has been appreciate a systemized as temperatures increase while the natural surface has been replaced by asphalt. The lack of urbanization, the presence of trees and vegetation zones mitigate the heat island effect in these spaces. This conclusion would open the possibilities for improving the urban climate of these areas with cold treatments pavement, revegetation, and buildings with external insulation (Santamouris 2007).

The expansion of these local studies of urban heat island to other areas of Madrid would identify those areas where the worst conditions and the surfaces that have a greater influence on microclimate modification. Thus, it could be established a program to improve the public space in Madrid adapting solutions to the conditions of each neighbourhood to attenuate the heat island effect and improve urban comfort and, by extension, the indoor comfort.

This data could be applied to other populations with similar geographical conditions, which were not able to carry out specific studies and for future situations arising from changes in urban planning.

Acknowledgements This research is funded by Programa de I+D+I orientada a los retos de la sociedad 'Retos Investigación' of the Ministry of Economy and Competitiveness. Grant code BIA2013-41732-R, MODIFICA Project: Predictive Model For Dwellings Energy Performance Under The Urban Heat Island Effect.Authors would also like to thank the State Meteorological Agency (AEMET) and Madrid Town Hall for weather data provided to carry out this research.

References

Comunidad de Madrid (2016) Información cartográfica y territorial. http://www.madrid.org/cartografia/planea/cartografia/html/web/index.htm (Fecha de consulta 10 de agosto de 2016)

Fernández García F (2002) El clima urbano de Madrid y su influencia sobre el confort térmico. Boletín de La Real Sociedad Geográfica 137–138:12

Gómez Muñoz G (2014) Método de análisis diacrónico para la intervención en el alojamiento con criterios ecológicos. El caso de Madrid 1940–2100. Tesis Doctoral. Universidad Politécnica de Madrid

López Gómez A, López Gómez J, Fernández García F, Arroyo Ilera F (1988) El Clima urbano de Madrid: La isla de calor. CSIC, Madrid

López Moreno H (2015) Hacia una evaluación energética de la influencia de la isla de calor urbana en la morfológica urbana de Madrid. Universidad Politécnica de Madrid Tesina final, Master de Medio Ambiente y Arquitectura Bioclimática, Madrid

Naredo JM, García R (Coord.) (2008) Estudio sobre la ocupación de suelo por usos urbano-industriales, aplicado a la Comunidad de Madrid. Ministerio de Medio Ambiente. http://habitat.aq.upm.es/oscam/

Núñez Peiró M (2015) La isla de calor urbana de Madrid. Principios de integración en simulación energética edificatoria. Tesina final, Master de Medio Ambiente y Arquitectura Bioclimática. Universidad Politécnica de Madrid

Santamouris M (2007) Heat island research in Europe: the state of the art. Adv Build Ener Res 1 (1):123–150. doi:10.1080/17512549.2007.9687272

Tapia C, Abajo B, Feliu E, Fernández JG, Padró A, Castaño J (2015) Análisis de vulnerabilidad ante el cambio climático en el municipio de Madrid. Ayuntamiento de Madrid

Threshold Values for Energy Loss in Building Façades Using Infrared Thermography

Juan José Moyano Campos, Daniel Antón García, Fernando Rico Delgado and David Marín García

1 Introduction

Recent research has shown that a significant percentage of primary energy usage, as well as greenhouse gas emissions (GHG), derive from the building sector. In the United States (according to the *Buildings Energy Data Book* issued by the US Department of Energy), around 38% of CO_2 emissions, 52% of SO_2, and 20% of NO_X derive from building-related energy usage (Fesanghary et al. 2012). In the European Union (EU), nearly half of the energy consumed is in the building sector. In addition, the commitment to complying with the *Kyoto Protocol* as regards energy usage and CO_2 emissions and other greenhouse gases (GHG) is leading to changes in regulatory standards in this sector to achieve buildings with nearly zero energy usage.

Likewise, there are significant changes in the standards regulating the building sector in different countries. In Spain, since Royal Decree 235/2013 (Ministerio de la Presidencia 2013) of 1 June 2013, it is mandatory to provide buyers with the energy efficiency certificate of any building being renovated, sold, or leased. This regulation aims to promote the use of more efficient buildings and the proposal of a series of measures to improve their energy rating in line with the recommendations of European Directive 2010/31/UE (European Parliament 2010). The calculation of this efficiency leads to its energy certification and ranking, which are used to consider the most significant factors in a building's energy consumption.

One of the most important aspects in energy savings is an envelope analysis of existing buildings since thermal losses or gains occur through it. Al-Habaibeh and Anderson (2010) state that the upgrading of existing dwellings could significantly reduce energy consumption up to an average of 25%. Thermal insulation plays a

J.J. Moyano Campos (✉) · D. Antón García · F. Rico Delgado · D. Marín García
Higher Technical School of Building Engineering, University of Seville,
4A Reina Mercedes Avenue, 41012 Seville, Spain
e-mail: jmoyano@us.es

P. Mercader-Moyano (ed.), *Sustainable Development and Renovation in Architecture, Urbanism and Engineering*, DOI 10.1007/978-3-319-51442-0_35

crucial role in reducing energy consumption. In order to compensate for the lack of insulation, the Technical Building Code (Código Técnico de la Edificación, hereafter CTE) (Ministerio de la Vivienda 2006) require increasing it in opaque areas (Castro 2009), then obtaining an average value below the thermal transmittance (*U*) threshold established. In addition, there are other low-cost measures such as installing double-glazed windows and measures to minimize thermal bridges during the building phase. Errors during building pose energy-loss problems, and it is therefore necessary to establish protocols for building interventions or upgrades. The latest studies by Taylor et al. (2013) reveal the possibility of infrared audits to identify building envelope defects during construction by 'active' thermography. This system requires an additional heating method to generate a difference in temperature between the inside and outside of the enclosure.

According to data from the Spanish Ministry of Housing, since the CTE came into force in 2006, around 715,000 dwellings were being built per year. The housing stock has increased by 1.8 million new dwellings in the last six years, of which 64% remain unsold. All of them are subject to the current CTE standard. Improving the thermal conditions and energy efficiency of this volume of buildings would reduce overall energy consumption and carbon dioxide emissions in the building sector (a commitment set by the EU in Horizon 2020) once they have been amortised as regards the energy consumption for their construction.

For this application, the infrared thermovision technique (ITT) is a non-invasive method of determining envelope energy losses, problems with electrical installations, and construction anomalies in buildings, as explained by Balaras and Argiriou (2002). In the scientific literature, few studies have examined the thermal power of the opaque portions of envelopes with the infrared thermovision technique. Albatici and Tonelli (2010) have demonstrated good results for infrared thermovision compared to the heat flow metre (HFM) method. Their results show ITT is faster and has both less restriction and more accurate results than the heat flow metre.

2 Objective

This work proposes a new methodology to detect the most significant sectors in energy transmission through erected building's envelope by means of infrared thermography and the thermal intensity assessment of each opaque area, building element or thermal bridge, taking into account different values of wind speed. As a result, selective interventions in façades and therefore optimised energy efficient improvement can be made.

3 Background

The assessment of thermal transmittance (*U* value) is present in every building audit. The theoretical calculation method is the most used among diverse options, considering the physical features of the construction from the standards (CTE). However, occasionally is not possible to obtain the technical information of the buildings' envelopes; then the calculation is based on an organoleptic inspection. Authors such as Li Francis et al. (2015) and Biddulph et al. (2014) assure that this procedure may involve significant errors as regards other experimental methods.

However, following the requirements of the European Standard EN 13187:1998 (1998) and adhering to its procedure, infrared energy audits entail proceeding under optimal climate conditions. During the data-gathering and field work, one must avoid sunshine, rainfall, a falling temperature gradient, and wind speed. Studies by Palyvos (2008) show that the heat energy budget of a building's surface depends in large part upon wind-induced convection. For a thermographic analysis outside, where atmospheric radiation may affect it, wind speed is recommended to be below 2 m/s.

The overall energy budget depends on the envelope's pathologies. Infrared thermovision identifies these troublesome zones as the temperature differences recorded by the thermograms, which provide a visual map of material changes. In addition, one must also have exhaustive graphic documentation describing the materials, finishes, and construction organization. One clear example of the protocol before the inspection is a review of the emittance values of the different envelope materials (Rico and Moyano 2012). Besides, moisture in walls can cause errors in the interpretation of infrared thermovision images (Dall'O' et al. 2013) in the same way as direct solar radiation.

4 Experimental Methodology

For the study, two sample buildings were used with very similar single-family construction types, with three exterior walls and one shared wall. The third sample is a residential block with a northern façade and another facing south, being considered as two independent audits. Therefore, they were named as Models 1 and 2, 3 and 4, respectively in this paper. The buildings are located in a neighbourhood in the southern part of Seville, then placing the envelopes in climate zone B4 (see CTE, DB HE-1 in Appendix B).

A portable weather station used recorded exterior and interior temperatures and wind speed for 24 h before the tests. The thermal camera was placed between 80° and 88° with respect to the surface plane and never perpendicular to the building element, thus avoiding the identification of sources of thermal reflection and low emissivity (Lezana 2011).

The measurement recorded in the thermographic images was taken by calibrating the device, consisting in determining the reflective surface temperature with a contact thermometer at a point on the façade surface and comparing it with the real-time temperature of the thermographic camera. This device calibration, which ensures that the temperature registered by the camera is the same as the temperature of the actual surface, was proposed by authors such as Hoyano et al. (1999) for when material surface emissivity is not low. This approach is corroborated by research such as that of Cerdeira et al. (2011), and Vidas and Moghadam (2013). One of the considerations confirming the complexity of the emissivity values used in thermographic surveys is issues such as material age, rugosity, and energy-absorption capacity in environmental exposure, as discussed by Avdelidis and Maropoulou (2003). The scientific literature questions the use of emissivity tables for infrared thermovision in buildings since, as mentioned above, values vary for a measurement under conditions of non-steady heat transfer (Fokaides and Kalogirou 2011). The procedure serves to ensure that the temperature difference (interior and exterior) for the two samples analysed is greater than 10 °C or 10 K. Model 1, taken on 30 May 2013, shows higher temperatures than Model 2, taken on 3 February 2014. Models 3 and 4 were also performed in winter, on 28 January 2016 and 16 February 2016, respectively. Figure 1 shows the architectural reality of Model 2.

In order to obtain real-magnitude images for each sector of the opaque area, the thermograms for the building surface are transformed via specific applications (Herráez and Navarro 1997). This procedure also used CAD software to calculate the area of envelope sectors from the prior thermograms with a different type of construction and variations in temperature.

The next step is to mark the zones delimited in the thermal image (Fig. 2). Areas are classified in accordance with variations in heat transfer in standard outer walls and areas such as thermal bridges. A distinction is made between the pillars for the

Fig. 1 Building for Model 2 with thermogram overlap

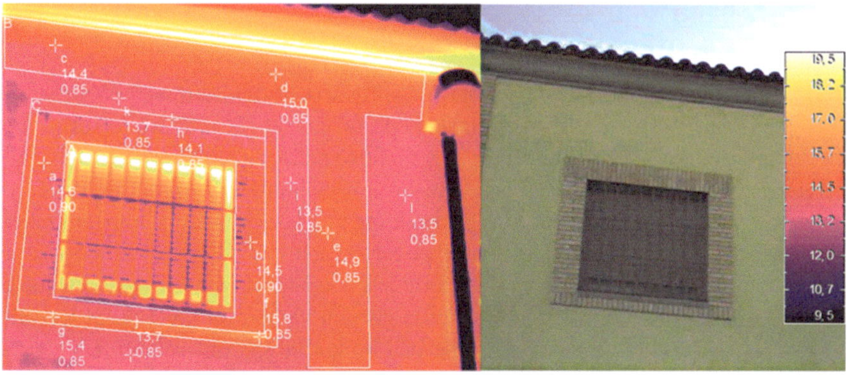

Fig. 2 Thermogram and photograph of Model 1

floor structures and elements comprising the voids (such as the relief and lintels enclosing window openings) and wall elements, which have a different heat transfer (U value) from the rest. The regions were considered based on their temperature, material, shape, and location in the opaque parts of the envelope.

The preliminary analysis of the results was performed with both software InfReC Analyzer NS9500 and FLIR Tools. The final processing of the thermograms was done with the Image Processor Pro II application, obtaining the average surface temperature for each of the sectors. These values depend on the material surface emissivity and the ambient temperature registered during each thermographic inspection. Inside the boundary of each polygon marked, the average temperature is calculated with Eq. (1).

$$T_m = \Sigma(A \cdot T_i)/\Sigma A \tag{1}$$

where T_m is the weighted surface temperature of each building element or thermal bridge in Kelvin, T_i is the surface temperature of a sector in Kelvin, and A is the area of each sector in square metres. Thus, each sector's percentage of surface temperature is determined in accordance with its area with respect to the total so that it can be compared with its thermal transmittance and, therefore, with the thermal intensity, which is the same as thermal power per area unit.

The next step is to calculate U using the infrared thermovision technique (ITT) as per Albatici and Tonelli. The parameters in Eq. (2) concern the thermographic audit, weather factors, and emissivity of the different materials of the envelope as well as the construction parts to be analysed.

$$U = (5.67\varepsilon_{tot}((T_m/100)4 - (T_{out}/100)4) + 3.8054\,\mathrm{v}(T_m - T_{out}))/(T_{int} - T_{out}) \tag{2}$$

where T_{int} is the interior ambient temperature in Kelvin, T_{out} the exterior ambient temperature in Kelvin, v the wind speed in m/s and ε_{tot} the total emissivity (coefficient).

5 Threshold Values

To establish the threshold values of the U-value thermal transmittance (W/m^2 K) in the opaque sections of walls, the climate zone of the analysis models is taken into account. The U_{lim} value (W/m^2 K) in outer walls and inner partitions in the thermal envelope that the CTE-DB establishes in Spain (Section HE-1 on the limitation of energy demands) is 0.82 W/m^2 K of the building in question, and the maximum heat transfer $U_{lim\ max}$ value (W/m^2 K) should never exceed 1.07 W/m^2 K for a B4 zone.

In order to enable comparison, this study has considered the weighted U value of different sectors of an envelope to obtain a general value as a function of the area covered by each building element or thermal bridge in accordance with Eq. (3), as required by the standard.

$$U_{Mm} = \left(\Sigma\left(A_{(M)}U_{(M)}\right) + \Sigma\left(A_{(PF)}U_{(PF)}\right)\right)/\left(\Sigma A_{(M)} + \Sigma A_{(PF)}\right) \tag{3}$$

where U_{Mm} is the weighted thermal transmittance of the façade in W/m^2 K, A is the area of each sector (S) or thermal bridge (TB) in square metres, and U is the thermal transmittance of these elements in W/m^2 K.

This paper proposes the implementation of threshold values for energy loss for thermal intensity (thermal power per surface unit) from thermal transmittance values that the different areas may show according to the heat transfer threshold values corresponding to a given climate zone.

The thermal intensity thresholds from the heat transfer tolerance values for a climate zone established in Sect. 6 are calculated using an expression by Díaz and Tenorio (2005), according to Eq. (4).

$$\phi_t = UA(T_{int} - T_{out}) \tag{4}$$

where ϕ_t is the thermal power in W, and the degree centigrade °C for T can be used to ease comprehension, especially in Table 1.Where ΔT_m is the thermal difference between the element and the largest sector without thermal alterations —standard wall—; ϕ_t/A is the thermal intensity; and L_x is either the upper threshold value L_{sup} or the mean threshold value L_m for the thermal intensity, as applicable, that are emitted from the CTE.

6 Results and Discussion

As an example, the results for Model 1 according to this methodology are given in Table 1 in order to clarify the calculation procedure of energy loss and the comparison between this loss and the threshold values established. The results for all the Models will be gathered in Fig. 3 before the explanation of Table 1.

Table 1 Results of thermal intensity of Model 1

Element	A (m^2)	U (W/m^2 K)	T_m (°C)	ΔT_m (°C)	ϕ_t (W)	ϕ_t/A (W/m^2)	L_{sup} (W/m^2)	$\phi_t/A - L_x$ (W/m^2)
Standard wall	29.71	1.33	15.10	0.00	395.75	13.32	10.70	2.62
Bathroom wall	3.47	0.82	14.68	−0.42	28.58	8.23		−2.47
Basement wall	8.30	5.99	18.91	3.81	497.65	59.93		49.23
Lower brick skirting	0.98	4.54	17.73	2.63	44.45	45.42		34.72
Upper brick skirting	0.97	3.97	17.26	2.16	38.47	39.66		28.96
Pillar fronts 35 cm	1.75	1.41	15.16	0.06	24.53	14.05		3.35
Slab fronts	7.57	2.10	15.73	0.63	158.72	20.98		10.28
Pillar fronts 30 cm	3.97	1.77	15.46	0.36	70.29	17.70		7.00
Window casing 24 cm	5.23	1.51	15.25	0.15	79.26	15.14		4.44
Window jambs	2.67	1.53	15.26	0.16	40.75	15.27		4.57
Bathroom jambs	1.21	1.53	15.26	0.16	18.49	15.27		4.57
Brick sill	0.91	1.89	15.56	0.46	17.21	18.91		8.21
Roof wall	2.91	2.99	16.46	1.36	87.00	29.88		19.18
Total	69.65	2.16	15.77	0.67	1501.13	21.55	…	10.85
						L_m (W/m^2)	8.20	13.35

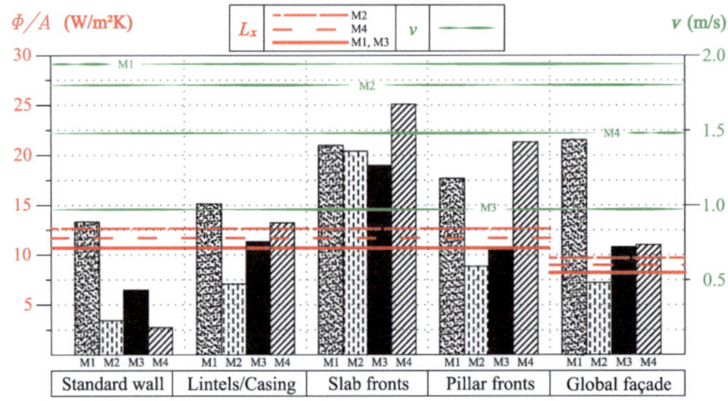

Fig. 3 Thermal intensity and wind speed results in Models 1, 2, 3 and 4

In the analysis in Table 1, the final row for overall outcomes refers to the weighted value of each magnitude as a function of the area of each element (not the arithmetic mean) except for parameters A and ϕ_t, which refer to the sum of the values in each column. Thus, the total value for U (W/m^2 K) corresponds to the result from Eq. (3) as the U_{Mm} average thermal transmittance of the façade.

This work analyses the ϕ_t/A values, which makes it necessary to convert the U-value transmittance threshold (W/m^2 K) from the national standards into this magnitude in order to directly compare it with the energy status of the overall envelope. Analytically relating the thermal intensity threshold values—deriving from the transmittance in the aforementioned standards—to the thermal intensity of each sector of the envelope, the result clearly shows which zones are more involved in the envelope's energy loss. These threshold values are obtained by solving for ϕ_t/A in Eq. (4), that is, finding the product of multiplying the U value by the temperature difference between the building's inside and outside. Consequently, the thermal intensity thresholds proposed will vary according to weather conditions and the interior temperature.

After applying the upper threshold L_{sup} in the column $\phi_t/A - L_x$, negative values can be obtained corresponding to the sectors that still show an increase in thermal intensity up to said threshold. In other words, their heat transfer is within current standards and so these zones have acceptable energy losses.

In contrast, the positive values calculated in the $\phi_t/A - L_x$ column derive from elements that have a ϕ_t/A above the established threshold L_{sup}, and they therefore require an intervention to address the energy losses.

By performing a weighted mean of the ϕ_t/A in its corresponding column as a function of the total area of the wall, its overall energy status is obtained. There is a threshold value of thermal intensity ϕ_t/A proposed as L_m that the building envelope must not exceed, considering its energy behaviour at a general level, based on U-value heat transfer for each climate zone. In Model 1, the mean ϕ_t/A value of the façade is 21.55 W/m^2 (see Table 1 or Fig. 3), showing excess in relation to the threshold of 8.20 W/m^2 (from the 0.82 W/m^2 K value for the CTE B4 climate zone). In contrast, in Model 2, when the mean ϕ_t/A value of 7.21 W/m^2 is compared with the threshold for its thermal gradients, the resulting values are negative and, therefore, favourable. A greater thickness in the thermal insulation of the envelope, together with building types that do not have or that minimise thermal bridges, can decrease energy losses through the envelope.

In addition, the importance of wind speed in energy loss has been analysed. Figure 3 gathers ϕ_t/A values of particular sectors within the envelopes of all Models and the average of their façades, the threshold values for that energy loss and the wind speed during each audit.

As displayed, the highest values of v in Model 1 increase thermal intensity and consequently thermal transmittance and energy loss. However, in Model 2 with higher thickness of insulation in its envelope, energy loss is clearly lower, except for slab fronts inappropriately solved. For their part, despite being the same building—see weighted thermal intensity of façades in Models 3 and 4-, in thermal bridges of the former v is reduced, causing ϕ_t/A to decrease significantly as regards the latter exposed to higher wind speed. The extreme case is the lack of wind, which causes the disappearance of convention and leaves radiation as the only mode of heat transfer (see the impact of a zero value for v on Eq. (2)).

In order to explain this, the theoretical weighted estimate of the façade is also compared with the empirical result. 0.60 W/m^2 K and 0.40 W/m^2 K for Models 1 and 2 respectively are the theoretical thermal transmittance values calculated using the technical information of the buildings and the regulations. It can be noticed that these theoretical U_{Mm} weighted values are lower than the transmittance calculated afterwards using this methodology with higher v (2.16 and 0.61 W/m^2 K). By contrast, U_{Mm} values for Models 3 and 4 are 1.15 W/m^2 K, which are slightly higher than the U obtained via infrared thermography with lower v (1.08 and 1.01 W/m^2 K).

7 Conclusions

This research is significant as few studies have analysed quantitatively the global thermal intensity and that of the components of existing building thermal envelopes. This protocol allows selective intervention on the shell, targeting detected thermal anomalies with a significant repercussion on the energy losses. Thus, indiscriminate actions are avoided and thermal retrofitting is optimised as regards material resources and labour. This lessens the investment and the environmental impact in both the action and the life cycle of the building itself. The results of this work should improve energy audit process and, therefore, good building codes.

It is necessary to take into consideration some variables in the procedure of this methodology in order to avoid significant errors. Infrared cameras with low sensitivity should not be used, since they increase the distortion error in readings. Furthermore, the location of many urban buildings makes it impossible to perform infrared audits (e.g. due to vegetation, other buildings, signs and other items on the façade). Additionally, the weather parameters must allow for a temperature gradient of at least 10 K, which usually occurs in summer and winter, or by means of heating or air conditioning.

Finally, it is worth mentioning the difference between the theoretical thermal transmittance value and the one obtained during the process of the thermographic audit. This denotes the great influence that meteorological conditions imply in the assessment of the energy efficiency. In spite of the fact that the results of this study clearly indicate the impact of wind speed on energy loss, additional audits could lead to quantify that loss when concrete wind speed increases take place.

Acknowledgements This work has been funded by V Research Plan (VPPI) of the University of Seville.

References

AENOR (1998) UNE-EN 13187:1998. Thermal performance of buildings. Qualitative detection of thermal irregularities in building envelopes. Infrared method (ISO 6781:1983 modified)

Albatici R, Tonelli AM (2010) Infrared thermovision technique for the assessment of thermal transmittance value of opaque building elements on site. Energy Build 42:2177–2183. doi:10.1016/j.enbuild.2010.07.010

Al-Habaibeh A, Anderson G (2010) Using infrared thermography for monitoring thermal efficiency of buildings—case studies from Nottingham Trent University. Paper presented at the 7th Jordanian international mechanical engineering conference (JIMEC'7), Amman-Jordan, 27–29 Sept 2010

Avdelidis NP, Maropoulou A (2003) Emissivity considerations in building thermography. Energy Build 35:663–667. doi:10.1016/S0378-7788(02)00210-4

Balaras CA, Argiriou AA (2002) Infrared thermography for building diagnostics. Energy Build 34:171–183. doi:10.1016/S0378-7788(01)00105-0

Biddulph P, Gori V, Clifford AE et al (2014) Inferring the thermal resistance and effective thermal mass of a wall using frequent temperature and heat flux measurements. Energy Build 78:10–16. doi:10.1016/j.enbuild.2014.04.004

Castro C (2009) Puentes térmicos en CTE HE1. In: Departamento Técnico de Dow Building Solutions. Available via DIALOG. http://www.andimat.es/wp-content/uploads/articulo-ccastro-puentes-termicos-en-cte-he1. Accessed 23 Sept 2015

Cerdeira F, Vázquez ME, Collazo J, Granada G (2011) Applicability of infrared thermography to the study of the behaviour of stone panels as building envelopes. Energy Build 43:1845–1851. doi:10.1016/j.enbuild.2011.03.029

Dall'O' G, Sarto L, Panza A (2013) Infrared screening of residential buildings for energy audit purposes: results of a Field Test. Energies 6(8):3859–3878. doi:10.3390/en6083859

Díaz MI, Tenorio JA (2005) Pérdidas de calor y formación de condensaciones en los puentes térmicos de los edificios. Paper presented at the 1st jornadas de investigación en construcción, Instituto de Ciencias de la Construcción "Eduardo Torroja", Madrid, 2–4 June 2005

Fesanghary M, Asadi S, Geem Zong Woo (2012) Desing of low-emission and energy-efficient residential buildings using a multi-objective optimization algorithm. Build Environ 49:245–250. doi:10.1016/j.buildenv.2011.09.030

Fokaides PA, Kalogirou SA (2011) Application of infrared thermography for the determination of the overall heat transfer coefficient (U-value) in building envelopes. Appl Energy 88:4358–4365. doi:10.1016/j.apenergy.2011.05.014

Herráez J, Navarro P (1997) Una práctica de fotogrametría desde principio a fin. Fotogrametría aplicada a la arquitectura. In: Abstracts of the 2nd jornadas de fotogrametría arquitectónica, Escuela Técnica Superior de Arquitectura de Valladolid, Valladolid, 25–28 Nov 1997
Hoyano A, Asano K, Kanamaru T (1999) Analysis of the sensible heat flux from the exterior surface of building using time sequential thermography. Atmos Environ 33:3941–3951. doi:10.1016/S1352-2310(99)00136-3
Lezana A (2011) Infrared thermography applied to construction. Eco-construction 29:109
Li Francis GN, Smith AZP, Biddulph P et al (2015) Solid-wall U-values: heat flux measurements compared with standard assumptions. Build Res Inf 43(2):238–252. doi:10.1080/09613218.2014.967977
Ministerio de la Presidencia (2013) Real Decreto 235/2013, de 5 de abril, por el que se aprueba el procedimiento básico para la certificación de la eficiencia energética de edificios. In: Boletín Oficial del Estado. Available via DIALOG. https://www.boe.es/boe/dias/2013/04/13/pdfs/BOE-A-2013-3904.pdf. Accessed 13 Apr 2013
Ministerio de la Vivienda (2006) Real Decreto 314/2006, de 17 de marzo, por el que se aprueba el Código Técnico de la Edificación. In: Boletín Oficial del Estado. Available via DIALOG. https://www.boe.es/boe/dias/2006/03/28/pdfs/A11816-11831.pdf. Accessed 13 Apr 2013
Palyvos JA (2008) A survey of wind convection coefficient correlations for building envelope energy systems' modeling. Appl Therm Eng 28:801–808. doi:10.1016/j.applthermaleng.2007.12.005
Parliament European (2010) Directive 2010/31/EU of the European parliament and of the council of 19 May 2010 on the energy performance of buildings (recast). Off J Eur Union L153:13–35
Rico F, Moyano JJ (2012) Tratamiento gráfico de imágenes generadas con cámaras termográficas con tecnología de infrarrojo (IR) al servicio del mapeado de anomalías en fachadas de edificios históricos. Paper presented at the XI congreso internacional de expresión gráfica aplicada a la edificación, Universidad de Valencia, Valencia, 29-1 Dec 2012
Taylor T, Counsell J, Gill S (2013) Energy efficiency is more than skin deep: Improving construction quality control in new-build housing using thermography. Energy Build 66:222–231. doi:10.1016/j.enbuild.2013.07.051
Vidas S, Moghadam P (2013) HeatWave: a handheld 3D thermography system for energy auditing. Energy Build 66:445–460. doi:10.1016/j.enbuild.2013.07.030

Assessment of the Energy Efficiency of a "Cool Roof" for Passive Cooling. Comparative Study of a Case of Tropical Climate and a Case of Southern Spanish Climate

Carlos A. Domínguez Torres and Antonio Domínguez Delgado

1 Introduction

Over the last decades, extensive research in the field of energy efficiency has focused on the need to reduce energetic costs in the thermal conditioning of buildings by using techniques that take advantage of environmental and climatic resources in a suitable form.

In hot climates, the high energy consumption needed to get internal comfort in buildings during the hot season poses a major problem. Hence the convenience of implementing techniques that minimize the cooling load.

Among the various existing passive cooling techniques, in the present work we focus on a specific case of Cool Roof.

A Cool Roof is a roofing system able to reject solar heat and keep roof surfaces cooler under sunny weather.

Its performance is due to the properties of the materials used, which reflect the solar radiation and release the heat they have absorbed.

Usually a Cool Roof is a roof system with a low absorptivity coefficient for short-wave radiation from solar origin and a large infrared emittance coefficient. This way a remarkable reduction in the temperature of the roof surface is achieved, thereby decreasing the heat flow into the building.

This effect is achieved thanks to the properties of the so-called "cold materials". characterized by high solar reflectance and high emissivity for longwave radiation.

In our case, the cold effect is achieved by applying on the roof external surface a white elastomer layer whose radiative properties have those requirements.

C.A. Domínguez Torres · A. Domínguez Delgado (✉)
Escuela Técnica Superior de Arquitectura de Sevilla, Seville, Spain
e-mail: domdel@us.es

P. Mercader-Moyano (ed.), *Sustainable Development and Renovation in Architecture, Urbanism and Engineering*, DOI 10.1007/978-3-319-51442-0_36

In this chapter the energy efficiency of the considered Cool Roof is analyzed in both the geographical and climatic contexts mentioned above, that is, Mediterranean climate (southern Spain, namely Seville city) and tropical equatorial climate (northern Colombia, namely Santiago de Cali city).

Due to the high levels of solar radiation and high temperatures frequent in southern Spain, the use of Cool Roofs in this area might turn out to be suitable to decrease the cooling load. However, its use penalizes the solar heat gain during the cold season by increasing the flow of heat from the roof towards the sky. Therefore its use should be assessed with the greatest possible precision in order to determine its suitability.

For the considered tropical equatorial climate, the weather is usually cloudy and characterized by very homogeneous levels of solar radiation throughout the year. This entails that the diffuse solar radiation and the flow of longwave radiation from the sky are quite high when compared to direct solar radiation. This determines the behavior of the Cool Roof, resulting in a much more favorable global energy balance than the obtained for the Mediterranean climate.

In this chapter, by using computational numerical simulation techniques applied to the air fluid thermal dynamic, to the heat transfer through the roof and to the radiative exchanges, we draw conclusions about the patterns displayed by the thermodynamic behavior of the considered Cool Roof when compared with a standard ceramic tiles roof in the two analyzed geographic and climatic contexts.

2 Case Studies

For our study a detached house with two floors has been considered. Since the influence of the Cool Roof affects basically the part of the building directly under it, its efficiency for tall buildings is limited to upper floors. However, the usefulness of Cool Roofs for low-rise housing and more collective use buildings, schools, office buildings, libraries, etc., is evident in the case studies included in the report "Cool Roof case studies in EU level" from the European Cool Roofs Council (Zinzi and Romeo 2010), and in a large number of recent publications such as those of Synnefa et al. (2012) and Pisello and Cotana (2014).

The studied roofs are flat type and have the same configuration, which complies with the specifications of the Spanish Technical Code. Such a configuration is described in Table 1 in which dimensioning and thermophysical characteristics of the various components of the roofs are described.

For the outer surface of ceramic tiles it is considered an absorptivity solar radiation coefficient of 0.75, while the emissivity coefficient of this layer is taken as equal to 0.83, as referenced by Gozalbo et al. (2008).

In order to transform the roof in a Cool Roof, it has been deemed the application of a white elastomer layer on the surface of the outer layer of the roof. On its technical specifications this layer provides a solar reflectance coefficient equal to 0.89 and an emissivity in the range of infrared radiation equal to 0.89.

Table 1 Thermophysical characteristics of the roofs

Layer	Description	Thickness (m)	Density (kg/m^3)	Specific heat (J/kg K)	Conductivity (W/m K))
1 (Ext.)	Ceramic tiles	0.01	2000	800	1.00
2	Mortar	0.02	2000	1000	1.40
3	Protective layer	0.02	1900	1000	1.80
4	Waterproofing (bitominous layer)	0.2	2100	1000	0.7
5	Separating layer	0.2	1900	1000	1.80
6	Insulation (extruded polystyrene)	0.05	35	1400	0.038
7	Lightweight concrete	0.15	1200	1000	0.57
8	Ceramic pieces	0.30	650	1000	1.58
9 (Int.)	Plastering	0.015	1000	1000	0.57

3 Roof Energy Balance

In this section a generic description of the physical problem involving the heat transfer and the movement of the air mass is set out. Specifically, the heat transfer in the roofs are determined by:

- Heat gain on the outer slab due to solar irradiation.
- Heat exchange by radiation between the outer surface and the sky.
- Heat exchange by convection between the outer surface and the ambient air.
- Heat transfer by conduction through the layers of the roof.
- Heat exchange by convection and radiation between the internal surface of the roof and the interior of the building.

Other possible factors such as radiative exchange between the roof and adjacent buildings or vegetable masses, even terrain, higher than the studied building, have not been taken into account in this study for the sake of brevity.

In order to establish the physical model it must be taken into account the fact that the equations describing the air flow, the equation for energy transport by the air flow and the heat transfer equations through the different layers of the roof must be solved in every time step. Likewise the radiative exchanges must be computed in each time step in order to adequately approximate the heat transfer through the roof.

4 Convective Heat Transfer Coefficients

Convective heat transfer coefficient for external building surfaces is essential in order to calculate heat gains and losses from the roof to the ambiance air. Following the recommendations of Hagishima and Tanimoto (2003) for low-rise buildings, we have considered the coefficient value equal to

$$h_c = 8.18 + 2.28\,U_R\,\mathrm{W/m^2K} \quad (1)$$

where U_R is the velocity in m/s measured 0.6 m away from the external roof surface. Since according to the authors this value is calculated specifically for convection heat transfer on the flat roof of a building with two floors, it adequately matches our case study.

This correlation allows a more accurate calculation of the convective coefficient h_c because it takes into account the wind speed calculated next to the roof surface by solving the Navier-Stokes equations instead of employing the wind speeds provided by weather stations as it is usual in energy simulation programs.

Internal roof surface energy balance calculation is made by taking a comfort room temperature T_{room} constant for each season. Then we calculate the exchange of heat by convection and radiation between the internal roof surface and the room by using a mixed convective-radiative transfer coefficient given by $h_i = 8\,\mathrm{W/m^2K}$, which is usually recommended in the technical specifications for this kind of indoor heat exchange.

5 Climatic Conditions

As mentioned above, passive systems are highly sensitive to local environmental conditions.

In Southern Spain, and specifically in the city of Seville, climatic characteristics are determined by a high seasonality. In summer daytime temperatures often reach high values, while in winter the values are significantly lower but moderate for what can be considered usual in Europe.

By contrast, in equatorial tropical climates, such as the case of Cali, temperatures are fairly uniform throughout the year, being the daily variation range quite narrow. The daily ambiance temperature close to the indoor temperature comfort is considered in this work, namely 25° for Cali.

Different behaviors of the temperatures can be seen in Fig. 1, which clearly illustrates the above.

In both cases winds are not usually strong but can range from total calm to relatively moderate winds that can significantly increase heat transfer by convection between the roof and the ambiance.

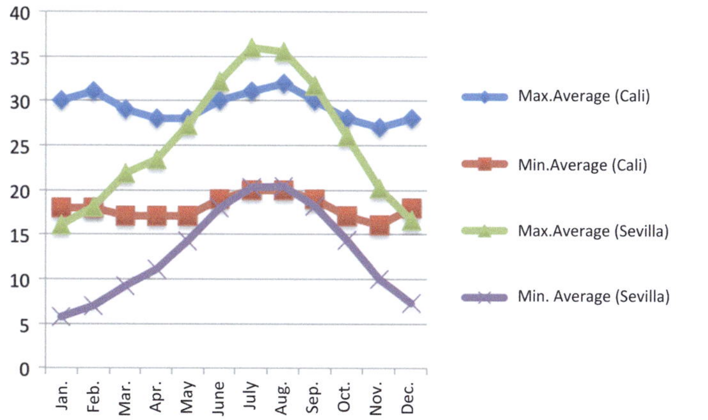

Fig. 1 Averages of maximum and minimum temperatures in Sevilla and Cali

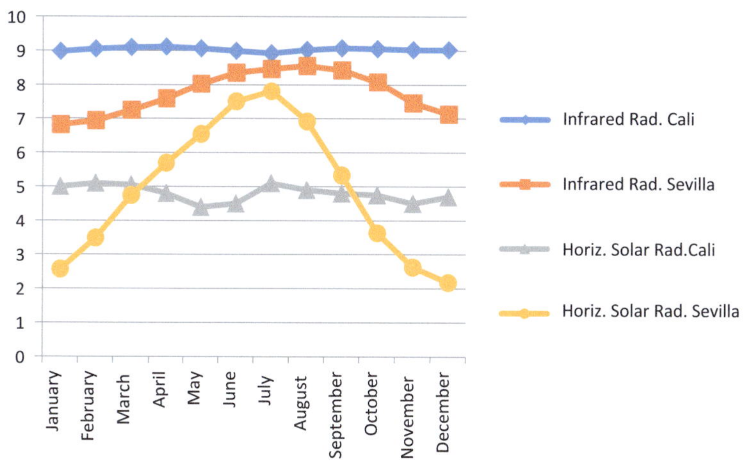

Fig. 2 Monthly averaged insolation and infrared radiadion incidents on a horizontal surface (kWh/m^2/day)

The studied velocities for the wind entering in the inlet of the computational domain are equal to 1, 4 and 8 m/s, which are considered sufficiently representative of the average values given in the two considered climatological frameworks.

As it is shown in Fig. 2, in the case of Seville, incident solar radiation on horizontal surface also has a markedly seasonal character with high values in summer and moderate in winter. Also longwave radiation reaches higher values in summer, but with a lower variation throughout the year.

However, in the case of Cali, solar radiation is relatively homogeneous throughout the year and is affected by the usual presence of clouds. As it is shown in Fig. 2, despite its latitude, solar irradiation values reached in summer are lower than the values for the south of Spain. Also infrared radiation remains quite constant in Cali throughout the year and it reachs higher values than for Sevilla.

6 Estimation of the Downward Long-Wave Radiation of the Sky

An essential difference between the two types of radiation affecting the roofs is that solar short wave radiation takes place only during daylight hours, but thermal downwelling radiation is present throughout the whole day. So, although this radiation is usually named nocturnal radiation, it takes place even during daylight hours, although at night the thermal radiative exchange is usually negative, that is, more longwave radiation is emitted that received by the roof. Hence the need for the most accurate calculation of this type of radiation as possible.

The presence of clouds significantly affects this type of radiation, increasing its value when the clouds have a higher density and a lower height.

A correlation quite used in such situations is the one presented by Goforth et al. (2002). Their model is a variant of the Swinbank model and is given by the following expression:

$$Q_{sky} = \left(1 + KC^2\right) 8.7810^{-13} T_{amb}^{5.852} RH^{0.07195} \tag{2}$$

where K is a parameter that depends on the height of the cloud layer, C is the percentage of cloudiness, being 0 for fully clear skies and 1 for completely cloudy skies and RH is the relative humidity.

In this chapter it has been used the Goforth model. For this purpose we have used the values provided by different climatic organisms: the Spanish Meteorological Agency, the Meteorological Services of the National Aeronautics and Space Administration (Nasa) and the National Oceanic and Atmospheric Administration (NOAA).

The obtained monthly average values for shortwave and longwave radiation are showed in Fig. 2, for both studied geographical locations.

7 Numerical Simulation

7.1 Computational Domain and Meshing

For the numerical solution of the set of equations describing the thermodynamic air flow, we have started from a two-dimensional computational domain that includes both the building and a wide region outside it, in order to adequately simulate air flow in the front and top of the building containing the roof.

The computational domain has been meshed by using triangular elements to perform a discretization of the problem by the Finite Element Method (FEM).

The mesh built has a total of 25,000 triangles, which guarantees a good accuracy in the numerical results obtained.

Furthermore it has been built a two-dimensional mesh of the roof in order to carry out the numerical calculation of the heat conduction through the layers that form the roof. This calculation is made by applying the FEM to the heat conduction equation through the roof.

7.2 Numerical Resolution

The numerical simulation of the flow air has been made by using a FEM discretization of the thermodynamic equations of Navier-Stokes.

The free code FreeFem++ software, from the French "Institut National de Recherche en Informatique et en Automatique" (INRIA), has been used for the computational implementation of the considered discretizations.

The problem has been solved in a time period of one week for each month of the year, making use of the typical meteorological monthly data for every month.

The time dependent problem has been solved by using the non-stationary equations of Navier-Stokes for the fluid thermodynamic dynamic and the non-stationary equations of heat conduction through the different layers of the cover.

To solve the equations of fluid dynamics it has been used the $k - \varepsilon$ model for turbulence. This is necessary due to the characteristics of the problem in which the air flow over the roof of the building presents high Reynolds values that make necessary modelization of the turbulence.

At each time step the energy balance for radiation and convection in the interior and exterior surfaces of the roof are taken as boundary conditions.

8 Results

In order to calculate the heat flows through the roofs, constant temperatures inside the building have been considered for each calculation period. This constant temperature is the comfort temperature that has been taken equal to 25 °C for the whole year in the case of the Cali.

In the case of Southern Spain, the comfort temperatures have been taken equal to 21 °C for the winter months, 23 °C for spring and autumn and 25 °C for summer.

8.1 Monthly Heat Flux

In Figs. 3 and 4 monthly flows of heat in kWh/m^2 are shown for the Cool Roof and for the ceramic tiles roof for the cities of Seville and Cali for different wind speeds. Positive values represent heat flux into the building and negative values outward heat flux.

For Seville city it is observed that heat flux values range from values close to -1.5 kWh/m^2 in winter months to values close to 1.5 kWh/m^2 in summer.

This result is quite expectable due to the typical climate of Seville, which is characterized by not too cold temperatures in winter, hot weather in summer and springs and autumns that have daily average ambient temperatures close to the comfort temperature.

It is noteworthy that the curves of heat flux for the Cool Roof are lower that the ceramic tiles ones for the whole year and for all the wind velocities.

For the city of Cali Fig. 4 shows that the observed heat flux is positive for almost every month. The addition of moderate ambient temperatures with average values slightly higher that the comfort temperature and heat gains involving solar radiation and infrared radiation from the sky produces this gain of heat in the building. In this result, the role played by the infrared radiation is very strong because as we said above, its values are higher than the solar irradiation on horizontal surface.

For Cali, like for the city of Sevilla, the curves of heat flux for the Cool Roof are lower that the ceramic tiles ones for the whole year and for all the wind velocities.

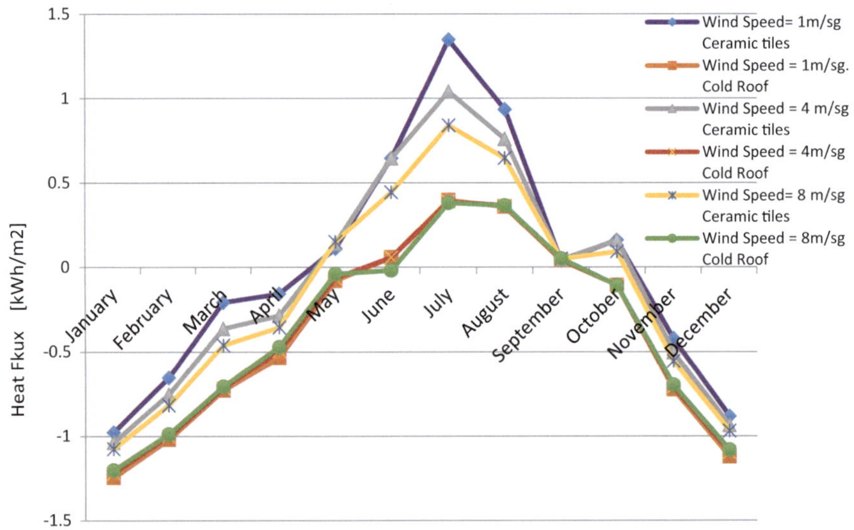

Fig. 3 Monthly heat flux (kWh/m^2) in Sevilla

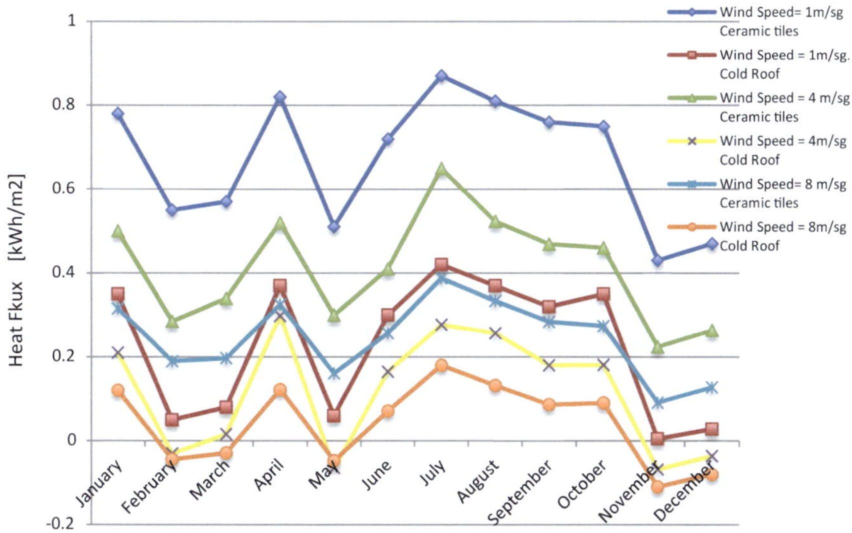

Fig. 4 Monthly heat flux (kWh/m^2) in Santiago de Cali

Moreover, for Cali it is showed that for higher wind speeds there is a greater cooling effect. This is because with increasing wind speed, convection between the roof and the air becomes greater than heat gain from solar radiation and longwave radiative exchange between the cover and the sky.

However, this effect is not so strong for the Cool Roof in the case of the dwe-lling located in Sevilla.

8.2 *Heat Flux Yearly Balance*

In Figs. 5 and 6 heat flux yearly balance for the Cool Roof and the ceramic tiles one are shown for the two analyzed cities and for the different wind velocities considered.

In general we can say that for both cities the calculations show a lower heat flux through the Cool Roof than through the classical ceramic tiles roof.

This heat flux reduction is more pronounced in the location of Santiago de Cali, where a sharp difference between the roofs is observed, especially for moderate wind speeds.

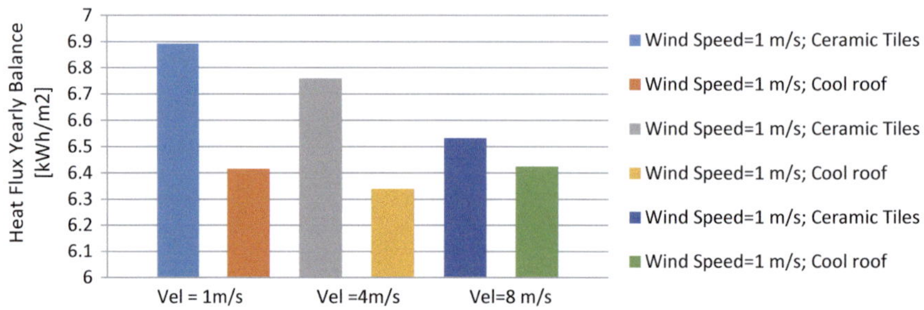

Fig. 5 Heat flux yearly balance (kWh/m^2) for Sevilla

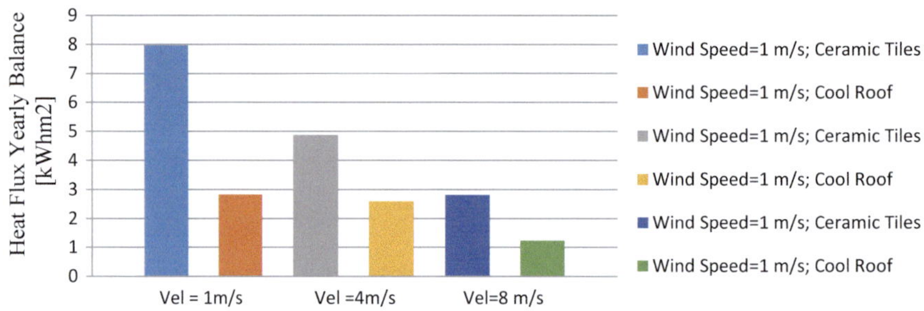

Fig. 6 Heat flux yearly balance (kWh/m^2) for Cali

9 Conclusions

The results seem to indicate that for the two analyzed locations the annual balance of heat flow is lower for the whole year when the Cool Roof is used instead of a typical ceramic tiles one. This fact is more evident for the city of Santiago de Cali than for the city of Seville that is related to the different climatic conditions of both cities.

Therefore, it can be established as a final conclusion that the analyzed Cool Roof provides greater energy savings in the geographic and climatic context of the present study.

However, its use in zones of colder climatology should be evaluated with caution due to the penalty in terms of the heat gain from a radiative source which is produced by this type of roof during the cold season.

References

Goforth MA, Gilcrest GW, Sirianni JD (2002) Cloud effects on thermal downwelling sky radiance. SPIE vol 4710

Gozalbo A, Orts MJ et al (2008) Study of the radiant properties of ceramic tiles. Instituto de Tecnología Cerámica (ITC). Asociación de Investigación de las Industrias Cerámicas (AICE). Universitat Jaume I. (Castellón, España) & KERABEN GRUPO, S.A

Hagishima Tanimoto J (2003) Field measurements for estimating the convective heat transfer coefficient at building surfaces. J. Building and Environ 38(7):873–881

Pisello A, Cotana F (2014) The thermal effect of an innovative cool roof on residential buildings in Italy: results from two years of continuous monitoring. Energy Build 69:154–164

Synnefa A, Saliari M, Santamouris M (2012) Experimental and numerical assess-ment of the impact of increased roof reflectance on a school building in Athens. Energy Build 55:7–15

Zinzi M, Romeo C (2010) Cool Roofs case studies in EU level. Intelligent Energy Europe program SAVE 2007

Author Index

A
Acosta, Ignacio, 381
Agudo Martínez, Andrés, 103
Aguirre-Maldonado, Eduardo, 253
Andújar Jose Manuel, 165
Antón García, Daniel, 427
Arriagada Sickinger, Carolina, 39

B
Baeriswyl Rada, Sergio, 39
Balcázar-Arciniega, Cristian, 227
Barrios-Padura, Angela, 117, 131, 289
Benito, Francisco J., 263
Blandón-González, Begoña, 131
Blasco Lucas, Irene, 351
Bobadilla Moreno, Ariel, 363
Bohórquez, M.A., 165
Bovea Edo, Mª Dolores, 15
Braulio-Gonzalo, Marta, 15

C
Cabello Montoro, Rafael, 63
Campano, Miguel ángel, 381
Carroquino Larraz, Santiago, 51
Cervero Sánchez, Noelia, 179
Coch, Helena, 77
Contreras Lopez, Christhopher, 325
Cordero, Sebastián, 277

D
de Luxán, Margarita, 203, 415
Domínguez Delgado, Antonio, 439
Domínguez Ruiz, Victoria, 153
Domínguez Torres, Carlos A., 439
Dueñas Ladrón de Guevara, Eduardo J., 311

G
Gómez de Terreros-Guardiola, Pedro, 131
Gómez, Gloria, 203, 415
Gomez Melgar, Sergio, 165
García Alvarado, Rodrigo, 363
González, Victoriano, 289
González-García de Velasco, Concepción, 103
González Vilchez, Miguel, 103
Guajardo, Alfonso, 27

H
Hernández-Olivares, Francisco, 227, 253
Herrera, Rafael, 277
Hidalgo, Pilar, 263

I
Isalgue, Antonio, 77

J
Jiménez-López, Isabel, 117

L
Ladrón de Guevara Muñoz, M. Carmen, 311
León Rodríguez, ángel Luis, 403
López-Alonso, Silvia, 191
Lopez-Besora, Judit, 77
López-Escamilla, Álvaro, 277
López-Martínez, Alicia, 301
López-Mesa, Belinda, 215

M
Marín García, David, 427
Martínez-Quesada, María del Carmen, 91
Martín-Mariscal, Amanda, 301
Martín Martínez, Luis, 311
Martín-Pastor, Andrés, 301
Mercader-Moyano, Pilar, 143
Miñano, Isabel, 263
Molina-Huelva, Marta, 117, 131, 289
Monzón, Marta, 215
Morales-Conde, Mª Jesús, 191
Moyano Campos, Juan José, 427

P. Mercader-Moyano (ed.), *Sustainable Development and Renovation in Architecture, Urbanism and Engineering*, DOI 10.1007/978-3-319-51442-0

N
Neila González, F. Javier, 3, 339
Nope Bernal, Alberto, 363
Núñez Peiró, Miguel, 3, 339

O
Ortega Rodríguez, Marcos, 311

P
Parra, Carlos J., 263
Pérez-Fargallo, Alexis, 393
Pérez-Gálvez, Filomena, 191
Pérez Téllez, Juan Diego, 403
Pineda, Paloma, 277
Pinto, Armando, 381
Prieto-Thomas, Ana, 239
Pulido-Arcas, Jesús A., 393

R
Ramos Martín, Manuel, 143
Rico Delgado, Fernando, 427
Roa, Jorge, 277
Rodríguez-Liñán, Carmen, 191
Román, Emilia, 203, 415
Ruá Aguilar, Mª José, 15
Rubio-Bellido, Carlos, 393
Rubio-de Hita, Paloma, 191

S
Sánchez-Guevara Sánchez, Carmen, 3, 339
Sendra, Juan J., 381
Suárez Medina, Rafael, 403

T
Tumini, Irina, 39

Subject Index

A

Advanced geometry, 308
Aphandra natalia, 227, 228, 230, 232–236
Architectural design techniques, 95, 309
Architectural recycling, 51, 54, 58, 60, 61
Architecture, 36, 51, 54, 56, 70, 78, 91, 93, 94, 101, 103, 114, 131, 140, 151, 153–155, 158, 159, 162, 175, 183, 186, 191, 193, 241, 248, 277, 282, 290, 301–307, 309, 334, 365, 369, 373, 379, 415
Assessment, 15, 23, 24, 47, 118, 120–123, 125, 143, 144, 146–148, 150, 151, 153, 179–182, 184, 186, 189, 191, 192, 195, 196, 254, 259, 290, 291, 294, 298, 352, 360, 363, 366, 372, 374, 378, 403, 404, 408, 411, 427–429, 436

B

Building envelopes, 18, 22, 150, 208, 213, 370, 372, 378, 379, 400, 428, 434
Building Information Modelling (BIM), 143, 147, 150, 151, 154, 170
Building simulation, 406
Built heritage, 153, 155, 160, 162

C

Calibration, 339, 403, 405, 406, 410, 430
Carbon fibre, 289, 291–294
Cement, 198, 253, 255, 256, 266, 268
City, 18, 21, 24, 55, 56, 63, 66, 72, 182, 239, 326, 422
Classrooms, 381–387, 389
Clevedon, 103, 107, 110–114
Climate, 77, 78, 144, 146, 166, 206, 207, 213, 277, 278, 285, 286, 339, 349, 351–353, 367, 371, 383–385, 393–395, 397, 400, 412, 417, 424, 427, 429, 432, 434, 439, 440, 442, 446
Climate adaption, 395
Climate change, 3, 4, 40, 143, 144, 203, 204, 206, 242, 311, 393, 394, 396, 397, 400, 404, 416, 417, 420
Collective creativity, 309
Comfort, 3, 4, 119, 125, 126, 131, 133, 140, 147, 165–167, 171, 174, 175, 203, 204, 206, 207, 209, 213, 277, 280, 282, 323, 339, 351, 353, 358, 362, 381, 388, 403, 404, 406, 408, 410, 415, 424, 439, 442, 445, 446
Complex morphologies, 363
Composite materials, 227, 236, 289, 290, 292, 293, 297
Compressive strength, 253, 255, 257–260, 264, 267, 268
Computational Fluid Dynamics (CFD), 381, 382, 385, 387
Concrete waste, 253, 254, 256, 257, 259, 261
Construction, 15, 18, 24, 27, 35, 52, 53, 56–58, 61, 63, 73, 92, 93, 95, 101, 103–105, 108, 117, 121, 135, 143, 147–151, 153, 155, 157–159, 165, 166, 171, 179, 180, 182, 184, 185, 187, 189, 191–196, 201, 204, 212, 218, 247, 248, 253–256, 263, 279, 283, 284, 289–291, 295, 301, 302, 305, 306, 308, 314, 325, 326, 360, 362–364, 367, 372, 394, 395, 397, 403, 405, 406, 408, 421, 428–431
Contamination of soil, 329, 332
Contrast, 10, 20, 31, 47, 53, 57, 58, 77, 78, 86, 104, 146, 183, 194, 197, 302, 385, 434, 435, 442
Cool roofs, 439, 440, 446–448
Courtyard, 20, 21, 24, 28, 31–36, 63, 66, 67, 69–71, 73, 91, 94–101, 134, 193–195, 204, 206–208, 210, 212

D

Daylight, 77, 87, 139, 444

P. Mercader-Moyano (ed.), *Sustainable Development and Renovation in Architecture, Urbanism and Engineering*, DOI 10.1007/978-3-319-51442-0

Diagnosis, 17, 91, 117, 131–134, 140, 181, 183, 216, 290, 356, 359, 362
Digital fabrication, 303–309

E
Eco-efficient, 63, 194, 277
Eco-efficient technologies, 242
Efficiency, 41, 43, 44, 86, 124, 145, 169, 187, 188, 191, 192, 194, 242, 243, 278, 289, 290, 303, 307–309, 312, 316, 373, 377, 381, 382, 384, 400, 427, 440
Embedded energy, 53
Energy certification, 133, 144, 146, 147, 427
Energy demand, 3, 15, 17, 18, 20, 21, 23, 24, 86, 146, 166, 208, 211–213, 222, 223, 363, 366, 369–371, 378, 393–396, 398–400, 406, 410, 411, 432
Energy efficiency, 15, 17, 24, 27, 51, 54, 140, 143–148, 150, 165, 180, 182, 188, 192, 197, 222, 351, 353, 407, 411, 427, 428, 436, 439, 440
Energy-efficient buildings, 427
Energy generation, 368
Energy instrumentation, 165, 171
Energy poverty, 165, 166, 169, 175, 203
Energy rehabilitation, 143, 150, 165, 166, 175, 211
Energy retrofit, 4, 403, 405, 406, 410, 411
Energy simulation, 20, 22, 165, 171, 339, 371, 382
Energy wreck, 000
Environment, 24, 40, 42, 51, 54, 55, 57, 75, 77–82, 86, 87, 101, 118–120, 122, 123, 125, 126, 131, 133, 143, 148, 151, 153, 182–184, 191, 192, 201, 216, 218, 248, 257, 263, 277, 286, 287, 289, 295, 297, 329, 387, 388, 415
Environmental assessment, 195, 196, 411
Ephemeral architecture, 302–304, 309
Evaluation report, 131–133, 140

F
Fluid dynamics, 381, 382, 445
Fuel poverty, 204, 207, 213

G
Geographic Information System (GIS), 7, 45, 153, 154, 156, 157, 339, 417, 420
Green areas, 45, 47, 57, 125, 332, 333, 325, 326, 329, 422

H
Habitability, 52, 54, 117, 119–121, 123, 217, 301, 325, 326, 333
Health facilities, 363, 364, 368, 370, 379
Heat island, 3–5, 12, 206, 339, 340, 344, 346–349, 415, 417–423, 418, 424
Heat wave, 4, 5, 416, 424
Heritage, 63–66, 68, 72, 74, 92, 93, 109, 110, 114, 153–156, 158, 159, 161, 186, 334
High performance, 263, 270–272
Housing, 29, 32, 36, 71–73, 131, 166, 175, 182, 183, 203, 206, 215, 217, 280, 281, 290
H-plan blocks, 28, 32, 35, 36
HVAC design, 381, 383
Hydro-efficiency, 239, 242

I
Indicators, 4, 6, 7, 9–12, 39, 40, 42–45, 51, 55–58, 61, 121, 124–126, 145, 148, 197, 207, 215–219, 221–223, 278, 318, 411
Indoor facilities, 169
Industrial waste, 263
Infrared thermography, 297, 408, 427, 428, 435
Innovative device, 323
Integral method, 369
Integrated method, 180
Integrated Nested Laplace Approximation (INLA), 15–17

L
Land reclamation, 000
Lightweight concrete, 197, 198, 263, 268, 441
Luminance, 77–79, 82–86

M
Madrid, 3–5, 8–11, 34, 179, 183, 185, 211, 345, 348, 415, 417, 418, 420, 422, 424
Mediterranean city, 77
Micro-Economic Assessment, 360
Monitoring, 135, 137, 196, 204, 211, 306, 327, 330, 341, 352, 358, 403, 404, 406, 408, 409

N
Natural risks, 40, 43
Nearly zero energy buildings (NZEB), 167
Neighborhood refurbishment, 000

O
Obsolescence, 27, 28, 36, 52, 55, 56, 64
Obsolescence of Use, 56
Office buildings, 393, 394, 400, 403, 405, 440

P
Passive cooling, 439
Passive strategies, 146, 203, 204, 213, 411
Passivhaus, 148, 165, 175

Piassaba, 227, 229, 230
Pier, 103, 107–109, 111, 112, 114
Plant palette, 325, 329, 332
Popular, 63, 65, 67, 69, 71, 73, 110, 159, 254
Popular architecture, 93, 162
Population access, 53, 59, 62
Public space, 27, 28, 32, 33, 40, 44, 47, 54, 100, 117–125, 140, 183, 185, 215, 308, 422, 424

Q
Quality control, 289, 290, 296, 297
Quality of life, 43, 118–120, 122, 131, 180, 333

R
Radiative cooling, 439, 440, 447, 448
Recycled, 61, 197, 227, 253, 254, 259, 263, 293, 306, 307
Refurbishment, 203, 204, 206–208, 213, 215–218, 362, 404
Rehabilitation, 24, 110, 114, 120, 133, 140, 151, 171, 175, 179, 186, 188, 197, 201, 203, 216, 218, 219, 221–223, 289, 290, 296, 412
Residential building stock, 15, 24
Residential states, 35
Restoration, 112, 154

S
Social housing, 27, 36, 53, 203, 204, 213, 215, 218, 277, 282, 284
Society, 27, 53, 63, 94, 114, 120, 141, 155, 245, 246
Software validation, 383
Solar Decathlon, 277
Standards, 45, 69, 98, 119, 121, 131, 137, 141, 151, 182, 184, 187, 259, 404, 434
Structural reinforcement, 290, 294, 374
Sustainability, 39, 40, 43, 46, 47, 51, 54, 73, 120, 131, 140, 147, 148, 151, 180, 191, 192, 195, 197, 201, 204, 206, 216, 239, 277, 289, 291, 294, 295, 305, 311
Sustainable, 27, 39–42, 44, 53, 63, 153, 179, 194, 216, 227, 239, 263, 277, 290, 291, 306
Sustainable building, 75, 147
Sustainable development, 40, 121, 143, 144, 150, 160, 191, 227
Sustainable refurbishment, 203, 206, 216
Sustainable strategy, 155
Sustainable urban planning, 43
Systematization criteria, 242, 247

T
Technology and social Change, 204
Tensile strength, 227, 228, 230, 232, 235
Thermal-energy efficiency, 353
Thermal intensity, 427, 431, 432, 434, 435
Thermal transmittance, 18, 369, 400, 431, 432, 434, 435
Tourism, 45, 63, 64, 68, 103, 191
Tradition, 93, 101, 153, 183, 279

U
UNESCO codes, 68, 72
Update, 339, 340
Urban climate, 340, 348
Urban heat island, 3, 340, 345, 347, 415, 418, 420, 424
Urban indicators, 122, 278
Urban regeneration, 12, 63, 179, 182, 215, 218
Urban sustainability, 40

V
Vegetation, 209, 227, 325–328, 330, 331, 333, 418, 421, 424, 435
Vulnerability, 4, 6, 10–12, 41, 45, 56, 60, 61, 217
Vulnerable population, 3, 4, 10, 12

W
Water mills, 153, 155, 157, 159
Water saving, 243, 246, 312, 323
Weibull analysis, 227, 231, 232
Wind speed, 429, 431, 435, 442, 446, 447

MIX
Papier aus verantwortungsvollen Quellen
Paper from responsible sources
FSC® C105338

If you have any concerns about our products,
you can contact us on
ProductSafety@springernature.com

In case Publisher is established outside the EU,
the EU authorized representative is:
Springer Nature Customer Service Center GmbH
Europaplatz 3, 69115 Heidelberg, Germany

Printed by Libri Plureos GmbH
in Hamburg, Germany